Phosphorus in Agriculture: 100 % Zero

Ewald Schnug · Luit J. De Kok
Editors

Phosphorus in Agriculture: 100 % Zero

Springer

Editors
Ewald Schnug
Institute for Crop and Soil Science
Julius Kühn Institute
Braunschweig
Germany

Luit J. De Kok
Laboratory of Plant Physiology
University of Groningen
Groningen
The Netherlands

ISBN 978-94-024-1394-6 ISBN 978-94-017-7612-7 (eBook)
DOI 10.1007/978-94-017-7612-7

Softcover reprint of the hardcover 1st edition 2016

Printed on acid-free paper

This Springer imprint is published by Springer Nature
The registered company is Springer Science+Business Media B.V. Dordrecht

Ewald Schnug (* 7 September 1954)

The contributing authors have decided to dedicate this book to Ewald Schnug for various reasons. One is his 60th anniversary, yet another one is his outright devotion to fertilizer research that started almost 40 years ago. It was the work of Armi Kaila (1920–2003) on the apparent utilization of phosphorus that inspired him and triggered his ambition to deliver a benchmark on phosphorus fertilization. The outcome is a standard work, which summarizes cutting-edge knowledge on phosphorus research in different disciplines commendable for a mixed spectrum of readers including scientists, farmers, and politicians. It is the comprehension of these different perspectives into decision-making rules for fertilization, which are the core for a truly balanced use of phosphorus, which is agronomically adequate, environmentally sound, and socio-economically just.

Preface

Nutrients in phosphorus (P) fertilizer which are completely (100 %) utilized by crop plants without (Zero) contaminating the soil with environmentally relevant substances is a future concept for P fertilization in agriculture. P scarcity is one of the major challenges for agriculture in the near future. The waste of the nonrenewable resource P, which is mainly consumed for fertilization is of major concern for a sustainable development and there is no doubt that P fertilization strategies require significant advances towards sustainability. Pessimistic assessments see world P reserves last only for another 50–150 years. In industrial countries agriculture uses more P for fertilization than it exports with agricultural products. Some uncertainty in practical P fertilization derives from the fact that there are no experimental means to estimate the true long-term utilization of fertilizer P. The traditional concepts (differential and radioisotope method) suggest that on an average only 75 % of the fertilizer P applied is finally used by plants, implicating that always more P has to be fertilized than will be recovered by agricultural products. Next to P scarcity, excessive P loads on soils with intensive livestock farming and use of P sources, which are not completely available for plant result inevitably in a dissipation of this nonrenewable resource. Yet, another problem is the guidelines for site-specific P fertilizer recommendations in relation to land use systems.

The breakthrough towards an understanding for the prerequisites for a complete P utilization and thus also a complete recycling of fertilizer P came with an empirical approach published in 1965 by the Finnish scientist Armi Kaila (1920–2003), for which the term “apparent utilization” has been coined. Based on Kaila’s hypothesis, P fertilization concepts can be designed, which warrant a full utilization of fertilizer P in agriculture. This book agglomerates the basic knowledge required to target a 100 % P utilization which comprises the basic concept of “apparent utilization”, the fundamental principles of P turnover in soils and plants, the evaluation of the P nutritional status, and characteristics of fertilizers and fertilization concepts whereby the latter is addressed on national level.

Second to the goal of a 100 % utilization of fertilizer P is the objective to avoid or at least to limit the soil contamination and accumulation of hazardous inorganic and organic substances coming with the various types of P fertilizers. What seems like cutting the Gordian knot may be a challenge for future development in fertilizer manufacturing technology, energy neutral P fertilizer production using high-temperature nuclear reactors, where the energy required to make a clean fertilizer comes with the raw material.

Last but not least, this book addresses one of the first not only basic issues of sustainability of P fertilization but also its socio-economic aspects and proposed governance instruments.

Braunschweig, Germany — Ewald Schnug
Groningen, The Netherlands — Luit J. De Kok

“Phosphora”

A hand-sewn textile collage from cotton and silk (44 × 42 cm) by Bettina Franckenberg, Bodrum, Turkey (2016)

In search for the elusive Philosopher's stone, which turns ordinary metal into gold, Henning Brand discovered in 1669 the element phosphorus. Alchemists of that time often used light as a symbol of the spirit. Thus they were especially interested in light that seemed to be trapped in matter and their vision was coupled into the idea of a perpetual, eternally burning light. The light emanated from phosphorus inspired a few artists: in E.T.A. Hoffmann's fairy tale "The Golden Pot" (1814), the lucent, obviously male, "Phosphorus" fights with a black dragon and in Bettina Franckenberg's "Phosphora" his female sibling debuts in form of a sphere of radiating light and luminosity in all directions. The stitched threads direct the way, the soft fabric nestles to flowing vibes, and the beads symbolize notions clear as crystals. The interplay of these elements reflects Phosphora's impression as spirited light. Children and open-minded are close to her heart, she empowers the will to live and her spiritual strength promises healing.

About the Artist

Bettina Franckenberg was born in February 3, 1956 in Essen, Germany. Handcraft and artistic abilities were encouraged early in her parent's house. After graduating from college in 1975 she continued with her education in the field of occupational therapy, graduated in 1979, and started to specialize in psychiatry, supplementary qualification as psychodrama assistant—at the same time initiation to "patchwork". In 1982 she discovered the Mediterranean (living the first 2 years in Greece) as well as her "textile ambitions"—1986 she opened her first exhibition in the Castle of St. Peter in Bodrum. She went back to live in Berlin for some years—between 1989 and 1992 she worked as an occupational therapist in psychiatry and organized several "textile art" exhibitions in Germany, Switzerland, Latvia, and Turkey.

Since 1992, Bettina Franckenberg has been living permanently in Bodrum, Turkey and between 1994 and 2005 she had her own Atelier-Gallery in the holiday-village Sea-Garden in Yaliçiftlik. She continues working as a freelance textile artist living in Kizilagac near Bodrum.

Her first artwork on a scientific subject was "Uranyum", a hand-sewn textile collage from velvet and silk (38 × 37 cm) which appeared on the cover of the book "Loads and Fate of Fertilizer Derived Uranium". Edited by L.J. De Kok and E. Schnug (2007), Backhuys Publishers, Leiden, The Netherlands.

The Preparation of the Fosperus, or the Astrum lunare Microcosmicum.

William Y-Worth, *Chymicus Rationalis... In which is contained, A Philosophical Description of the Astrum Lunare Microcosmicum, or Phospheros.* London, 1692.

As is before shewed. take Urine well putrefied in a Tub. exposed to the Air for seven Weeks. all one as you do when you're to make a Spirit of it; the Spirit being drawn. or rather the whole of it being evaporated to the consistance of Honey. in which lies the Fosperus; but the Art is somewhat difficult to get it from thence. in two cases. the one is in making choice of a proper Agent to be mixed therewith. and the other is the exact regiment of the Fire. Now as to the Agent. it must not (as some have said) be Sand; but one that is free from Salt. and consequently from the glass-making Faculty. or any thing that will vitrifie. for if it doth that. you shall never obtain the Fosperus; but that you may. we direct you to make use of a Fire within and without your Vessel. which is thus to be understood; take the Sap of Urine. as before 'tis said. or that of Dung and Urine putrified together. and incorporate it with Charcoal-dust. and put it into a German Retort. and lute on a Receiver. that is filled half full with Water. that when the Fosperus comes over. it may immediately drop into the Water and be preserved thereby. Observe. That you are to begin with a very small Fire. but you are to increase it gradually into the highest degree that possibly can be made. and let it be in such a Furnace. that upon occasion the Bellows may be used. and that the Fire may play well about the Vessel. continuing so long at its greatest heighth. till no more comes over. and you'l see Flakes like Lightening. that will come over in the Receiver of two substances. the one like a thick muddy Water. which will sink to the bottom. the other swimming over it; separate by inclination these two substances. and keep each by it self in a Glass. the gummy Nature may be made into what form you please. by being dissolved in Spirit of Sal Armoniack or other Urinous Spirits; in B. 'tis thereby putrified.

Observations concerning it. It must always be kept in Water. for the Air sets it on fire. The thick Liquor. if rubbed upon the Hands. Cloaths. or Hair. they will appear in the dark. as if all in fire. but will not burn. and if you open the Cork of the Glass. all its vacancy will appear like the flame of Brimstone. If you put this thick Liquor. hardened by digestion into a Gum in Oyl of Cloves or of Cinnamon. and let it remain there 24 hours. and pour off the Oyl from it into a Glass. the same Oyl will as you pull out the Cork. and set it in the Air. appear so bright. that in the dark. if it's in a large and clear Vial. you may see it to read distinctly. If you write upon the Palm of your Hand. or upon Paper with the said Gum. what ever you write will appear all on fire. and the Letters may be read a long time after; but you must have a great care. that you do it softly. and to put it into Water. as soon as you have done. for if it happen to fire 'twill burn the place most dreadfully.

Cut a little piece of the said Gum. and lay it upon Paper. and rub it on with the point of a Knife. and 'twill set the Paper on fire Put a little piece of the said Gum to steep 24 hours in Spirit of Wine. then taking it out again. if you pour this Spirit of Wine by drops upon a Bason of Water. you will see Flashes like Lightening. somewhat like the flame of Brimstone. and somewhat of a purplish colour. Rx. Of the aforesaid Gum one drachm. and in a white earthen Vessel expose it to the Air. and the Air will set it on fire. and when it hath done burning. you will have four drachms of Spirit of Salt. which it attracts out of the Air. If the Privy Parts be therewith rubb'd. they will be inflamed and burning for a good while after.

Contents

Urgent Need for Action in the Baltic Sea Area

Leena-Marja Kauranne and Marjatta Kemppainen

Abstract Phosphorus has a central role in eutrophication of the Baltic Sea. Severe algal bloomings are the visible sign of the excess of nutrients for all living around the sea. Agriculture is responsible for more than a half of the waterborne nutrient loads in the Baltic Sea. It is also the most difficult sector to tackle in curbing the nutrient emissions. Since 1974 the Helsinki Commission (HELCOM) has taken a major role in joint efforts to improve the state of the Baltic Sea. The HELCOM (Helsinki Commission) Baltic Sea Action Plan aims at reaching a good environmental status of the Baltic Sea before 2021.

Keywords Baltic sea action plan · Diffuse source · EU strategy for the Baltic Sea region · Eutrophication · Helsinki commission · Manure · Nutrient load · Point source · Water framework directive

Introduction

The Baltic Sea is an extraordinary and unique, almost land-locked, shallow water basin with brackish water and an average depth of only 54 m. Very specific hydrographic, chemical and physical conditions make the Baltic Sea extremely sensitive to environmental impacts. Its catchment area covers 1,720,000 km^2 with a population of 85 million people. The state of the Baltic Sea has drastically changed over the recent decades. Human activities both at sea and throughout its catchment area are placing rapidly increasing pressures on the marine ecosystem. Of the many

---

Retired from the Ministry of Agriculture and Forestry of Finland in 2013.

L.-M. Kauranne (✉)
Ministry of the Environment, P.O. Box 35, 00023 Helsinki, Finland
e-mail: leena-marja.kauranne@ymparisto.fi

M. Kemppainen
Ministry of Agriculture and Forestry of Finland, Helsinki, Finland

E. Schnug and L.J. De Kok (eds.), *Phosphorus in Agriculture: 100 % Zero*,
DOI 10.1007/978-94-017-7612-7_1

environmental challenges, eutrophication is one of the major problems for the sea, for the lakes and the rivers in the catchment area, especially in the southern and eastern parts of the Baltic Sea. It is caused by excessive nitrogen and phosphorus inputs. This leads to problems like increased algal blooms, murky waters, oxygen depletion and lifeless sea bottom (HELCOM 2009).

Nutrients enter the Baltic Sea via rivers, through atmospheric deposition and in direct discharges from pollution sources located along the coastline. Phosphorus enters the sea mainly as waterborne input; only about 1–5 % of the total input is airborne. The riverine discharges originate both from point sources, such as industrial or municipal wastewater plants, and from diffuse sources, such as agriculture and forestry, scattered dwellings, traffic and atmospheric deposition within river basins. The Baltic Sea is connected to the North Sea and the Atlantic Ocean through the narrow Danish Straits and Sound areas. Inflows of fresh seawater occur rarely and poor oxygen conditions release phosphorus accumulated in the bottom sediments.

Agriculture is the main source of phosphorus and nitrogen inputs to the Baltic Sea accounting about 50 % of the total diffuse loads. Managing emissions from agricultural land and livestock operations is critical to restore the ecological balance of the Baltic Sea.

Putting an end to further destruction and avoiding an irreversible disaster of the Baltic Sea marine environment calls for immediate wide-scale coordinated action in all sectors. Failure to react now would undermine both the prospects for the future recovery of the sea and its capability to react to the projected stress by the climate change with the risk of increasing precipitation and thereby intensified run-off from agriculture as well as changes in water temperature and salinity. Furthermore, inaction will affect vital resources for the future economic prosperity of the whole region and would cost much more than the cost of action. Efforts to combat eutrophication have already been and are taken on many different fora at different levels but it is a complicated and long-term task to curb loads from diffuse sources.

Action Already Taken

The Baltic Sea is a target of actions at several different levels. Many tools to protect and restore the Baltic Sea have been and are currently taken to improve its state: the EU Water Framework Directive (WFD), the Marine Strategy Framework Directive (MSFD) as well as the EU Strategy for the Baltic Sea Region (EUSBSR) at the European level, the HELCOM Baltic Sea Action Plan (BSAP) at the regional level and numerous projects and programmes at national and local levels. The impact of the EU Common Agricultural Policy on the state of the Baltic Sea is also significant. Moreover, the EU Nitrates Directive sets a frame for prevention of nitrate pollution from agricultural sources but no equivalent legislation exists for phosphorus at the EU level.

HELCOM Shows Way for Baltic Sea Protection

The Baltic Marine Environment Protection Commission, more usually referred to as the Helsinki Commission or HELCOM, is an international organisation made up of the nine Baltic Sea coastal countries and the European Community, working to protect the marine environment of the Baltic Sea. HELCOM is the governing body of the "Convention on the Protection of the Marine Environment of the Baltic Sea Area"—also known as the Helsinki Convention, which was signed by all the Baltic Sea countries in 1974 and it came into force in 1980. A new updated convention was signed in 1992 by all the states bordering on the Baltic Sea and the European Community, and the Convention entered into force in 2000. The Convention covers not only the Baltic Sea, but also the surrounding catchment area within the coastal countries.

The Baltic Sea Action Plan Sets the Protection Targets

Since the establishment of the Helsinki Convention, HELCOM has, with the help of several programmes, been working to reduce the nutrient loads to the sea. The latest HELCOM programme with new nutrient reduction targets is the Baltic Sea Action Plan (BSAP), which was adopted by all nine Contracting Parties (eight EU Member Countries and Russia) and the European Community at the Ministerial Meeting on 15 November 2007 in Krakow, Poland (HELCOM 2007).

HELCOM follows the progress towards reaching politically agreed goals by collecting and reporting load data. The latest Baltic Sea Pollution Load Compilation (PLC-5) was published in late 2011 whereas the first one (PLC-1) in 1987. A project called PLC-5.5 provided updated and corrected data and more complete information on pollution loads including data for 2009 and 2010 for the HELCOM Ministerial Meeting in 2013 (HELCOM 2011).

The ambitious target of the BSAP is to restore the good ecological status of the Baltic marine environment by 2021. The cross-sectoral Plan identifies the specific actions needed to achieve the agreed targets within a given timeframe for the main environmental priorities: (1) combatting eutrophication, (2) curbing inputs of hazardous substances, (3) ensuring maritime safety and response capacity to accidents at sea, and (4) halting habitat destruction and the ongoing decline in biodiversity. The HELCOM Member States have developed their own national implementation plans. The BSAP has strong links to global legislative frameworks and is also seen, for those Parties being also EU Member States, as a contribution to the implementation of key EU directives, in particular, the Marine Strategy Framework Directive and the Water Framework Directive.

The objectives of the Baltic Sea Action Plan to combat eutrophication include: no excessive nutrient concentrations; clear water; no excessive algal blooms; natural oxygen levels; and natural distributions and abundance of plants and animals.

Actions are needed both for point and for diffuse sources in the following areas: (i) waste waters: municipalities, scattered settlements and single-family homes; (ii) agriculture; (iii) transboundary air- and waterborne pollution (HELCOM 2007).

HELCOM has estimated that for good environmental status to be achieved by the year 2021, the maximum total allowable annual nutrient inputs into the Baltic Sea would be 21,000 tonnes of phosphorus and about 600,000 tonnes of nitrogen. Over the period 1997–2003, average annual inputs amounted to 36,000 tonnes of phosphorus and 737,000 tonnes of nitrogen. Therefore, annual reductions of some 15,000 tonnes of phosphorus and 135,000 tonnes of nitrogen would be required to achieve the crucial 'clear water' objective set in the Baltic Sea Action Plan. The 2013 HELCOM Ministerial Meeting adopted the revised HELCOM nutrient reduction scheme. To decrease nutrient inputs into the Baltic Sea to the maximum allowable levels, the HELCOM countries have agreed to take actions not later than in 2016 to reduce nutrient loads in waterborne and airborne inputs, aiming to reach good environmental status of the Baltic Sea by 2021. Also the country-wise provisional nutrient reduction targets were agreed upon (Table 1). To reach these targets each Member Country is encouraged to choose the most appropriate and cost-effective measures for its special needs.

For addressing the problems of the large agro-industrial clusters, contracting states also agreed to identify individual hotspots such as major facilities of intensive rearing of cattle, poultry and pigs, where actions should be prioritised in order to comply with revised requirements for prevention of pollution from agriculture (Annex III of the 1992 Helsinki Convention).

Environmentally sound manure management has to be assured by construction of sufficiently large and watertight manure storages, proper agri-environmental

Table 1 Updated country-wise reduction targets for nitrogen (N) and phosphorus (P) (HELCOM 2013a)

	Phosphorus		Nitrogen	
	2007	2013	2007	2013
Denmark	16	38	17,210	2.890
Estonia	220	320	900	1800
Finland	150	330 (+26[a])	1200	2.430 (+600[a])
Germany	240	110 (+60[a])	5.620	7.170 (+500[a])
Latvia	300	220	2560	1.670
Lithuania	880	1470	11,750	8970
Poland	8760	7480	62,400	43,610
Russia	2500	3790	6970	10,380
Sweden	290	530	20,780	9240

2007: as in the HELCOM Baltic Sea Action Plan

2013: as adopted by the HELCOM Ministerial Meeting on 3 October 2013

[a]Loads originating from the country but being discharged to the Sea via another country; additional specific footnotes to the above table can be found in the text of the Ministerial Declaration

measures and advisory services and by identification of agricultural areas that are critical for nutrient pollution of the Baltic Sea. This implies designation of relevant parts of agricultural land as nitrate vulnerable zones and performing risk assessments of nutrient leaching from agricultural areas. The 2013 HELCOM Copenhagen Ministerial Meeting agreed (HELCOM 2013b) that measures such as annual nutrient accounting at farm level and environmentally sound utilization of nutrients in manure should be implemented in Member Countries by specific timeframes.

PLC-5 Shows Some Positive Signs

According to the data presented on total nutrient loads into the Baltic Sea until year 2008 in the Fifth Pollution Load Compilation (PLC-5) in 2011, it seems that there has been some progress in reduction of nutrients losses from agriculture within the Baltic Sea catchment area since 2000. In fact, losses from agriculture appear to have increased in some Contracting Countries although load figures for diffuse sources are uncertain due to the impact of climatic factors and different methodologies applied, making comparisons of loads between years difficult. The total input of phosphorus and nitrogen to the Baltic Sea in 2008 was 29,000 and 859,000 tonnes, respectively (HELCOM 2011).

Comparison of the flow normalized load during 1997–2003, which is the reference period for setting provisional reduction targets in the BSAP, with flow normalized input during 2006–2008 indicates a reduction with less than 1900 tonnes of nitrogen (2 %) and more than 3300 tonnes of phosphorus (10 %). The decrease since the inception of the Action Plan was 9 % for nitrogen and 10 % for phosphorus from figures of the BSAP baseline years (1997–2003). In the longer perspective, since 1994, the reductions have been even larger inputs of N and P to the Baltic Sea have been cut by 18 and 16 %, respectively (HELCOM 2013a).

Reductions in nutrient inputs have primarily been achieved through improvements at major point sources, such as municipal sewage treatment plants and industrial wastewater outlets. Achieving further reductions is a more difficult task. Reducing nutrient losses from agriculture is much more complicated than curbing loads from point sources and will involve several measures. Due to retention in soils, groundwater and inland surface waters, a reduction of nitrogen or phosphorus in the local emissions will result in less reduction to the Baltic Sea. Besides, there is a considerable time-lag between the implementation of agricultural water protection measures and any observed impacts on lakes and rivers, and even more so for marine water bodies. This means that reductions in the loads into marine water bodies will not be observable for years or even decades.

Around the Baltic Sea, the rise in the living standards of the eastern EU member countries and of Russia and Belarus will probably increase the proportion of animal-sourced food in people's diets. The increasing animal production may require an expansion of the cultivated area, increasing the leaching of nutrients.

Pressures should be progressively reduced since only significant reductions of phosphorus and nitrogen will result in improved ecosystem health of the Baltic Sea. To reduce phosphorus and nitrogen losses from countless individual field plots and animal production units is a huge task that calls for identification of high-risk areas and implementation of cost-efficient agri-environmental measures. Such measures could e.g. be catch crops, buffer zones, restored and constructed wetlands, reduced mechanical treatment of soils, lime application, application of more environmentally friendly cultivation and manure handling technology, higher storage capacity for manure and application of less than yield-optimal fertilizer amounts. Enlarged animal farms and their clustered locations require more effective methods in manure handling practices and in recycling of nutrients. Slurry could e.g. be processed chemically or mechanically in order to separate nitrogen and phosphorus rich components (liquid/fibre part) and be further used for energy production. To this end, there is a need for interdisciplinary research and advice and recognition of economic, social and political constraints.

For more information about the Baltic Sea, see http://www.helcom.fi and the HELCOM Fifth Baltic Sea Pollution Load Compilation 2011.

References

HELCOM (2007) HELCOM Baltic Sea action plan

HELCOM (2009) Eutrophication in the Baltic Sea—an integrated thematic assessment of the effects of nutrient enrichment and eutrophication in the Baltic Sea region: executive summary. Balt Sea Environ Proc 115A

HELCOM (2011) The fifth Baltic Sea pollution load compilation/PLC-5. Balt Sea Environ Proc 128

HELCOM (2013a) Overview of implementation of the HELCOM Baltic Sea action plan (BSAP)/ 2013 HELCOM ministerial meeting. www.helcom.fi/Documents/Ministerial2013/Associated%20documents/Supporting/BSAP_Overview_with%20cover.pdf

HELCOM (2013b) www.helcom.fi/Documents/Ministerial2013/Ministerial%20declaration/Adopted_endorsed%20documents/Revised%20palette%20of%20agri-environment%20measures.pdf

The Enigma of Fertilizer Phosphorus Utilization

Ewald Schnug and Silvia H. Haneklaus

Abstract The need for a sustainable use of the limited resource phosphorus in agriculture is generally acknowledged though scarcity is no pressing topic. There is no doubt that phosphorus losses from agricultural soils to the environment are a major contributor to eutrophication of water bodies. There is also no doubt that the excessive application of farmyard manure in the vicinity of big livestock enterprises yields an undesired accumulation of P in soils, while on numerous arable farms phosphorus mining can be observed. The premise for a sustainable P use in agriculture is a balanced P fertilization where inputs equal outputs. This chapter summarizes the chemical behavior of fertilizer-derived phosphorus in soils, and it presents a novel approach to assess the fertilizer phosphorus utilization completely and reliably as a tool to quantify the genuine phosphorus demand of agricultural crops.

Keywords Actual phosphorus utilization · Apparent phosphorus utilization · Balanced fertilization · Immobilization · Off-take · Speciation · Utilization coefficient

Introduction

Phosphorus (P) is an essential plant nutrient and a sufficient P supply is required in order to maintain a high level of crop productivity. Next, P has an eminent and well-known relevance as an ecological contaminant (Bai et al. 2013; Frede and Bach 2003; Perspectives Agriculture 2011). Thus it is vital for sustainable crop production to develop strategies and measures to close the agricultural P cycle (Perspectives Agriculture 2011).

E. Schnug (✉) · S.H. Haneklaus
Federal Research Centre for Cultivated Plants, Institute for Crop and Soil Science, Julius Kühn-Institut (JKI), Bundesallee 50, 38116 Brunswick, Germany
e-mail: ewald.schnug@julius-kuehn.de

E. Schnug and L.J. De Kok (eds.), *Phosphorus in Agriculture: 100 % Zero*,
DOI 10.1007/978-94-017-7612-7_2

The waste of the non-renewable resource P, which is mainly consumed for fertilizer purposes, is of major concern for sustainable development. Here, P fertilization strategies request significant advancements towards sustainability. Pessimistic assessments see world P reserves last only for another 50–150 years and the term 'peak phosphorus' (Cordell et al. 2009) has caused audible public uproar (Schwarz and Weingärtner 2013).

In industrial countries agriculture uses more P for fertilization than it exports with agricultural products. There is an ongoing discussion about the scarcity of P reserves that will affect agricultural production within the next decades. Ulrich and Frossard (2014), however, showed that available data do not support such apprehension. Meanwhile the problem stepped from scientific to public awareness and finally has reached the political level: in the coalition agreement for the 18th legislative period of the Lower House of German Parliament the P problem is addressed in the chapter "Water bodies and marine environment protection" with the phrase: "We will end the agricultural use of sewage sludge and retrieve P and other nutrients." (Anonymous 2014).

But retrieving P is only one part of the story, making it into suitable fertilizers, however, a complete different one. The core process of separating P from wastes is making it insoluble and undergoing precipitation. In contrast, the core process of making P-fertilizers is making it soluble and thus available for plant uptake. It had been one of the great breakthroughs in plant nutrition and fertilization when *John Bennet Lawes* showed more than 150 years ago "that bones, treated with sulfuric acid and to produce superphosphate increased the yields of turnips grown on the soils at Rothamsted" (e-RA 2014). A key in assessing the overall sustainability of P fertilization is the question to which extent soil-applied P is taken up by crops cumulatively over an infinite period of time. Common understanding is that if P is used for fertilization a certain amount of the nutrient will always be lost in the system requiring that more P needs to be fertilized than is removed by the crop plants (Djodjic et al. 2005; Finck 1979; Platzen and Munk 1975; Syers et al. 2008; Wiechmann and Werner 1976). It is intrinsic that this assumption postulates no global validity as otherwise no ecosystem would be able to survive over time without any external supply of P. Despite a number of such ecosystems (Vitousek et al. 2010) the perhaps most popular example and evidence are "ecospheres" which are materially closed ecological systems and which are self-sustaining over a period of years (Taub 1974; Sagan 1986; Wikipedia 2014).

P fertilizers are the most significant carriers of hazardous substances into agro-ecosystems. P fertilizers manufactured from rock sources are contaminated with heavy metals and radio-nuclides (Taylor et al. 2016; Schnug and Haneklaus 2015; De Kok and Schnug 2008) and those deriving from organic sources may carry significant amounts of pharmaceuticals and other organic xenobiotics (Bloem et al. 2016), both hampering the sustainability of P fertilization from the material point of view. Promising technology has been developed to destroy organic xenobiotics in sewage sludges by incineration combined with solubilization of P and heavy metal extraction which delivers recycled P fertilizer materials that are comparable to manufactured mineral P fertilizers (Adam et al. 2009). With view to the contamination of phosphate rock with uranium (U) Haneklaus and Schnug (2016b) developed an innovative concept to employ the extracted U for energy and fertilizer production. Sustainable P fertilization not only requires a complete utilization of the nutrient itself, but also needs to limit collateral damages to soil and food quality. The primary questions to be answered in this chapter are: "What are the factors governing the utilization of fertilizer P and what are the circumstances under which a complete (100 %) utilization of fertilizer applied P can be expected".

Fate of Fertilizer P in Agro-ecosystems

The following section summarizes the common understanding of the behavior of fertilizer-derived P in agro-ecosystems in the context of its utilization by plants.

Turnover of Inorganic P in Soils

A general overview about the turnover of P in soils is given in Fig. 1. On an average half of the total P in a soil is bound organically, the other half exists in form of inorganic compounds. A small amount of the inorganic P is to be considered as virtually inert. This implies for instance apatites in the parent material of the soil, and insoluble P-containing fertilizer materials.

Plants have access to the P resources of a soil only through P dissolved in the soil solution. This is usually orthophosphate, but also in distinctly lower amounts low molecular organic compounds, which are able to pass the *Caspary* strip in the roots (Barber 1980). The amount of P dissolved in the soil solution at any time is always only a fraction (approximately about one-tenth of a percent) of the total P content in the soil, and constitutes not more than half of the amount of P taken up by an average harvest of crops (20 kg/ha P). This implies that P in the soil solution has to be constantly replenished from soil resources. A site will become P deficient and respond to P fertilization if the P demand of the crop for optimum growth exceeds the amount of plant available P in the soil. Fertilizer P increases the P concentration in the soil solution, the amount of adsorbed P and the quantity of P precipitated as secondary minerals. The nature of the precipitates depends on the predominance of Ca, Fe or Al ions and thus the pH-value of the soil (Sample et al. 1980). The utilization coefficient of P in the year of application depends strongly on the amount and nature of available adsorption sites in the soil (Schwertmann and Taylor 1989).

The utilization coefficient of mineral P fertilizers decreases with an increasing number of adsorption sites relatively to the P concentration (Roemer 2003). Doering (1956) estimated that 30–55 % of adsorbed P is available to plants in the first year after application. The speciation of fertilizer P in secondary minerals is mainly governed by the soil pH: in acidic soils Fe- and Al-phosphates prevail, while

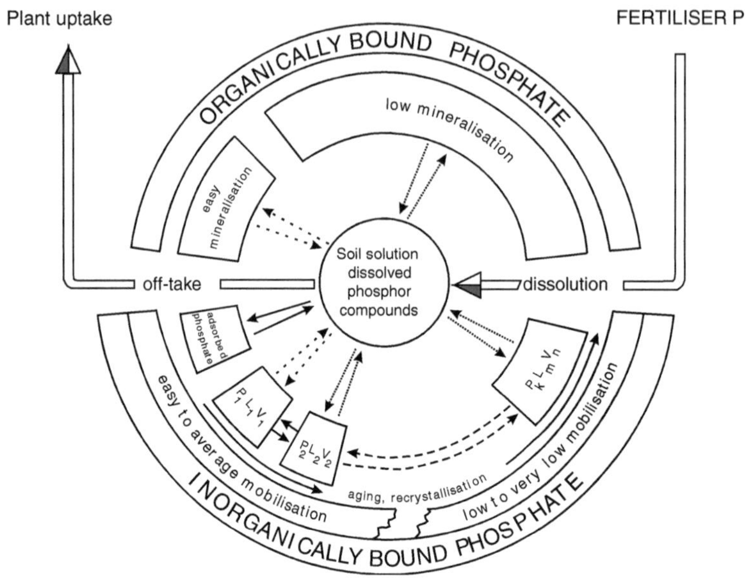

Fig. 1 The soil P engine. Plants take P up as dissolved orthophosphate from the soil solution. P replaced by fertilization first dissolves in the soil solution, some of it is taken up by plants but the majority transfers into the organic and inorganic storage pools. The latter undergoing aging over time: $P_1, P_2, \ldots, P_k$ = inorganic P species; $L_1, L_2, \ldots, L_m$ = solubility of P_1–P_k, decreasing from L_1 to L_m; $V_1, V_2, \ldots, V_n$ = rate of solubility of P_1–P_k, decreasing from V_1 to V_n

in neutral to alkaline soils Ca-phosphates are predominant (Munk and Gross 1975; Werner 1969). These initial, still labile precipitation products undergo further reactions with the soil environment. They change into more stable and less soluble, and thus less plant available P compounds over time (Sample et al. 1980). These processes follow thermodynamic principles, and thus are driven by an increase in entropy. In calcareous soils the first reaction product is brushite, whereby a continuous exchange of Ca^{2+} for H^+ in the crystal surface leads to the formation of hydroxyapatite. However, in temperate climates the formation of apatite has never been observed under field conditions (Werner 1970, 1971a; Lookman et al. 1995). This is highly significant as precipitation products deriving from P introduced through the soil solution by fertilization obviously never become completely inert.

Plants developed a number of mechanisms to revise the process of P immobilization in the soil through organic acids excreted by the root, i.e. citric, oxalic and galacturonic acid which form stabile complexes with metals, and thus are able to exchange and utilize even strongly bound P (Haneklaus and Schnug 2016a; Nagarajah et al. 1970; Oburger et al. 2013; Rovira 1965).

In acidic soils the aging of fertilizer P is characterized by a change from amorphous to more crystalline phosphate minerals, whereby orthophosphate displaces OH^- from coordination positions at the surface of the Fe- and Al-oxides (Werner 1971a; Werner and Wiechmann 1972). Although the initial reactions of

P fertilizers in the soil are rapid and result quickly in equilibrium between reaction products and soil solution, aging is a continuous process that yields the formation of more stable and less available P forms (Werner and Wiechmann 1972; Scharafat and Finck 1973; Sample et al. 1980).

In addition, thermodynamic processes of energy releasing phosphate transformations are accompanied by energy consuming microbiological processes that have a solubilizing effect on phosphate minerals (Sharma and Singh 1971; Mosse et al. 1976; Frossard et al. 2000).

Half or more of the total P in soils may be present in organic forms (Harrison 1987). Organically bound P originates from residues of crops grown on a site, or from organic fertilizers charging soils with additional P in organic form. Despite low molecular P containing organic compounds such as ATP organic P is less readily available for absorption by the plant than the inorganic P pool. More than 50 % of P in organic matter occurs as high molecular weight molecules such as esters of the phosphoric acid or inositol phosphates (Baeumer 1971; Schachtschabel 1960; Anderson 1980; Noack et al. 2013). Inositol phosphates form a number of insoluble salts and also form strong complexes with proteins and some metal ions, which cannot be utilized by the plant without prior mineralization by micro-organisms. Ivanov and Sauerbeck (1972) estimated an annual utilization efficiency of about 1.0–1.5 % for these P species. The rest of the organic P species occurs in form of easily soluble nucleic acids and phospholipids. The annual utilization efficiency is estimated to be 15–20 % (Ivanov and Sauerbeck 1971, 1972).

Turnover of Organic P in Soils

The release of organic P into the soil solution, where it is translocated before being adsorbed by the root, is controlled by the mineralization rate of organic matter (Turner et al. 2013). Generally, tillage of the soil promotes the decomposition of organic matter by stimulating the activity of the soil micro-flora (Harrison 1987). As with N, a high ratio of soil organic matter to organically bound P may result in P deficiency in the soil system. This so-called biological P-lock has been reported for soils with a C/P_{org} ratio >150 (Baeumer 1971; Finck 1976; Kaila 1949). In this process the fixation kinetics of the different organic P species follow the same principles as the aging of the inorganic P fraction. Thus in the first year of application, the utilization efficiency of P applied as organic fertilizers does not differ from those reported for P in inorganic fertilizers (Diekmann 1966; Debruk and Voemel 1970; Asmus et al. 1973; Isermann 1978; Latkowicz 1977; Sanderson and Jones 1997; Ibrikei et al. 1999; Anderson et al. 2001; Whalen and Chang 2001). Crop type and associated land management practices influence decomposition processes in the long run, and each management system will stabilize around a steady organic P level.

P-Losses from Agro-ecosystems

P is a macro-nutrient with no chemical or physical interferences with the atmosphere so that P will not evade from a soil through gaseous losses. For most agricultural systems P losses from soils by leaching or surface run-off are considered low with an average of 0.5 kg/ha year P except when extraordinary conditions favoring preferential flow occur (Boysen 1977; Hoffmann 1978; Svanbaeck et al. 2014). This amount seems like a ubiquitous background level which is little affected by agronomical or environmental factors. A much more significant factor for P losses from agro-ecosystems is erosion (Catt et al. 1997; Fuchs et al. 2009; Sharpley et al. 2000; Schroetter et al. 2016) which, based on averaging data provided by UBA (2015) would account to a mean value of 7–8 kg/ha year P in Germany. In contrast to leaching, losses through erosion can be controlled and kept on low levels by implementing codes of good agricultural practices (GAP) (Al-Kaisi 2015; BMVEL 2002). The expressed spatial persistence of P in soils was recognized long ago by European geographers and since then used in archaeology to trace back ancient settlements (Arrhenius 1931; Broadbendt 1981; Proudfoot 1976).

The Utilization Coefficient of P in Fertilizers and Its Significance for Fertilization

Phosphorus taken up by plants from soils originates either from native soil reserves or from P not taken up by previous crops. These residuals from former fertilizer applications complement the pool of native soil P reserves. Both P sources will never be utilized completely by crop plants so that a 'utilization coefficient' can be calculated for each nutrient source, which is required in order to determine the plant available amount of P. A defined amount of crop yield corresponds to an assigned amount of P removed from the soil. Assuming that P is yield limiting, the difference between the amount of P available from the soil pool and the amount of P removed with the targeted crop yield needs to be supplied by fertilization. In addition, it needs to be taken into account that P in a fertilizer product is not completely taken up by plants so that a utilization coefficient for this P source is necessary in order to determine the fertilizer rate. The previously mentioned parameters are summarized in the following equation:

$$D = \frac{E - \left(\frac{B \times b}{100} + \frac{C \times c}{100}\right)}{d} \times 100 \tag{1}$$

B P content in the ploughed soil layer originating from previous fertilizer applications in (kg/ha)

C Native P content in the ploughed soil layer in (kg/ha)

E P off-take by harvest products in (kg/ha) for target yield
b Utilization coefficient of accumulated P, fertilizer-derived in (%)
c Utilization coefficient of native P in (%)
d Utilization coefficient of fertilizer-derived P in the year of application in (%)
D Fertilizer rate expressed as amount of P in (kg/ha).

Equation 1 describes the factors, which determine the P utilization of the current crop. As mentioned before plants utilize fertilizer-derived P only partly in the year of application so that the residual amount restocks soil reserves. Figure 2 provides a graphical approach for a better understanding of the processes involved: the inner circle in the Fig. depicts the amount of nutrient found in the harvest products and its origins: native soil reserves (C), secondary soil reserves (B) and fertilizer (D). From the entity of each source only a faction is utilized by the plants and removed with harvest products, which implies that a time factor needs to be considered, too. Fertilizer P not taken up in the year of application is attributed to the pool of

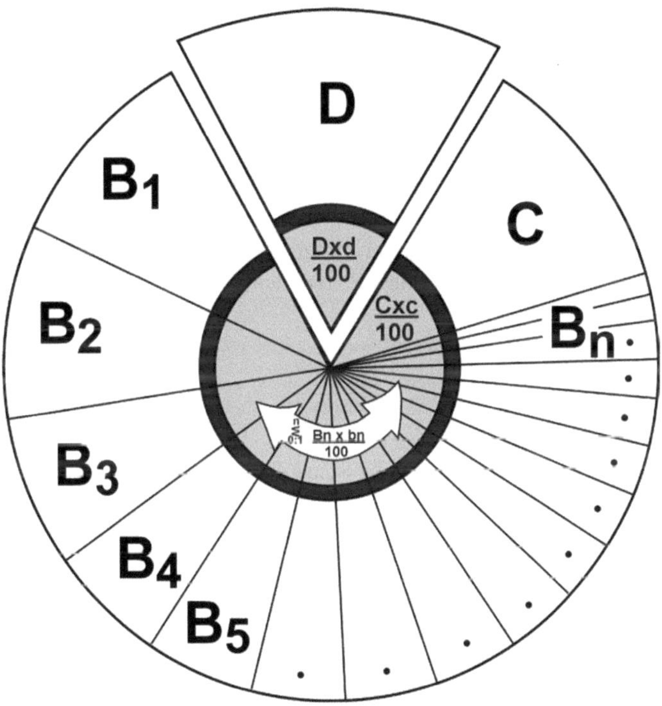

Fig. 2 Origin of P taken up by a crop. ● P off-take for target yield; D = P fertilizer rate; B = residual P in the soil from former fertilizer applications; C = native P content of the soil; E = P off-take; b = utilization efficiency of residual P; c = utilization efficiency of native P; d = utilization efficiency of P in fertilizers in year of application

secondary soil reserves in the P balance of the following crop. This process reiterates within each vegetation period whereby the residual amount of fertilizer-derived P declines over time. The remaining quantity of residual P is a fraction that has aged one more year and adds to the secondary soil reserves. This implies that the total amount of P taken up by plants constitutes of fractions from previous fertilizations which contribution decrease over time.

During the year of application plants utilize only a fraction from the total amount of P applied by fertilization. The amount of fertilizer that is not taken up by plants during the growth period will remain in the soil and is potentially plant available within the next vegetation period apart from losses by soil erosion.

The potential availability of P in following years depends on immobilization processes in the soil (see above). As this is a continuous process the term $\frac{B*b}{100}$ in Eq. 1 is actually a cumulative quantity:

$$\frac{B * b}{100} = \frac{B_1 * b_1}{100} + \cdots + \frac{B_n * b_n}{100} \tag{2}$$

with:

$B, B_1, \ldots, B_n$	residual P from fertilization
1	of previous year
2	of year the last but two years
n	of n years ago
$b_1, b_2, \ldots, b_n$	corresponding utilization efficiencies of $B_1, B_2, \ldots, B_n$.

On agricultural fields with regular P fertilizer applications it is necessary to consider the total efficacy of applied P. Besides the potential supply from native resources,

$$\left(\frac{C * c}{100}\right)$$

the utilization of P by fertilizers in the year of application

$$\left(\frac{D * d}{100}\right)$$

and their potential residual effect

$$\left(\frac{B * b}{100}\right)$$

need to be determined for calculating the required P fertilizer input. Due to the accumulation of residual P from fertilizers over time the term

$$E - \left(\frac{B * b}{100} + \frac{C * c}{100} \right) = 0 \tag{3}$$

equals zero if the utilization efficiency is calculated for the year of application in dependence on P off-take and supply by residual P so that it is sufficient to replace losses by off-take, leaching, erosion and fixation (Fig. 2). To do so the utilization in the first year of application and following years needs to be considered. Fertilizer rates, which maintain this *status quo* are economically optimum and ecologically balanced.

The utilization efficiency in the year of application represents the actual utilization, while the utilization over long periods of time with its cumulative efficacy expresses the effective (Karlovsky 1962), or apparent (Kaila 1965; Finck 1979) utilization efficiency (s). Actual utilization rate and the utilization rate from previous fertilizer applications are required for the calculation of the apparent utilization efficiency:

$$\frac{D * s}{100} = \frac{D * d}{100} + \frac{B * b}{100} \tag{4}$$

whereby the apparent utilization efficiency can be determined as follows

$$s = \frac{\frac{D * d}{100} + \frac{B * b}{100}}{D} * 100 \tag{5}$$

The turnover of P supplied by fertilization can be balanced as follows

$$D = E + V + F \tag{6}$$

with:

D Fertilizer rate
E P off-take by harvest products for target yield
V P losses by leaching and erosion
F Losses by P fixation.

and after considering a potential supply from native resources the following equation can be set up

$$\frac{E + V + F - \frac{C \times c}{100}}{D} = 1 \tag{7}$$

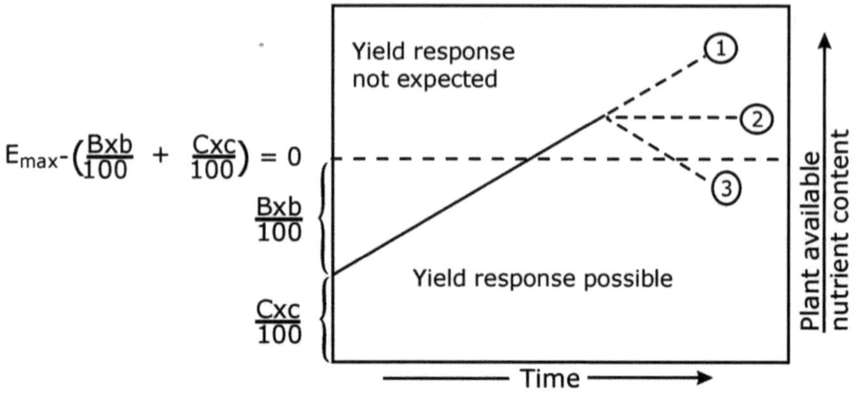

Fig. 3 Changes in soil nutrient concentration after reaching the optimum soil fertility level (based on fertilization considering the utilization efficiency of the fertilizer in the year of application (*1*), fertilization considering the effective or apparent utilization efficiency of the fertilizer (*2*), without fertilization (*3*) (Engelstad and Parks 1976, modified)

As a result of long-term P fertilization a level of residual P in the soil will adjust which is sufficiently large enough to replenish P in the soil solution required for the entire crop without adding supplementary fertilizer P after a given time (Fig. 3). At this point it would be sufficient in the future to replace only the amount of P which has been removed with the harvest products and in the form the plants have taken up this P from the soil solution which is water-soluble orthophosphate.

Assuming that P losses by leaching, erosion and fixation can be neglected the long-term cumulative utilization efficiency of P or 'apparent' utilization coefficient (s) is then simply deduced by balancing plant removal and supply by of fertilization:

$$s = \frac{E}{D} * 100 \tag{8}$$

with:

s apparent utilization coefficient for P fertilizer (%)
E P off-take by crop (kg/ha P)
D amount P fertilized (kg/ha P).

The validity of this approach demonstrates the experimentation of Engelstad and Parks (1976). On soils which are sufficiently supplied with P, P rates which equal the off-take by harvest products proved to be adequate in order to maintain the level of P supply. In contrast, P rates based on the actual P utilization rate resulted in a continuous increase of the plant available P pool (Fig. 3).

Empirical Methods to Determine the Fertilizer Nutrient Utilization Efficiency

It is one of the basic questions in fertilizer research: how much of a nutrient applied will be taken up by plants and removed by harvest products. The classic empirical approach to determine the utilization coefficient is the so-called differential method. Employing this approach the amount of nutrients taken up by plants from an unfertilized plot is subtracted from the amount of nutrients taken up from a fertilized plot, divided by the amount of fertilized nutrient:

$$(U_f - U_0)/D * 100 = d \tag{9}$$

with

U_f net nutrient off-take by harvest products in (kg/ha) from the fertilized plot
U_0 net nutrient off-take by harvest products in (kg/ha) from the unfertilized plot (control)
D fertilizer rate expressed as amount of nutrient (kg/ha)
d utilization coefficient of fertilizer-derived nutrient in the year of application in (%).

A major drawback of the differential method is that it is not feasible to distinguish between nutrients originating from the soil and fertilizer-derived nutrients. The method simply assumes that all nutrients taken up from an unfertilized plot must be soil-borne. Yet, another shortcoming of this approach can be attributed to so-called priming effect, which describes increasing plant vigor by fertilization associated with a higher nutrient uptake from soil resources if the nutrient has been a minimum factor (Lehne 1968; Rauhe and Bornhak 1968; Jacob et al. 1949). Here, the differential method tends to overestimate the utilization efficiency of a fertilized nutrient (Nethsinghe 1977; Michael and Machold 1957; Behrens 1954). Vice versa the method tends to underestimate the utilization efficiency if the initial soil supply is close to optimum and additional fertilization will only yield small increases in nutrient uptake (Nethsinghe 1977; Michael and Machold 1957; Behrens 1954).

The most important fault of the differential method is its inability to estimate the valid utilization over an infinite period of time. By means of the differential method it is impossible to verify whether a fertilized nutrient can be utilized completely or not. The reason is that although the amount might be marginal, plant biomass, even when little is produced on a control plot so that the denominator in Eq. 8 will always be higher than the enumerator. At the end of the day from a basically faulty experimental design the myth of occluded or fixed soil P accrued which justified among other reasons strong over-fertilization with P during the second half of the 20th century.

The experimental attempt to assess the utilization efficiency of residual fertilizer P by means of exhaustion trials was not successful as it was not possible to assess the potential amount of P that can be utilized by plants. In such experiments crops are grown consecutively on a soil subjected afore to long-term fertilization, however, without additional supplements of fresh fertilizers (Werner 1971b; Werner and Wiechmann 1972). The reason for the ineptness of the method is that the viability and thus the ability of plants to take up nutrients is negatively affected when the crop nutritional P level declines to the deficiency range. The common fallacy made on the basis of these experiments was to interpret the physiological disability of the plants to take up nutrients as a nutrient fixation in the soil. Consistent with the old view on an always incomplete utilization of fertilizer P Werner and Wiechmann (1972) reported an average, effective utilization efficiency of 48 % for residual fertilizer P in soils. They evaluated this low efficiency as an indication of P fixation in soils (Platzen and Munk 1975; Wiechmann and Werner 1976). Finally, a total P utilization efficiency of only 60 % has been calculated by adding the previously made estimations for the utilization efficiency of residual P and the reported initial utilization efficiency of fertilizer P in the application year of 15 % (Arbeitskreis Phosphat 1978).

Quite from the beginning of research in fertilizer nutrient utilization isotopes of the nutritive elements in question have been used for labeling and tracking (Hevesy 1923; Larsen 1952). Isotopes are any of two or more forms of a chemical element, having the same number of protons in the nucleus, or the same atomic number, but having different numbers of neutrons in the nucleus, or different atomic weights. Because all isotopes of an element have the same atomic shell they show all the same chemical characteristics. Fertilizer nutrients with altered isotope composition have the same physiological features than those with natural isotope composition. Depending on availability, modifying the isotope composition can be practiced either with stable or radioactive isotopes of the particular element. In the first case, the altered isotope composition is detected by an altered atomic weight of the element or by the radiation signature of the instable isotope.

The utilization coefficient is then calculated according to:

$$\left(\frac{A * \frac{B-C}{D}}{E}\right) * 100 = \text{Utilization rate } (\%) \text{ of fertilizer-derived nutrient} \qquad (10)$$

A total uptake of element by plants
B isotopic abundance (or specific activity in case of radioactive isotopes) in plants from fertilized plots
C isotopic abundance (or specific activity in case of radioactive isotopes) in plants from unfertilized plots
D isotopic abundance (or specific activity in case of radioactive isotopes) in fertilizer
E amount of fertilizer.

Stable isotopes of macro-nutrients are available only for nitrogen (^{15}N), sulfur (^{33}S), potassium (^{41}K), calcium (^{43}Ca) and magnesium (^{25}Mg). Expensive ^{15}N (which costs approximately 100 times more than the price for the element in natural isotope configuration) still is the most affordable of all stable isotopes as all other nutrients yield prices that are several thousand times higher. Radioactive isotopes are available for studies on phosphorus (^{32}P), sulfur (^{35}S), potassium (^{42}K) and calcium (^{45}Ca). Radioactive isotopes are distinctly cheaper than stable ones, but in case of P and S only applicable for short-term studies because of their short half-life time (^{32}P: 14.3 days; ^{35}S: 87.5 days).

There are only a few trials, which are able to proof the hypothesis that water-soluble fertilizer P can be utilized completely over infinite times. Shortcomings of traditional experimental designs such as the differential method and exhaustion trials have been outlined before.

Unlike C and N, which can be added to the soil systems from the atmosphere, the P status of natural systems is essentially controlled by the occurrence of primary apatite minerals (Walker and Syers 1976; Bowman et al. 1998). While this is true for most tropical soils, in temperate soils the importance of primary minerals is negligible and subsequently the aging of applied fertilizer essentially controls the P cycle. Thus, in temperate agro-ecosystems the utilization efficiency of fertilizers can be expressed as a percentage of the plant uptake (Koehnlein and Knauer 1965). Karlovsky (1962) introduced the so-called balance method, whereby under the condition of constant yield and fertilizer application rates the effective utilization efficiency can be calculated by balancing fertilizer input and plant uptake.

The non-equilibrium behavior of phosphate sorption reactions, known as hysteresis, whereby the desorption reaction usually does not take place as fast as the sorption reaction (Uehara and Gillman 1981), led to the prevailing opinion that P immobilization is higher than the dissolution of adsorbed fertilizer P and thus justifies a fertilizer rate higher than plant uptake (Arbeitskreis Phosphat 1978; Johnston 1976). However, in a seven years field trial Schachtschabel (1976) demonstrated that an application rate of water-soluble P fertilizer equivalent to plant uptake did not change the soil P level of the soils. As the natural supply of P by the soil did not have an effect on the results he concluded that a fertilization rate higher than the uptake was not economically sound. An application rate higher than the uptake will result in an increase of the soil P level which equals an accumulation of P, whereby a rate lower than the uptake would result in a decrease of the soil P level, thus in mining of P (Fig. 3). Bergmann and Witter (1965) obtained similar results (Fig. 4). Both field trials clearly demonstrate that the dissolution of initial fertilizer reaction products during the cropping season was sufficient to keep the concentration of P in the soil solution at the level of the plant requirement. Thus in these trials the apparent utilization efficiency of fertilizer P was 100 %.

These results not only demonstrate that plants are capable of utilizing aging P fertilizer products despite their decreasing solubility, but also that immobilization and mobilization processes are in an equilibrium. In the light of the above discourse the intrinsic hypothesis is derived that in soils where P fertilization does not result in a surplus in crop yield (Fig. 2) as the P supply is already optimum, the fertilizer rate should be equivalent to the plant uptake, as a utilization efficiency of 100 % can be assumed. For the validation of this hypothesis it is important to note that the assumption of 100 % utilization efficiency together with a balanced fertilizer

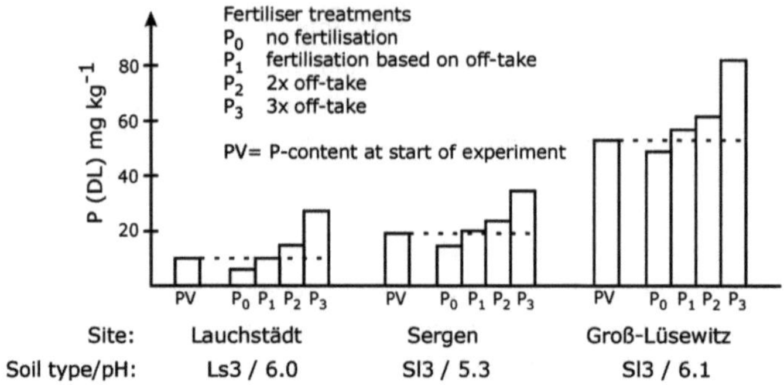

Fig. 4 Changes in the concentration of double-lactate-soluble P in the surface soil during a 9-year field trial with increasing P fertilization using super-phosphate (extracted from Bergmann and Witter 1965)

strategy is only feasible for P species, which will undergo total dissolution within one cropping season. For the majority of agricultural soils this effective percentage of P fertilizers corresponds to their water-soluble P concentration on soils with a pH between 5.6 and 7.5, and the citrate-soluble P concentration on soils with a pH < 5.8, respectively. For organic fertilizers it is fair to assume that in the long run all P undergoes mineralization and will be converted into orthophosphate. Thus, the effective percentage of organic fertilizers corresponds with the total P concentration.

Following the concept of balanced fertilization as it is proposed by environmentalists (Sharpley and Withers 1994; Higgs et al. 2000), in particular by a variety of conventions for the protection of marine bodies (Vermeulen et al. 1998; Schnug et al. 2001), it is sufficient for agricultural production to replenish the soil P withdrawal by crops by the equivalent amount of water-soluble P.

Summarizing the significance of the information given in this section about the premises for a complete P utilization by plants, Table 1 shows the results of a meta data analysis of references on experiments determining P utilization.

According to the data presented in Table 1 utilization rates for fertilizer P in the year of application are higher when based on pot trials and employing the differential method. Empirical assessments are based consistently on differential trials

Table 1 Utilization rates (%) of fertilizer P determined by different methods (results from a meta data analysis provided by Schnug (1978))

Time scale	Year of application				Long-term differential	
Method	Differential	Differential	Isotope	Isotope		
	Field	Pot	Field	Pot	Field	Pot
Mean	17.5	19.9	7.9	12.7	66	70
Median	11.8	14.0	8.0	11.8	70	75
Minimum	8.0	11.3	3.5	5.0	32	36
Maximum	37.0	45.5	10.5	21.5	83	89

because of the lack of stable isotopes other than the natural isotope ^{31}P, or alternatively radioactive isotopes with a long half-life time. The experiments indicate constantly that about a quarter of the fertilized P has not been used by the plants.

It is widely accepted that more than 80 % of the applied P is immobilized in the soil due to precipitation and sorption processes, whereby the limiting step to furnish crop requirements is the dissolution of initial reaction products during the cropping season (Frossard et al. 2000). However, it has been demonstrated conclusively in this section that there is virtually no experimental set-up available to assess the amount of P finally used by crops.

In this context it appears more than doubtful to draw conclusions on the effectiveness of a P fertilizer from any growth experiment. There was a lot of speculation in literature on stocking up soil P reserves with non-water soluble P forms. The greatest efforts in this context were done by the fertilizer industry marketing partially acidulated rock phosphates in order to save costs for sulfuric acid in P fertilizer production (Hagin and Katz 1984; Zapata and Roy 2004). Recently, the effectiveness of new P fertilizer products derived from recycling processes is compared by means of growth experiments in pot and field for one vegetation period and year, respectively (Roemer 2003, 2013; Cabeza et al. 2011; Siebers and Leinweber 2013; Severin et al. 2013). However, differences in efficiency in such trials reflect only the initial speed of dissolution of the product and are no measure for the long-term utilization rate.

Conclusions

The compilation of data since the 1920s elucidated some long-known, however, generally ignored facts about the utilization of fertilizer-derived P. Firstly, water-soluble and citrate-soluble inorganic P fully contribute on a long-term basis to the plant available P pool on agricultural soils in temperate climates, while less soluble P forms need to be discarded as their contribution to the P supply of crops is strictly limited. This aspect is of prime relevance for mineral P fertilizers and recycled, secondary P fertilizers. Secondly, the determination of the actual P utilization conceals the long-term effectiveness of fertilization and fertilizer product, and is no suitable indictor for assessing the optimum fertilizer rate. Thirdly, for a truly balanced P fertilization where input equals output, it is important to determine the apparent P utilization, which is not feasible with common experimental set-ups. However, limited proper data show that regular P fertilization yields a level of residual P in the soil, which is sufficiently high to replenish P in the soil solution which is required for optimum crop growth (Fig. 4). At this point it is sufficient to replace only the amount of P, which has been removed with the harvest products in water-soluble orthophosphate form.

References

Adam C, Peplinski B, Michaelis M, Kley G, Simon FG (2009) Thermochemical treatment of sewage sludge ashes for phosphorus recovery. Waste Manage 29:1122–1128

Al-Kaisi M (2015) Soil erosion: an agricultural production challenge. http://www.ipm.iastate.edu/ipm/icm/2000/7-24-2000/erosion.html; http://en.wikipedia.org/wiki/Erosion_control

Anderson G (1980) Assessing organic phosphorus in soils. In: Khasawneh FE, Sample EC, Kamprath EJ (eds) Proceedings symposium the role of phosphorus in agriculture, 1–3 June 1976, National Fertilizer Development Center, Tennessee Valley Authority, Muscle Shoals, Alabama, pp 411–431

Anderson G et al (2001) In: Pfeffer E, Hristov AN (eds) Nitrogen and phosphorus nutrition of cattle. CABI Publishing

Anonymus (2014) Deutschlands Zukunft gestalten—Koalitionsvertrag zwischen CDU, CSU und SPD 18. Legislaturperiode http://www.cdu.de/sites/default/files/media/dokumente/koalitionsvertrag.pdf; http://www.bundesregierung.de/Content/DE/_Anlagen/2013/2013-12-17-koalitionsvertrag.pdf;jsessionid=33D94E2F55707D1A79CB57C0A900149C.s3t1?__blob=publicationFile&v=2

Arbeitskreis Phosphat der deutschen Düngerindustrie (1978) P_2O_5; Phosphatduengung—unentbehrlich für Boden und Pflanze. Information fuer die Beratung

Arrhenius O (1931) Soil analysis in the service of archaeology. Zeitschrift Pflanzenernaehrung, Duengung, Bodenkunde 10b:427–439

Asmus F, Hermann V, Lange M, Specht G (1973) Wirkung und Ausnutzung des Stickstoffs aus Gülle. Archiv für Acker-und Pflanzenbau 17:927–934

Baeumer K (1971) Allgemeiner Pflanzenbau. UTB 18, Ulmer Verlag, Stuttgart, pp 99–137

Bai Z, Li H, Yang X, Zhou B, Shi X, Wang B, Li D, Shen J, Chen Q, Qin W, Oenema O, Zhang F (2013) The critical soil P levels for crop yield, soil fertility and environmental safety in different soil types. Plant Soil 372:27–37

Barber SA (1980) Soil-plant interactions in the phosphorous nutrition of plants. In: Khasawneh FE, Sample EC, Kamprath, EJ (eds) The role of phosphorus in agriculture. ASA CSSA and SSSA, Madison, pp 591–613. ISBN 89118 062 1

Behrens WU (1954) Ueber die Verwendung von radioaktivem Phosphor zum Studium von Problemen der Pflanzenernaehrung und Duengung. Landwirtschaftliche Forschung 6:1–5

Bergmann W, Witter B (1965) Die Wirkung der Phosphorsaeure in statischen P-Steigerungsversuchen und der Verbleib der Restphosphorsaeure im Boden. Albrecht-Thaer-Archiv 9:901–922

Bloem E, Kratz S (2016) Organic xenobiotics. This volume

BMVEL (2002) Gute fachliche Praxis zur Vorsorge gegen Bodenschadverdichtungen und Bodenerosion. Bundesministerium für Verbraucherschutz, Ernaehrung und Landwirtschaft (BMVEL) Referat 516, Postfach, 53107 Bonn http://www.umwelt.sachsen.de/umwelt/download/boden/Broschuere_GfP_Bodenschutz.pdf

Bowman RA, Rodriguez JB, Self JR (1998) Comparison of methods to estimate occluded and resistant soil phosphorus. Soil Sci Soc Am J 62:338–342

Boysen P (1977) Nährstoffauswaschung aus gedüngten und ungeduengten Boeden in Abhaengigkeit von Standorteigenschaften und Nutzung der Moraenen—und Sandergebiete Schleswig-Holstein. PhD thesis, University of Kiel, pp 176

Broadbent N (1981) Phosphate analysis in archeology. Anthropological uses of an old method. Soc Am Archeol Proc 42:1–16

Cabeza R, Steingrob B, Romer W, Claassen N (2011) Effectiveness of recycled P products as P fertilizers, as evaluated in pot experiments. Nutrient Cycling Agroecosyst 91:173–184

Catt JA, Johnston AE, Quinton JN (1997) Phosphate losses in the Woburn erosion reference experiment. In: Tunney H, Carton OT, Brookes PC, Johnston AE (eds) Phosphorus loss from soil to water. CAB International, pp 374–377

Cordell D, Drangert JF, White S (2009) The story of phosphorus: Global food security and food for thought. Global Environ Change (Elsevier) 19:292–305
Debruk J, Voemel A (1970) Die Verwertung der Naehrstoffe in Abwasserschlamm. Landwirtschaftliche Forschung 27/I SH 39-60
De Kok LJ, Schnug E (2008) Loads and fate of fertilizer-derived uranium. Backhuys Publishers, Leiden, p 229
Diekmann K (1966) Unser Acker, 9th edn. Parey Verlag, Hamburg
Djodjic F, Bergström L, Grant C (2005) Phosphorus management in balanced agricultural systems. Soil Use Manage 21:94–101
Doering H (1956) Untersuchungen über die Ausnutzbarkeit der an Kolloiden sorbierten Phosphorsäure durch die Pflanze. Zeitschrift Pflanzenernaehrung, Duengung, Bodenkunde 73:235–245
Engelstad OP, Parks WL (1976) Buildup of P and K in soils and effective use of these reserves. In: TVA Fertilizer Conference 27–28 July 1976, Cincinatti, Ohio. TVA Bull 016:50–58
e-RA (2014) John Bennet Lawes 200 years. http://www.rothamsted.ac.uk/sites/default/files/Sir%20John%20Bennet%20Lawes.pdf
Finck A (1976) Pflanzenernaehrung in Stichworten. Hirt Verlag, Kiel
Finck A (1979) Fertilizers and fertilization. Verlag Chemie, Weinheim
Frede HG, Bach M (2003) Heutiger und kuenftiger Beitrag der Guelle bei der Phosphorversorgung der Böden. In: Procceedings of symposium on Rueckgewinnung von Phosphor in der Landwirtschaft und aus Abwasser und Abfall, Berlin, 6–7 February 2003, Umweltbundesamt, Institut für Siedlungswasserwirtschaft der RWTH Aachen, ISBN 3-932590-99-6
Frossard E, Condron LM, Oberson A, Sinaj S, Fardeau JC (2000) Processes governing phosphorus availability in temperate soils. J Environ Qual 29:15–23
Fuchs JW, Fox GA, Storm DE, Penn CJ, Brown GO (2009) Subsurface transport of phosphorus in riparian floodplains: influence of preferential flow paths. J Environ Qual 38:473–484
Hagin J, Katz S (1984) Effectiveness of partially acidulated phosphate rock as a source to plants in calcareous soils. Fertilizer Res 8:117–127
Haneklaus SH, Schnug E (2016a). Assessing the plant phosphorus status. This volume
Haneklaus N, Schnug E (2016b). Energy neutral phosphate fertilizer production using high temperature reactors. This volume
Harrison AF (1987) Soil organic phosphorus: a review of world literature. CAB International, Wallingford, UK
Hevesy G (1923) The absorption and translocation of lead by plants. A contribution to the application of the method of radioactive indicators in the investigation of the change of substance in plants. Biochem J 17:439–445
Higgs B, Johnston AE, Salter JL, Dawson CJ (2000) Some aspects of achieving sustainable phosphorous use in agriculture. J Environ Qual 29:80–87
Hoffmann W (1978) Phosphor-und Stickstoffzufuhr aus der Landwirtschaft in die Ostsee, insbesondere durch die Schwebstoffe der Gewässer. Umweltforschungsplan des Bundesministers des Inneren—Wasser—Forschungsbericht 120 04 015 Kiel
Ibrikei H, Hanlon EA, Rechigl JE (1999) Inorganic phosphorous and manure effects on bahiagrass on a Spodosol. Nutrient Cycling Agroecosyst 54:259–266
Isermann K (1978) Die Phosphat und Kali Bilanz von viehstarken Gruenland-und Ackerbaubetrieben. Limburgerhof 1978
Ivanov P, Sauerbeck D (1971) Festlegung, Umwandlung und Aufnehmbarkeit von Phytin-Phosphor im Boden. Zeitschrift Pflanzenernaehrung, Duengung, Bodenkunde 129:113–123
Ivanov P, Sauerbeck D (1972) Die Pflanzenverfügbarkeit des Phosphors aus verschiedenen organischen Substanzen. Landwirtschaftliche Forschung 25:216–225
Jacob CW, van Middelem CH, Nelson CD, Welch CD, Hall NS (1949) Utilisation of phosphorus by potatoes. Soil Sci 68:113–120

Johnston AE (1976) Additions and removals of nitrogen and phosphorous in long-term experiments at Rothamsted and Woburn and the effect of the residues on total soil nitrogen and phosphorous. Techn Bull Min Agric Food UK 32:101–119
Kaila A (1949) Biological absorption of phosphorous. Soil Sci 68:279–289
Kaila A (1965) Apparent recovery of fertilizer nitrogen. J Sci Agric Soc 37:163–174
Karlovsky J (1962) Method of assessing the utilization of phosphorous on permanent pastures. In: Internatiuonal Soil Conference on New Zealand Rukuhia Soil Research Station, Hamilton, New Zealand, pp 726–730
Koehnlein J, Knauer N (1965) Ergebnisse der Kieler Dauerdüngungsversuche mit Phosphat und Kali. Schriftenreihe der Landwirtschaftlichen Fakultät der Christian-Albrecht-Universitaet Kiel: Heft 39
Larsen S (1952) The use of ^{32}P in studies on the uptake of phosphorus by plants. Plant and Soil 4:1–10
Latkowiczs I (1977) Wirkung von organischen und Mineraldüngern in Mais-Monokulturen. Agrochimica 11:67–73
Lehne J (1968) Zur Quantifizierung der Gesamtwirkung von Düngermaßnahmen (Pflanze und Boden) auf einem Standort mit Hilfe der Differenzmethode. Albrecht-Thaer-Archiv 12:717–731
Lookman R, Freese D, Merckx R, Vlassak K, van Riemsdijk H (1995) Long-term kinetics of phosphate release from soil. Environ Sci Technol 29:1569–1575
Michael G, Machold O (1957) Untersuchungen ueber die Verwendung von 32P zur Ermittlung der Ausnutzung von Phosphorsaeureduengemitteln. Zeitschrift Pflanzenernaehrung, Duengung, Bodenkunde 77:1–18
Mosse B, Powell CL, Haymann D (1976) Plant growth responses to vesicular arbuscular mycorrhiza. IX. Interactions between V.A. mycorrhiza, rock phosphate and symbiotic nitrogen fixation. New Phytologist 76:331–342
Munk H, Gross K (1975) Phosphatfraktionierung an langjaehrig differenziert geduengten Boeden. Landwirtschaftliche Forschung 31/I:96–108
Nagarajah S, Posner AM, Quirk JP (1970) Competitive adsorption of phosphate with polygalacturonate and other organic anions on kaolinite and oxide surfaces. Nature 228:83–90
Nethsinghe DA (1977) The use of isotopes and radiation in studies on the efficient use of fertilizers. In: Improved use of plant nutrients; report of the expert consultation on better exploitation of plant nutrients, FAO Soils Bull 37:125–134
Noack SR, McLaughlin MJ, Smernik RJ, McBeath TM, Armstrong RD (2013) Crop residue phosphorus: speciation and potential bio-availability. Plant Soil 359:375–385
Oburger E, Jones DL, Wenzel WW (2013) Phosphorus saturation and pH differentially regulate the efficiency of organic acid anion-mediated P solubilization mechanisms in soils. Plant Soil 341:363–382
Perspectives Agriculture (2011) Sustainable agriculture in the Baltic Sea region in times of peak phosphorus and global change. Baltic 21 series No. 4/2011
Platzen H, Munk H (1975) Zur Frage nach der optimalen P-Zufuhr auf Boeden mit ausreichendem Naehrstoffpotential. Landwirtschaftliche Forschung 31/I:258–269
Proudfoot N (1976) The analysis and interpretation of soil phosphorus in archeological contexts. In: Davidson DA, Shackley ML (eds) Geoarchaeology. Duckworth, London, pp 93–113
Rauhe K, Bornhak H (1968) Quantifizierung der Gesamtwirkung von Mineralstickstoff auf Pflanze und Boden mit Hilfe von ^{15}N im Feldversuch. Albrecht-Thaer-Archiv 12:701–716
Roemer W (2003) Pflanzenverfuegbarkeit von Phosphor bei verschiedenen Duengematerialien—insbesondere bei Klärschlamm. In: Proceedings of symposium on Rueckgewinnung von Phosphor in der Landwirtschaft und aus Abwasser und Abfall. Berlin, 06–07 Feb 2003, Umweltbundesamt, Institut für Siedlungswasserwirtschaft der RWTH Aachen, ISBN 3-932590-99-6
Roemer W (2013) Phosphor fertilization effects of phosphorus recycling products. Berichte ueber Landwirtschaft 91(1):202–215

Rovira AD (1965) Ecology of soil-borne pathogens. In: Baker KF, Smyder WC (eds) University of California Press Berkeley
Sagan C (1986) The world that came in the mail. Parade Magazine
Sample EC, Soper RJ, Racz GJ (1980) Reaction of phosphate fertilizers in soils. In: Khasawneh FE, Sample EC, Kamprath EJ (eds) The role of phosphorus in agriculture. ASA-CSSA-SSSA, Madison, pp 263–304
Sanderson MA, Jones RM (1997) Forage yields, nutrients uptake, soil chemical changes and nitrogen volatilization from Bermuda grass treated with dairy manure. J Prod Agric 10:266–271
Schachtschabel P (1960) Umwandlung der Duengerphosphate im Boden und Verfuegbarkeit des Bodenphosphors. Landwirtschaftliche Forschung (special issue) 17:30–36
Schachtschabel P (1976) In: Scheffer-Schachtschnabel: Lehrbuch der Bodenkunde, Kap. XI, Nährstoffe. Enke Verlag, Stuttgart, pp 196–264
Scharafat I, Finck A (1973) Umwandlung und Verfügbarkeit von Rophosphaten (Hyperphos) verglichen mit aufgeschlossenen Phosphaten unter kontrollierten Temperaturen-und Feuchtebedingungen. Zeitschrift Pflanzenernaehrung, Duengung, Bodenkunde 134:257–267
Schnug E (1978) Die Ausnutzung von Duengernährstoffen und die Bedeutung des Ausnutzungsgrades für die Hoehe der Duengung. Diplomarbeit, Kiel. http://books.google.de/books/about/Die_Ausnutzung_von_D%C3%BCngern%C3%A4hrstoffen_u.html?id=DbC0PgAACAAJ&redir_esc=y
Schnug E, Haneklaus N (2015) Uranium in phosphate fertilizers—review and outlook. In: Merkel BJ, Arab A (eds) Uranium—past and future challenges, pp 123–130
Schnug E, Sparovek B, Storck W, Jering A, Volkgenannt U (2001) The HELCOM working group on agriculture (WGA). Landbauforschung-Voelkenrode 51:5–9
Schroetter S, Rogasik J, Schnug E (2016) Light particle erosion—a significant pathway for nutrient losses? J Cultivated Plants (in press)
Schwarz C, Weingaertner M (2013) Die Phosphorkrise. http://www.arte.tv/guide/de/046557-000/die-phosphor-krise; https://www.youtube.com/watch?v=q7BSLOoUU9o
Schwertmann U, Taylor RM (1989) Iron oxides. In: Dixon JB, Weed SB (eds) Minerals in soil environments. SSSA, Madison, pp 379–427
Severin M, Ahl C, Kucke M, van den Weghe H, Greef JM (2013) Phosphate solubility and phosphate fertilizer effect of substances from sludge treatment—analysis of different treatment methods. Landbauforschung-Voelkenrode 63:235–243
Sharma HC, Singh RM (1971) Effect of phosphobacteria culture on the efficiency of phosphatic fertilizers and on the yield of maize. Indian J Agron 16:422–424
Sharpley A, Withers P (1994) The environmentally-sound management of agricultural phosphorus. Fertilizer Res 39:133–146
Sharpley A, Foy B, Withers P (2000) Practical and innovative measures for the control of agricultural phosphorus losses to water. An overview. J Environ Qual 29:1–9
Siebers N, Leinweber P (2013) Bone char: a clean and renewable phosphorus fertilizer with cadmium immobilization capability. J Environ Qual 42:405–411
Svanbaeck A, Barbro U, Ararso E (2014) Mitigation of phosphorus leaching losses via subsurface drains from a cracking marine clay soil. Agric Ecosyst Environ 184:124–134
Syers JK, Johnston AE, Curtin D (2008) Efficiency of soil and fertilizer phosphorus use: reconciling changing concepts of soil phosphorus behviour with agronomic information. FAO Fertilizer Plant Nutrition Bull 18:xi + 108
Taub FB (1974) Closed ecological systems. Annu Rev Ecol Syst 5:139–160
Taylor M, Kim N, Smidt G, Busby C, McNally S, Robinson B, Kratz S, Schnug E (2016) Trace element contaminants and radioactivity from phosphate fertilizers – development of a risk assessment model. This volume
Turner BL, Lambers H, Condron LM, Cramer MD, Leake JR, Richardson AE, Smith SE (2013) Soil microbial biomass and the fate of phosphorus during long-term ecosystem development. Plant Soil 367:225–234

UBA (2015) Erosion. http://www.umweltbundesamt.de/themen/boden-landwirtschaft/bodenbelastungen/erosion. Webauftritt 14.05.2015
Uehara G, Gillman G (1981) The mineralogy chemistry and physics of tropical soils with variable charge clays. Westview Press, Boulder
Ulrich AE, Frossard E (2014) On the history of a reoccurring concept: phosphorus scarcity. Sci Total Environ 490:694–707
Vermeulen S, Steen I, Schnug E (1998) Nutrient balances at the farm level. In: Proceedings of 11th international symposium of CIEC Codes of good fertilizer practice and balanced fertilization, pp 108–123
Vitousek PM, Porder S, Houlton BZ, Chadwick OA (2010) Terrestrial phosphorus limitation: mechanisms, implications, and nitrogen-phosphorus interactions. Ecol Appl 20:5–15
Walker TW, Syers JK (1976) The fate of phosphorus during pedogenesis. Geoderma 15:1–19
Werner W (1969) Kennzeichnung des pflanzenverfügbaren Phosphats nach mehrjähriger Düngung mit verschiedenen Phosphaten. Zeitschrift Pflanzenernaehrung, Duengung, Bodenkunde 122:19–32
Werner W (1970) Untersuchungen zur Pflanzenverfügbarkeit durch langjaehrige Phosphatduengung angereicherten Bodenphosphats. 1. Mitt.: Die Verfuegbarkeit der Umwandlungsprodukte in sauren Boeden. Zeitschrift Pflanzenernaehrung, Duengung, Bodenkunde 126:135–150
Werner W (1971a) Phosphatdynamik in carbonathaltigen Böden. Der Deutsche Weinbau: 30–31
Werner W (1971b) Untersuchungen zur Pflanzenverfügbarkeit durch langjaehrige Phosphatduengung angereicherten Bodenphosphats. 2. Mitt.: Die Verfuegbarkeit von Umwandlungsprodukten mit defekter Apatitstruktur in neutralen und carbonathaltigen Boeden. Zeitschrift Pflanzenernaehrung, Duengung, Bodenkunde 129:30–42
Werner W, Wiechmann H (1972) Untersuchungen zur Pflanzenverfuegbarkeit des durch langjaehrige Phosphatduengung angereicherten Bodenphosphats. 3. Mitt.: Die Verfuegbarkeit der Umwandlungsprodukte in neutralen und kalkhaltigen Boeden. Zeitschrift Pflanzenernaehrung, Duengung, Bodenkunde 133:3–17
Whalen JK, Chang C (2001) Phosphorous accumulation in cultivated soils from long-term annual applications of cattle feedlot manure. J Environ Qual 30:229–237
Wiechmann H, Werner W (1976) Umwandlung von Phosphaten im Boden. DLG-Mitteilungen 91:1147–1150
Wikipedia 2014. http://de.wikipedia.org/wiki/Ecosphere_%28Aquarium%29
Zapata F, Roy NR (2004) Use of phosphate rocks for sustainable agriculture. FAO Agric Bull, vol 13, Rome

Fate of Fertilizer P in Soils: Inorganic Pathway

Markku Yli-Halla

Abstract Soils never fertilized with P have usually a low P concentration (<0.02 mg l^{-1}) in soil solution, which is below the crop needs. In acid soils, added P is adsorbed on the surfaces of iron and aluminum (hydr)oxides or, in calcareous soils, predominantly precipitated as calcium phosphates. Abundant organic and inorganic P fertilization in the industrial world has elevated the P status of agricultural soils and the large P surplus remains at least partly available to plants. In acidic soils *Degree of P Saturation* (*DPS*) can be used to indicate to which extent the potential sites for P sorption are already occupied. Repeated P applications increase the DPS and decreases P retention, and in calcareous soils, relatively easily soluble calcium phosphates are accumulated. Abundant residual fertilizer P results in a higher P concentration in soil solution. P supplying power in such soils is sufficient to meet the crop needs for several years, even decades, with P additions equaling to or smaller than the crop P uptake and apparently 100 % utilization of fresh fertilizer P can be achieved.

Keywords Soil solution · Inorganic phosphorus · Fertilizer phosphorus · Degree of phosphorus saturation

Inorganic Phosphorus in Soil Solution

The original source of P in soil is the rock from which the soil has been formed. Igneous and metamorphic rocks contain small amounts of P mainly in the form of different apatites, which has a general formula of $(Ca, Sr, Pb)_5[(P, As, V, Si)O_4]_3$ (F, Cl, OH), fluorapatite, $Ca_5(PO_4)_3F$, being the most common. In these minerals, P is originally unavailable to plants or to other organisms. The minerals are, however, gradually weathered, meaning that they break down in the soil environment.

M. Yli-Halla (✉)
Department of Food and Environmental Sciences, University of Helsinki,
P.O. Box 27, 00014 Helsinki, Finland
e-mail: markku.yli-halla@helsinki.fi

E. Schnug and L.J. De Kok (eds.), *Phosphorus in Agriculture: 100 % Zero*,
DOI 10.1007/978-94-017-7612-7_3

Consequently, P is released into the surrounding solution in the form of phosphate anion ($H_2PO_4^-$, HPO_4^{2-}, PO_4^{3-}), the charge being dependent on the solution pH.

Plants take up P from soil solution in the form of $H_2PO_4^-$ and HPO_4^{2-} at different efficiencies. Plants adapted to low P concentrations, such as sweet potato and even some maize varieties, can take up some P from concentrations as low as 0.003 mg l^{-1} (Fox et al. 1974; Bhadoria et al. 2009), but according to an early solution culture investigation (Olsen 1950), the phosphate concentration had to be 0.009–0.03 mg l^{-1} before most cultivated plants can take up appreciable amounts of P. Together with shortage of N, this low P concentration is commonly the limiting factor for growth in native ecosystems and in agricultural soils that have received little P fertilization.

The P concentration needed for abundant growth varies from plant species and cultivar to another and also at different growth stages (Fox et al. 1974; Cassman et al. 1981; Nielsen and Schjørring 1983). When solution P concentration is elevated from the deficiency level, P uptake is increased. A threshold value for rye has been established at about 0.3 mg l^{-1} (Adams and Odom 1985), which is also typical for many other crops. In the classical paper by Fox et al. (1974), it was reported that 95 % of crop yield in volcanic soils of Hawaii was produced at the soil solution P concentration of 0.4 mg l^{-1} for lettuce and 0.1 mg l^{-1} for sweet potato. Recently, adequate soil solution phosphate concentrations of 0.23–0.86 mg l^{-1} were reported for barley (Matula 2011). Elevation of solution P concentration beyond the threshold value does not contribute to increased yield.

Concentrations ranging from 0.1 to 0.3 mg l^{-1} are commonly reported in agricultural soils of Europe (Table 1), but both lower and higher concentrations are often measured, as well. In this respect the study of Delgado and Torrent (1997) is particularly interesting, since the 12 soils investigated were called "phosphate-rich soils". Some of them exhibited soil solution concentrations less than 0.1 mg l^{-1} while in three soils values above 2 mg l^{-1} were measured. Thus, solution P concentrations in European agricultural soils seem to range from deficient to excessive in regard to the crop need.

Table 1 Soil solution P concentrations in various studies

Crop	P in soil solution (mg l^{-1})	Reference
Silage grass, barley, 7 fields after harvest; basaltic soils, UK	0.13–0.28	Benians et al. (1977)
Several soils, UK	0.06–0.16	Webber and Mattingly (1970)
Sugar beet, 3 unfertilized fields in spring, Sweden	0.14–0.15	Wiklander and Andersson (1974)
Various crops, 12 soils, Germany, Spain, Scotland, Italy	0.04–3.7 median 0.12	Delgado and Torrent (1997)

The objective of P fertilization is to increase the P concentration of soil solution to a level sufficient to crop so that this limitation for growth is removed. However, a simple calculation presented below shows that the sufficiency of P supply to a crop cannot be estimated solely on the basis of soil solution P concentration. A rough estimate about the content of P in soil solution in a 22-cm deep plough layer can be made by assuming 50 % pore space in soil and 40 % of that space being filled with water (i.e., soil solution occupying 20 % of soil volume). This amount of soil solution, having a phosphate-P concentration of 0.1 mg l^{-1}, would contain 44 g P ha^{-1}. Even at the rather high solution P concentration of 1 mg l^{-1}, soil solution P content would be mere 440 g P ha^{-1}. These are small amounts, compared to the typical P uptake of 20–30 kg ha^{-1} by a 8000 kg grain yield of wheat. In order to provide the crop with sufficient P, the amount of P dissolved in soil solution has to be continuously replenished from the solid phase of the soil. Phosphorus supply to the crop cannot be understood without knowing the reactions taking place between P in the solid phase and that in the soil solution, which is the immediate source of P to plants.

Reactions Between P and the Inorganic Soil Constituents

Phosphate ions released from the lattices of weathering minerals do not, for the most part, remain in the solution. In acidic soils, most P is readily adsorbed on the surfaces of iron (Fe) and aluminum (Al) (hydr)oxides, which are common weathering products in soil. Like from the inorganic minerals, P contained in the organisms living in the soil environment is released into the soil solution after they have completed their life cycle. Also this phosphate reacts with the solid phases in a similar manner. Phosphate adsorption takes place by ligand exchange reactions (Fig. 1), oxygen bridging between Fe and Al and P. Phosphate can be adsorbed by one bond (monodentate binding), like in Fig. 1, or it can also be adsorbed more strongly with bidentate binding, where two oxygen bridges are formed with the same Fe/Al atom or binuclear binding, with oxygen bridges formed between P and two different Fe/Al atoms.

Calcareous soils also contain oxide surfaces that serve as agents for P sorption (Holford and Mattingly 1975; Ryan et al. 1985; Hammad et al. 1992). However, in calcareous soils P is predominantly adsorbed on calcite surfaces or precipitated as calcium phosphates (Havlin et al. 2005). Dicalciumphosphate ($CaHPO_4 \cdot H_2O$) is

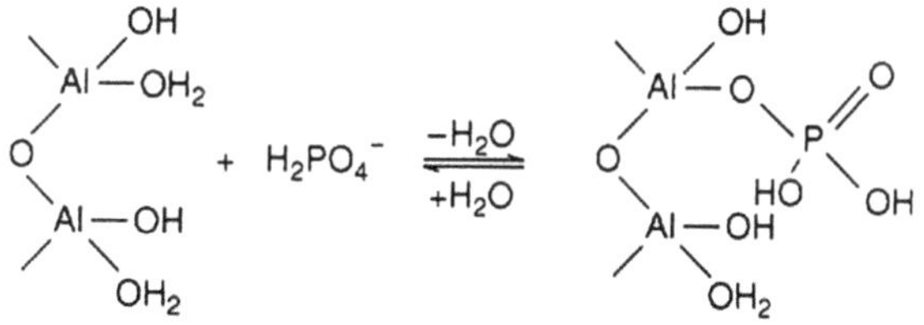

Fig. 1 Schematic presentation of phosphate adsorbing on an aluminium hydroxide surface by ligand exchange (Havlin et al. 2005)

precipitated, and further converted gradually to more sparingly soluble compounds such as octacalcium phosphate, $Ca_8(HPO_4)_2(OH)_2$, and hydroxyapatite, $Ca_{10}(PO_4)_6(OH)_2$ (Lindsay et al. 1989).

In soils never fertilized with P, the solid faces maintain a very low P concentration (<0.02 mg l^{-1}) in soil solution. In order to elevate the solution P concentration, agricultural soils are amended with phosphate fertilizers. Substantial part of P in manure, is in the form of inorganic phosphate, and more of it is released by mineralization of organic P compounds. Most P in common mineral P fertilizers is readily soluble. For example, the traditional P fertilizer "Superphosphate", containing $Ca(H_2PO_4)_2{\cdot}H_2O$ and $CaSO_4{\cdot}2H_2O$, is manufactured by dissolving apatite in sulfuric acid, 93 % of its P being water-soluble. Presently, the NPK fertilizers contain usually 60–70 % P in the water-soluble form, the rest being mostly dicalciumphosphate, $CaHPO_4$, which is soluble in dilute acids. After application into the soil the fertilizer granule starts to dissolve, and a concentrated P solution diffuses into the surrounding soil. Different calcium phosphates, predominantly dicalciumphosphate, are precipitated in that zone as their solubility products are exceeded (Havlin et al. 2005). Later, upon dilution of the soil solution, these precipitates are gradually dissolved and fertilizer P enters the different pools of soil P.

Inorganic Fractions of P in Soil

Inorganic P is held by various constituents of the soil matrix. This distribution of P can be investigated with soil analyses. In the traditional Chang and Jackson's (1957) procedure, soil samples are sequentially extracted with (1) ammonium chloride (NH_4Cl) to remove the most easily soluble P, (2) ammonium fluoride (NH_4F) to solubilize P retained by aluminum (hydr)oxides, "Al-P", (3) sodium hydroxide (NaOH) to desorb P held by iron (hydr)oxides, "Fe-P" and (4) sulfuric acid (H_2SO_4) to dissolve calcium phosphates, "Ca-P". Part of calcium phosphate precipitates are likely dissolved already with NH_4F (Bromfield 1967) and the fraction dissolved with H_2SO_4 represents more sparingly soluble forms. Phosphorus occluded within the oxides can be extracted with citrate—bicarbonate—dithionite treatment. Several modifications, compiled by Olsen and Sommers (1982) have been made to the original protocol, such as extracting the "Al-P" and "Fe-P" with NaOH as a single fraction, without a previous extraction with NH_4F, and using hydrochloric acid (HCl) instead of H_2SO_4 to extract apatite.

In cold desert soils of Antarctica, soil minerals have undergone only minimal chemical weathering, and upon fractionation of P as much as 300–400 mg kg^{-1} (95–99 % of P) was found as "Ca-P" extractable with H_2SO_4 (Blecker et al. 2006). Kaila (1964) investigated P reserves of rather young (<10,000 years) Finnish soils according to the Chang and Jackson procedure and identified "Ca-P" as the most abundant in virgin mineral soils (Table 2). In another Finnish study, Peltovuori et al. (2002) observed that this fraction increased with depth, i.e. in horizons that are less exposed to weathering.

Table 2 Fractions of inorganic phosphorus (mg kg^{-1}) in sandy soils of Finland

Soil group	Horizon	Samples	"Al-P"	"Fe-P"	"Ca-P"
Cultivated sand	Topsoil	19	135 ± 55	130 ± 50	125 ± 35
Cultivated fine sand	Topsoil	38	120 ± 35	160 ± 35	155 ± 30
Virgin sand and fine sand	Topsoil	16	15 ± 5	50 ± 15	85 ± 55
Cultivated sand and fine sand	Subsoil	25	40 ± 10	65 ± 15	120 ± 25
Virgin sand and fine sand	Subsoil	11	50 ± 30	80 ± 35	155 ± 105

Mean values are presented with the confidence limits at 95 % level. Selected results from Kaila (1964)

Hartikainen (1989) showed that the "Ca-P" fraction was also significantly decreased in several soils in field experiments when cropping without P fertilization for 7 years, indicating the gradual depletion of this fraction upon intensive plant P uptake. This fraction is indeed gradually decreasing in acid soils over time. Studying 120 soils at different stages of weathering from different countries (USA, Indonesia, Malaysia, Papua New Guinea, Philippines, Sudan), Sharpley et al. (1987) reported that in slightly weathered soils ($n = 40$) there was on average 17 % Ca-bound P while in highly weathered soils ($n = 39$) this form comprised of 4 % of soil P, reflecting the depletion of native apatite in acidic soils upon pedogenesis.

The effect of acidification on the "Ca-P" fraction was demonstrated by the present author (unpublished data) in a two-month incubation experiment where HCl was added into three mineral soils to decrease soil pH by about one unit. In a subsequent fractionation of P, it was observed that the "Ca-P" fraction decreased while the "Fe-P" and often also "Al-P" and the most easily soluble NH_4Cl extractable P increased (Table 3). These data indicate that apatitic P is depleted

Table 3 Impact of acidification in the distribution of inorganic P in three soils in a 2-month incubation study

Treatment	pH ($CaCl_2$)	Phosphorus (mg kg^{-1}) extracted with				Sum of fractions (mg kg^{-1})
		NH_4Cl	NH_4F	NaOH	H_2SO_4	
0	6.0	7.8	145	184	398	734
HCl (I)	5.2	+1.1	+14	+15	−36	728
HCl (II)	4.8	+4.1	+35	+27	−77	723
Fine sand						
0	5.2	3.2	101	291	319	715
HCl	4.3	+1.5	+17	+22	−35	719
Silty clay						
0	4.9	1.2	20	156	171	348
HCl	3.9	+0.6	+1	+13	−22	340

The changes are presented in comparison to the soil incubated without addition of acid. All the changes were statistically significant ($P = 0.05$) except the NH_4F-extractable fraction in silty clay

from soil quite rapidly upon severe acidification and after dissolution retained by Al and Fe hydroxides.

Calcareous soils, with pH values normally above 7.5, differ from acidic soils in P distribution, "Ca-P" dominating in alkaline environments, even though they can contain appreciable amounts of P bound by Fe oxides (Saavedra and Delgado 2005). In the international material of 120 soils mentioned above (Sharpley et al. 1987), "Ca-P" comprised 55 % of total P in calcareous soils (n = 41) with minor amounts of P associated with Fe and Al. The same trend was observed in 5 calcareous soils from UK, Italy, Austria and Hungary, where 89 % of P was extractable with H_2SO_4 as compared to 46 % in 19 non-calcareous soils, dominated by P forms extracted with NH_4Cl, NH_4F and NaOH (Hartikainen et al. 2010). Higher percentages of "Ca-P" suggest that this soil material was in general at a less advanced state of pedogenesis than the soils of Sharpley et al. (1987).

Where Does the Fertilizer P Go in the Soil?

In intensive agriculture P applications have for decades been much higher than P uptake by the crop with positive P balances (e.g. Tunney et al. 2003). Therefore fertilizer residues have accumulated in the soil. Already nearly 50 years ago, Kaila (1964) showed that acidic agricultural mineral surface soils of Finland were much higher in "Al-P" and "Fe-P" than the respective subsoils and virgin topsoils while there was no difference in the abundance of P fractions between the deeper horizons of cultivated and virgin soils (Table 2). In another study (Kaila 1967), P addition (100 mg kg^{-1}) in a water-soluble form to 12 acidic mineral soils was rather completely (95 %) recovered by the Chang and Jackson procedure. In that study 46 and 48 % of the recovered P was retained as "Al-P" and "Fe-P", respectively, only 2 and 4 % found in the fraction extracted with NH_4Cl and as "Ca-P".

In soil samples taken from a field experiment where different amounts of P had been applied for 11 years in a silty clay (pH 6.0) in Finland, the fractions of "Al-P" and "Fe-P" were significantly higher in the fertilized plots while there was practically no difference in the "Ca-P" fraction between the different treatments (Yli-Halla 1989; Table 4). Assuming a 25-cm deep plough layer and the bulk density of 1.2 kg dm^{-3}, the recovery of residual fertilizer P in the soil was 42–85 %. Also in other field experiments in Finland (Hartikainen 1989; Jaakkola et al. 1997; Turtola and Yli-Halla 1999) residual fertilizer P has been predominantly recovered from the "Al-P" and "Fe-P" fractions with little change in the "Ca-P" fraction. This outcome about the changes of inorganic P fractions is principally in agreement with results obtained in various acidic soils with different fractionation procedures (Shelton and Coleman 1968; Tran and N'dayegamiye 1995; Motavalli and Miles 2002; Saleque et al. 2004; Zhang et al. 2004; Kuo et al. 2005).

In calcareous soils added P is partly adsorbed on Fe and Al oxides but mostly on calcite surfaces and precipitated as calcium phosphates (Castro and Torrent 1998; Delgado and Torrent 2000). Results of detailed sequential fractionation schemes

Table 4 Accumulation of P in a silty clay in a field experiment after 11 years of repeated P applications

Annual P application	Total P applied	P in the grain	P balance	"Al-P"	"Fe-P"	"Ca-P"	Sum of fractions
(kg ha^{-1})		(mg kg^{-1})					
0	0	153	−137	42	150	230	422
13–16	154	162	−8	+4	+8	+6	440
26–32	309	171	138	+35	+45	−14	494
47–56	541	162	379	+53	+48	+6	559
60-72	696	174	522	+78	+91	+12	605

Results are presented as differences to the control not receiving P (adapted from Yli-Halla 1989)

show that calcareous soils contain both lithogenic apatite and secondary calcium phosphates (Delgado and Torrent 2000). These fractions differ substantially in their solubility, the secondary calcium phosphates maintaining a much higher P concentration in the soil solution. In a 21-year field experiment with annual P applications up to 79 kg ha^{-1} as superphosphate (i.e. mostly water-soluble P), octacalcium phosphate-like compounds increased most both in absolute and relative amounts, with some increase in the fractions representing "Al-P", "Fe-P", occluded P and hydroxyapatite-like compounds (Wang et al. 2010). There was also increase in the fraction representing monocalcium phosphate and dicalciumphosphate but the results clearly demonstrate that water-soluble P added to a calcareous soil is by far predominantly converted to less soluble calcium phosphates. It has however been shown that in spite of lower solubility compared to dicalciumphosphate, octacalcium phosphate serves as a source of plant-available P (Olsen et al. 1983; Wang et al. 2010).

Applied P thus enters the P cycle of soil, as shown in Fig. 2. After dissolution from the fertilizer into the soil solution, P is retained by solid soil constituents. When P is taken up from the soil solution, the chemical (semi)equilibrium between the solid phases and the soil solution results in desorption/dissolution of P from the solid surfaces, attempting to maintain the established solution P concentration. Abundant P fertilization, with large stock accumulated on the adsorbing surfaces or

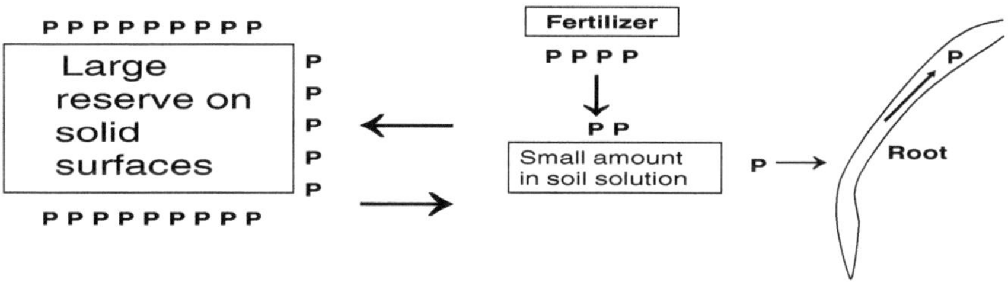

Fig. 2 Fertilizer P equilibrates with the stock of soil P accumulated in soil in previous fertilizer applications

as easily dissolved precipitates, increases the ability of the soil to maintain a solution P concentration that is adequate for the crop. P supplying power of the soil is thus increased.

Phosphorus Sorption Capacity of Soil

It has been well established that in acidic soils Al and Fe (hydr)oxides are by far the most important constituents of P retention in soil. Already in the 1920s it was reported that removing these constituents decreases the ability of soil to retain P (Mattson 1927). In principle, these (hydr)oxides have a certain number of sites that can bind P (see Fig. 1). The amount of poorly crystalline Al and Fe (hydr)oxides, extracted from soil with acid ammonium oxalate solution, has been considered a measure for phosphate sorption capacity (PSC) of soil (e.g., van der Zee and van Riemsdijk 1988):

$$\mathrm{PSC} = \alpha \cdot (\mathrm{Al_{ox}} + \mathrm{Fe_{ox}})$$

In the equation, $Al_{ox} + Fe_{ox}$ stands for the sum of Al and Fe extracted with oxalate, all expressed in mmol kg^{-1} of soil. The coefficient α stands for the fraction of Al and Fe that is accessible for the external solution containing phosphate. For example, values of 0.61 (van der Zee and van Riemsdijk 1988) and 0.11–0.14 (Uusitalo and Tuhkanen 2000) have been reported, and the value of 0.5 has often been used in practice (Lookman et al. 1995; Peltovuori et al. 2002). In a large material of 301 soil samples of various textures from northern Belgium (Lookman et al. 1995), the PSC most commonly ranged from 30 to 50 mmol kg^{-1}, corresponding to about 900–1600 mg P kg^{-1}. According to Peltovuori et al. (2002), four agricultural topsoils of Finland had PSC's of 66–169 mmol kg^{-1}. PSC can also be estimated by saturating soil with a concentrated P solution (e.g. 150–250 mg l^{-1}) and actually measuring the decrease of P concentration of the solution, which is assumed to be attributed to P sorption to soil. With this method, the median of PSC in 102 mineral soils of Finland was 27 mmol kg^{-1} (840 mg kg^{-1}), individual observations up to 87 mmol kg^{-1} (2700 mg kg^{-1}; Niskanen 1990).

Another approach to estimate the fertilization need of a soil is to determine how much P has to be added to soil in order to elevate the soil solution concentration to a level sufficient to a crop (Fox and Kamprath 1970). For example, Fox et al. (1974) reported that in Hawaii two Oxisols with a very high P sorption tendency required 400 and 800 mg kg^{-1} of P to increase the soil solution concentration to 0.2 mg l^{-1}. If we estimate that 20 % of soil volume is filled with soil solution, there would be 0.04 mg P dm^{-3} in the solution phase, suggesting that 10,000–20,000 units of P had to be added to have one unit remaining in the solution phase. This may be an extreme case but it reflects the situation in soils never or scarcely fertilized with

P. In such a soil almost all sites capable of adsorbing P are free for capturing P from the solution.

Degree of P Saturation

In a virgin soil reclaimed for agriculture, even a large P dose is nearly completely adsorbed (fixed), almost all added P being removed from solution and retained in the solid phase. However, abundant P fertilization with manure and mineral fertilizers that result in strongly positive P balances has continued in large parts of the industrial world for decades with a consequent accumulation of P in soil. The tendency of soil to fix added P has decreased as an increasing number of the potential sites for P sorption become occupied. Gradual filling of sorption sites has contributed to higher P concentrations in solution. On the basis of Chang and Jackson fractionation of soil P and acid oxalate extraction of Al and Fe in 103 mineral soils of Finland, Hartikainen (1982) calculated the ratio of the NH_4F extractable P ("Al-P") and oxalate-extractable Al, and showed that this ratio very closely predicted the concentration of P extracted with deionized water (1:60), all concentrations presented in mmol kg^{-1}:

$$P_w = -0.209 + 9.792 \times \text{"Al–P"}/\text{Al} \left(R^2 = 0.87\right)$$

Phosphate is adsorbed also by the Fe hydroxides but the corresponding ratio between the NaOH extractable P ("Fe-P") and oxalate extractable Fe seemed to be less important in explaining the immediate level of water extractable P in soil.

The concept Degree of P Saturation (DPS) was originally launched by van der Zee et al. (ref. Lookman et al. 1995) and has been widely adopted ever since. The concept is valid in soils where Al and Fe (hydr)oxides are mainly responsible for the sorption of added P. It is assumed that there is a certain number of sites capable of adsorbing P. The DPS was originally operationally defined as the molar ratio of oxalate-extractable P to the sum of oxalate-extractable Al + Fe:

$$\text{DPS}\ (\%) = 100 \times P_{ox}/0.5 \times (Al_{ox} + Fe_{ox}) = 100 \times P_{ox}/\text{PSC}$$

As mentioned earlier, the coefficient 0.5 stands for the assumption that half of the potential sorption components is actually active and in contact with soil solution. PSC stands for Phosphate Sorption Capacity. The lower the DPS, the more readily and more strongly added P is adsorbed. Upon repeated applications and positive P balances, increasing number of the potential sorption sites become occupied with P. At the same time, the bonding strength of successively adsorbed P decreases, and a higher P concentration is maintained in soil solution. A DPS value of 24 % has been set for a soil where from the environmental point of view solution P concentration becomes excessive (>0.1 mg l^{-1} in groundwater), allowing P leaching in dissolved form or abundant P release from eroded soil material.

The DPS concept is easy to comprehend. Another advantage is that all the three elements are determined in the same oxalate extract. This simple procedure has however been criticized (Uusitalo and Tuhkanen 2000; Peltovuori et al. 2002; Hartikainen et al. 2010) because the acid oxalate solution dissolves at least partly also Ca-bound apatitic P which does not actively contribute to solution P concentration. Pedologically young soils are rich in this form of P, and therefore an alternative, and a more laborious, approach to calculate the DPS has been proposed. Peltovuori et al. (2002) calculated the DPS as the following molar ratio:

$$\mathrm{DPS} = 100 \times (\mathrm{P\ extracted\ with\ NH_4Cl,\ NH_4F\ and\ NaOH})/(\mathrm{Fe_{ox}} + \mathrm{Al_{ox}})$$

where the P fractions stand for those of the Chang and Jackson procedure and Fe_{ox} and Al_{ox} are extracted with acid oxalate solution. When applied to the soils taken from an 11-year field experiment (Yli-Halla 1989), there was a linear relationship between the DPS and P extracted with water (Fig. 3). Still another way to estimate the DPS was used by Börling et al. (2004) in Sweden. They calculated the ratio of P extracted with acid ammonium lactate, the method used in soil testing in Sweden, and P sorption capacity determined by a saturation method. The ratio of P extracted with $NaHCO_3$ at pH 8.5 (Olsen method) and the sorption capacity was also calculated. In spite of the differences in methodology, all approaches aim at the same objective, i.e., they attempt to give an estimate for the filling of sorption sites with P.

Even though the DPS concept was primarily developed for environmental assessment, it can to a certain extent be utilized roughly in estimating the sufficiency of P supply to a crop. In many parts of intensive agriculture, with a long history of abundant P fertilization with mineral fertilizers or manure and highly positive P balances, the DPS values are considerably high. In a large soil material from Northern Belgium (Lookman et al. 1995), the average DPS of 59 % in the 0–30 cm soil as compared to 20 % at 30–60 cm and 9 % at 60–90 cm

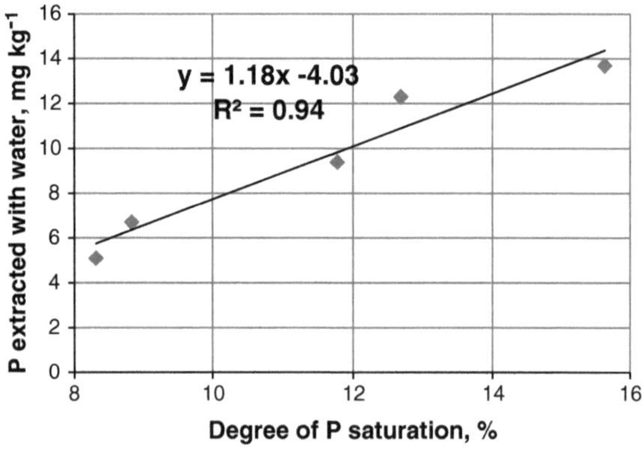

Fig. 3 Relationship between the degree of P saturation of aluminum and iron hydroxides and P extracted with water at the end of an 11-year field experiment (calculated from Yli-Halla 1989)

demonstrating a substantial accumulation of P in the plough layer of many agricultural soils.

P Stock and P Supply to Plants

Over time, P initially adsorbed on easily accessible surfaces diffuses further into the soil pores (Rajan and Fox 1972) and gets out of the reach of soil organisms. Upon ageing, bonding of P onto the oxide surfaces also becomes stronger (Kaila 1965) as monodentate bonds are converted to bidentate and binuclear ones (Kafkafi et al. 1967). Continuous weathering of soil minerals produces new Al and Fe hydroxide faces and new sites for P sorption are generated (Rajan et al. 1974). Phosphorus can even be embedded (occluded) in the oxides and lose contact with the soil solution. Recently precipitated calcium phosphates are also converted to more sparingly soluble forms. These processes, competing with plant P uptake, slowly decrease the availability of residual fertilizer P.

In soils with a low DPS the processes mentioned above result in the consumption of P added to soil with a consequent need of P fertilization much exceeding the crop uptake. Abundant P fertilization has increased the DPS and accumulation of the P stock, increasing the P supplying power of soil. On the basis of a pot experiment with 12 "phosphate-rich" soils from Germany, Spain, Scotland and Italy, the total amount of plant-available P was on average 169 mg kg^{-1} (range 23–392 mg kg^{-1}) (Delgado and Torrent 1997), corresponding to 500 kg ha^{-1} (range 69–1170 kg ha^{-1}) in a plough layer of 23 cm with a bulk density of 1.3 kg dm^{-3}. In another pot experiment with an acidic (pH 6.0) clay soil of Finland (Yli-Halla 1991), 21–25 % of the "Al-P" + "Fe-P" was taken up by ryegrass, corresponding to 40–90 mg kg^{-1} (130–293 kg ha^{-1}) at the various levels of previous P fertilization (up to 72 kg ha^{-1} annually in 11 years). Novais and Kamprath (1978) measured a 82–108 kg ha^{-1} release to nine successive crops of pearl millet in a pot experiment conducted with acidic (pH 5.3–6.3) heavily fertilized soils from the eastern coast of USA varying in clay contents. These results indicate that a surplus of P accumulated in soil due to abundant P fertilization remains to a large extent available to plants.

Often it is estimated that only 10–20 % of P contained in the crop originates from the most recent fertilization, the rest 90–80 % coming from the reserves accumulated in the soil in earlier fertilizer applications (Sharpley 1986; McLaughlin et al. 1988). Small immediate responses of crop to added P are measured (Valkama et al. 2009) because in many soils that have a history of abundant P fertilization the P supplying power meets or exceeds the need of the crop for optimum growth. Even though P is continuously subjected to processes that slowly decrease its solubility also in soils of high P status, such soils can provide the crop with adequate amounts of P for several years, even decades, with P applications that are equal to or less than P removal by the crop yields. A prerequisite for apparently 100 % utilization of fertilizer P is thus a rather high DPS or large reserves of easily soluble calcium phosphates, both resulting from residual P of manure or chemical fertilizers. These soil characteristics are conducive to an abundant P supplying power of the soil.

References

Adams JF, Odom JW (1985) Effects of pH and phosphorus rates on soil solution phosphorus and phosphorus availability. Soil Sci 140:202–205

Benians G, Scullion P, Fitzhugh GR (1977) Concentrations and activities of ions in solutions displaced from basaltic soils. J Soil Sci 28:454–461

Bhadoria PS, Samal D, Steingrobe B, Claassen N (2009) Phosphorus uptake kinetics, root-shoot relations and P-flux of groundnut and maize grown in solution culture. In: Proceedings of the XVI international plant nutrition colloquium. UC Davis, USA. http://escolarship.org/uc/item/9p6432sv. (visited 31 May 2012)

Blecker SW, Ippolito JA, Barrett JE, Wall DH, Virginia RA, Norvell KL (2006) Phosphorus fractions in soils of Taylor Valley, Antarctica. Soil Sci Soc Am J 70:806–815

Börling K, Barberis E, Otabbong E (2004) Impact of long-term phosphorus fertilization on accumulation, sorption and release of phosphorus in five Swedish soil profiles. Nutr Cycl Agroecosyst 69:11–21

Bromfield SM (1967) An examination of the use of ammonium fluoride as a selective extractant for aluminium bound soil phosphate in partially phosphate systems. Aust J Soil Res 15:225–234

Cassman KG, Whitney AS, Fox RL (1981) Phosphorus requirements of soybean and cowpea as affected by mode of N nutrition. Agron J 73:17–22

Castro B, Torrent J (1998) Phosphate sorption by calcareous Vertisols and Inceptisols as evaluated from extended P-sorption curves. Eur J Soil Sci 49:661–667

Chang SC, Jackson ML (1957) Fractionation of soil phosphorus. Soil Sci 84:133–134

Delgado A, Torrent J (1997) Phosphate-rich soils in the European Union: estimating total plant-available phosphorus. Eur J Agron 6:205–214

Delgado A, Torrent J (2000) Phosphorus forms and desorption patterns in heavily fertilized calcareous and limed acidic soils. Soil Sci Soc Am J 64:2031–2037

Fox RL, Kamprath EJ (1970) Phosphate sorption isotherms for evaluating the phosphate requirements of soils. Soil Sci Soc Am J 34:902–907

Fox RL, Nishimoto RK, Thompson JR, de la Pena RS (1974) Comparative external phosphorus requirements of plants growing in tropical soils. Hawaii agricultural experiment station technical paper no 1517. Electronic publication (visited 18 May, 2012)

Hammad ME, Rimmer DL, Syers JK (1992) Effect of iron oxide on phosphate sorption by calcite and calcareous soils. J Soil Sci 43:273–281

Hartikainen H (1982) Water soluble phosphorus in Finnish mineral soils and its dependence on soil properties. J Sci Agric Soc Finland 54:89–98

Hartikainen H (1989) Effect of cumulative fertilizer dressings on the phosphorus status of mineral soils. I changes in inorganic phosphorus fractions. J Agric Sci Finland 61:55–59

Hartikainen H, Rasa K, Withers P (2010) Phosphorus exchange properties of European soils and sediments derived from them. Eur J Soil Sci 61:1033–1042

Havlin JL, Beaton JD, Tisdale SL, Nelson WL (2005) Soil fertility and nutrient management: an introduction to nutrient management, 7th edn. Pearson/Prentice Hall, Upper Saddle River 515 pp

Holford ICR, Mattingly GEG (1975) The high- and low-energy phosphate adsorbing surfaces in calcareous soils. J Soil Sci 26:407–417
Jaakkola A, Hartikainen H, Lemola R (1997) Effect of fertilization on soil phosphorus in a long-term field experiment in southern Finland. Agric Food Sci Finland 6:313–322
Kafkafi U, Posner AM, Quirk JP (1967) Desorption of phosphate from kaolinite. Soil Sci Soc Am Proc 31:348–353
Kaila A (1964) Fractions of inorganic phosphorus in Finnish mineral soils. J Sci Agric Soc Finland 36:1–13
Kaila A (1965) The fate of water-soluble phosphate applied to some mineral soils. J Sci Agric Soc Finland 37:104–115
Kaila A (1967) Effect of liming on the fate of applied superphosphate phosphorus in some mineral soils. J Sci Agric Soc Finland 39:5–13
Kuo S, Huang B, Bembenek R (2005) Effects of long-term phosphorus fertilization and winter cover cropping on soil phosphorus transformations in less weathered soils. Soil Biol Fert 41:116–123
Lindsay WL, Vlek PLG, Chien, SH (1989) Phosphate minerals. In: Dixon JB, Weeds SB (eds) Minerals in soil environment, 2nd edn. Soil Sci Soc Am, Madison, Wisconsin, USA, pp 1089–1130
Lookman R, Vandeveert N, Merckx R, Vlassak K (1995) Geostatistical assessment of the regional distribution of phosphate sorption capacity parameters (Fe_{ox} and Al_{ox}) in northern Belgium. Geoderma 66:285–296
Mattson S (1927) Anionic and cationic adsorption by soil colloidal materials of varying SiO_2/ $Al_2O_3 + Fe_2O_3$ ratio. In: Proceedings 1st international congress of soil science II, pp 199–211
Matula J (2011) Relationship between phosphorus concentration in soil solution and phosphorus in shoots of barley. Plant Soil Environ 57:307–314
McLaughlin MJ, Alston AM, Martin JK (1988) Phosphorus cycling in wheat-pasture rotations. I. The source of phosphorus taken up by wheat. Aust J Soil Res 26:323–331
Motavalli P, Miles R (2002) Soil phosphorus fractions after 111 years of animal manure and fertilizer applications. Biol Fert Soils 36:35–42
Nielsen NE, Schjørring JK (1983) Efficiency and kinetics of phosphorus uptake from soil by various barley genotypes. Plant Soil 72:225–230
Niskanen R (1990) Sorption capacity of phosphate in mineral soils. II. Dependence of sorption capacity on soil properties. J Agric Sci Finland 62:9–15
Novais R, Kamprath EJ (1978) Phosphorus supplying capacities of previously heavily fertilized soils. Soil Sci Soc Am J 42:931–935
Olsen C (1950) The significance of concentration on the rate of ion adsorption by higher plants. Physiol Plant 3:152–164
Olsen SR, Sommers LE (1982) Phosphorus. In: Page AL, Miller, RH, Keeney DR (eds) Methods of soil analysis, part 2. Agronomy monographs no. 9. Am Soc Agron and Soil Sci Soc Am. Madison, Wisconsin, USA, pp 403–430
Olsen SR, Watanabe FS, Bowman RA (1983) Evaluation of fertilizer phosphate residues by plant uptake and extractable phosphorus. Soil Sci Soc Am J 47:952–958
Peltovuori T, Uusitalo R, Kauppila T (2002) Phosphorus reserves and apparent phosphorus saturation in four weakly developed cultivated pedons. Geoderma 110:35–47
Rajan SSS, Fox RL (1972) Phosphate adsorption by soils. I. Influence of time and ionic environment on phosphate adsorption. Commun Soil Sci Plant Anal 3:493–504
Rajan SSS, Perrott KW, Saunders WMH (1974) Identification of phosphate-reactive sites of hydrous alumina from proton consumption during phosphate adsorption at constant pH values. J Soil Sci 25:438–447
Ryan J, Curtin D, Cheema MA (1985) Significance of iron oxides and calcium carbonate particle size in phosphorus sorption by calcareous soils. Soil Sci Soc Am J 49:74–76

Saavedra C, Delgado A (2005) Phosphorus fractions and release patterns in typical Mediterranean soils. Soil Sci Soc Am J 69:607–615

Saleque MA, Naher UA, Islam A, Pathan ABMPU, Hossain ATMS, Meisner CA (2004) Inorganic and organic phosphorus fertilizer effects on the phosphorus fractionation in wetland rice soils. Soil Sci Soc Am J 68:1635–1644

Sharpley A (1986) Disposition of fertilizer phosphorus applied to winter wheat. Soil Sci Soc Am J 50:953–958

Sharpley AN, Tiessen H, Cole CV (1987) Soil phosphorus forms extracted by soil tests as a function of pedogenesis. Soil Sci Soc Am J 51:362–365

Shelton JE, Coleman NT (1968) Inorganic phosphorus fractions and their relationship to residual value of large applications of phosphorus on high phosphorus fixing soils. Soil Sci Soc Am J 32:91–94

Tran TS, N'dayegamiye A (1995) Long-term effects of fertilizers and manure application on the forms and availability of soil phosphorus. Can J Soil Sci 75:281–285

Tunney H, Csathó P, Ehlert P (2003) Approaches to calculating P balance at the field-scale in Europe. J Plant Nutr Soil Sci 166:438–446

Turtola E, Yli-Halla M (1999) Fate of phosphorus applied in slurry and mineral fertilizer: accumulation in soil and release into surface runoff water. Nutr Cycl Agroecosyst 55:165–174

Uusitalo R, Tuhkanen H-R (2000) Phosphorus saturation of Finnish soils: evaluating an easy oxalate extraction method. Agric Food Sci Finland 9:61–70

Valkama E, Uusitalo R, Ylivainio K, Virkajärvi P, Turtola E (2009) Phosphorus fertilization: a meta-analysis of 80 years of research in Finland. Agric Ecosyst Environ 130:75–85

Van der Zee SEATM, van Riemsdijk WH (1988) Model for long-term phosphate reaction kinetics in soil. J Environ Qual 17:35–41

Wang J, Liu W-Z, Mu H-F, Dang T-H (2010) Inorganic phosphorus fractions and phosphorus availability in a calcareous soil receiving 21-year superphosphate application. Pedosphere 20:304–310

Webber MD, Mattingly GEG (1970) Inorganic soil phosphorus. II. Changes in monocalcium phosphate and lime potentials on mixing and liming soils. J Soil Sci 21:121–126

Wiklander L, Andersson A (1974) The composition of the soil solution as influenced by fertilization and nutrient uptake. Geoderma 11:157–166

Yli-Halla M (1989) Effect of different rates of P fertilization on the yield and P status of the soil in two long-term field experiments. J Agric Sci Finland 61:361–370

Yli-Halla M (1991) Phosphorus supplying capacities of soils previously fertilized with different rates of P. J Agric Sci Finland 63:75–83

Zhang TQ, MacKenzie AF, Liang BC, Drurya CF (2004) Soil test phosphorus and phosphorus fractions with long-term phosphorus addition and depletion. Soil Sci Soc Am J 68:519–528

Fate of Fertilizer P in Soils—The Organic Pathway

Emmanuel Frossard, Else K. Bünemann, Lucie Gunst, Astrid Oberson, Michael Schärer and Federica Tamburini

Abstract This chapter addresses the following questions: (i) is phosphorus (P) that passes through the "organic pathway", i.e. that is found at some point in time in an organic form or within soil organisms or plant residues completely used by plants (100 % use by plants) and (ii) does P that passes through the "organic pathway" fully remain in the soil/plant system (zero losses to the environment). Positive answers to these questions would lead to the conclusion that P management could be optimized by increasing soil organic P content and soil biological activity. The chapter begins with a short presentation of the actors of the "organic pathway", and how their dynamics influence soil P transformations. Then we show how the "organic pathway" controls soil P availability in natural systems and how soils under native vegetation can lose significant amounts of P during their pedogenesis. Afterwards we discuss the role of the "organic pathway" in determining P availability and limiting P losses in agricultural systems. The analysis of exemplary cropping systems studied in the Swiss DOK field experiment (bio-Dynamic, bio-Organic and conventional (K) cropping systems) shows that it is not possible to reach 100 % apparent P recovery and that the recovery is negatively correlated with soil available inorganic P (Pi). A high apparent P fertilizer recovery is observed in the system in which the soil shows the highest biological activity. This high recovery can be explained by the low P inputs in this treatment and by the transfer of P through the soil biota. However, this high recovery can also be interpreted as an artifact in the calculation of the apparent recovery rate, concealing a higher use

E. Frossard (✉) · E.K. Bünemann · A. Oberson · F. Tamburini
Group of Plant Nutrition, Institute of Agricultural Sciences, ETH Zurich, Eschikon 33, 8315 Lindau, Switzerland
e-mail: emmanuel.frossard@usys.ethz.ch

L. Gunst
Agroscope Reckenholz-Tänikon ART, Reckenholzstrasse 191, 8046 Zurich, Switzerland

M. Schärer
Abteilung Wasser Sektion Oberflächengewässer Qualität, Bundesamt für Umwelt BAFU, CH 3003 Bern, Switzerland

E. Schnug and L.J. De Kok (eds.), *Phosphorus in Agriculture: 100 % Zero*,
DOI 10.1007/978-94-017-7612-7_4

of P derived from the soil in this treatment. The higher use of soil P by crops can be explained by the higher root colonization by mycorrhizal fungi and by the good soil structure observed in this treatment. In permanent grasslands, the very high microbial P content observed in the upper soil horizon suggests that the "organic pathway" plays a major role in taking up and releasing P to the plants. Finally, the importance of this pathway for limiting P losses from agricultural systems is discussed. Whereas it might have a role for conserving P in the soil profile, it is probably not the most important point to address for limiting P losses from agricultural systems.

Keywords Fertilizer use · Immobilization · Losses · Mineralization · Organic P · P availability

Introduction

Phosphorus (P) is essential to all organisms and cannot be substituted by any other element. P is therefore essential for agricultural production. As agricultural systems are open, soil P must be managed at a level that ensures optimal plant growth (Sanchez et al. 1997; Follain et al. 2009). Long-term P inputs in excess to plant needs increase soil available P, with the often associated release of P towater resulting in eutrophication of water bodies (Granger et al. 2010). The abundance of rare terrestrial plant species declines when systems change from P-limited to non-P-limited conditions (Wassen et al. 2005; Ceulemans et al. 2011). In Western and Northern European agriculture many soils have reached a sufficient or a high P level. Nevertheless, more P is still applied in Europe than removed (Lemercier et al. 2008; Schröder et al. 2010; Spiess 2011). Whereas a fertilization strategy leading to an enrichment in soil available P can make sense in P-poor agricultural systems to increase and stabilize crop production (Sanchez et al. 1997), it is not relevant for Western and Northern Europe. Applying good agricultural practices (FAO 2003) and integrated nutrient management schemes (Frossard et al. 2009) have the potential to increase P use efficiency and decrease P losses to the environment as these strategies aim at applying P according both to the needs of crops and to the soil available P content, and at the same time aim at replenishing soil organic matter content, optimizing soil biological activity, and minimizing erosion and water runoff.

This chapter addresses the following questions: (i) is phosphorus (P) that passes through the "organic pathway", i.e. that is found at some point in time in an organic form or within soil organisms or plant residues completely used by plants (100 % use by plants) and (ii) does P that passes through the "organic pathway" fully remain in the soil/plant system (zero losses to the environment). Positive answers to these questions would lead to the conclusion that P management could be optimized by increasing soil organic P content and soil biological activity.

Before starting this chapter, the limits of the system to be considered must be clarified. Except P from igneous origin, most of the phosphate on earth has been going through an "organic pathway". Indeed P found in sediments has been metabolized by plankton, algae, bacteria or fish before being precipitated as apatite (Benitez-Nelson 2000; Filippelli 2011). As most of the fertilizers applied in agriculture are made from sedimentary P, the added P has had at some point in time a biological origin. The geological cycle of P, however, runs over millions of years (Follmi 1996) and is not considered in this review. If we focus on the farm level, P is being metabolized by soil organisms, but also by plants and animals. Manure that is being applied contains P from urine and faeces mixed with P present in bedding material. Although manure does not exclusively contain organic P forms, all its P has gone through an "organic pathway". We argue therefore that at farm level it does not make sense to consider solely the importance of P going through the soil "organic pathway", but indeed we should consider the soil-plant-animal "organic pathway". We therefore need to include in the "organic pathway" P fertilizers derived from a biological process (compost, manure). The metabolism of P in farm animals will not be discussed in this chapter, for more information the reader is referred to Sims and Sharpley (2005).

The chapter is constructed as follows. First we present the actors of the "organic pathway" and how their dynamics affect P transformations; then the importance of the "organic pathway" is discussed in natural systems in relation to P availability and losses; finally the relevance of the "organic pathway" is addressed for Western and Northern European agricultural production systems, also in relation to P availability and losses. In this last part we focus on agricultural systems managed according to good agricultural practices. Systems that are not managed according to these practices, e.g. where excessive amounts of P are applied, where long term mono-cropping is the rule or where erosion or runoff are not controlled, are not discussed.

The Actors of the "Organic Pathway" and Their Dynamics

The actors of the "organic pathway" are plants, soil macro and micro fauna and the soil microbial biomass (bacteria, archea, and fungi; Paul 2007). All of them influence P transformations. The biological reactions controlling the P cycle in the soil/plant system have been reviewed recently (see e.g. Bünemann et al. 2011; Richardson et al. 2011; Frossard et al. 2012). The reader is referred to these reviews for more information. We provide here a summary of reactions involving plants and microorganisms in order to allow a good understanding of the following sections of this chapter. The P cycle in the soil/plant system is sketched in Fig. 1 with the relevant pools and fluxes.

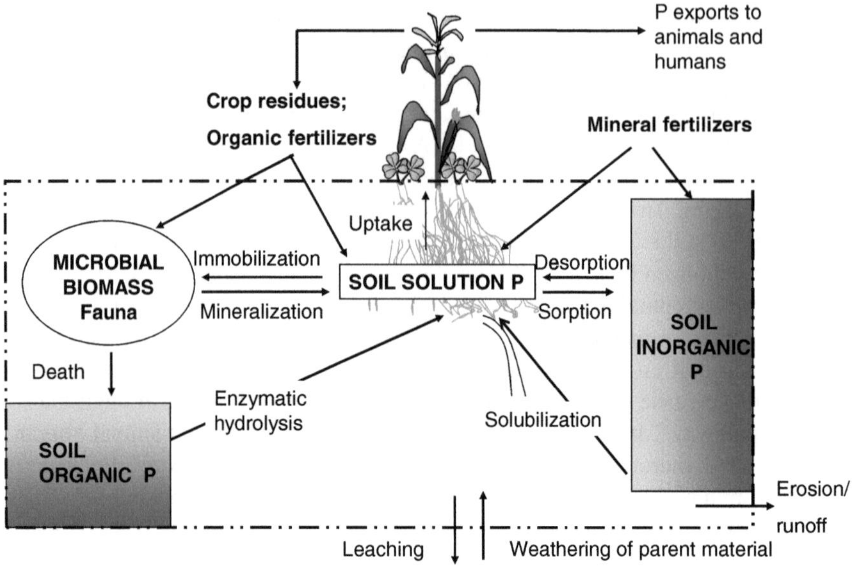

Fig. 1 The phosphorus cycle in the soil/plant system

The Role of Plants

Plants take up P as orthophosphate ion (Pi) from the soil solution through roots and hyphae of associated mycorrhizal fungi. In agricultural systems, the amount of P annually exported by plants vary from less than 10 kg P ha^{-1} $year^{-1}$ for a little intensively managed grassland, to 50 kg P ha^{-1} $year^{-1}$ for sugar beet used as forage (Sinaj et al. 2009). Plants can adapt to low Pi availability by different mechanisms. They can activate high affinity Pi transporters for taking up Pi at low concentrations, they can release organic acids and protons into the external environment to solubilize Pi bound to calcium, and they can release phosphatase enzymes to hydrolyze organic P compounds (George et al. 2011; Richardson et al. 2011). They can adapt their root system to access a larger soil volume by increasing specific root length and decreasing root diameter, by increasing root hair length, or by establishing a symbiosis with mycorrhizal fungi (George et al. 2011; Jansa et al. 2011; Richardson et al. 2011). In agricultural systems of Western and Northern Europe there are only a few crops such as rapeseed or sugar beet that are not mycorrhized and that are able to access scarcely soluble P forms through root exudation (Hoffland et al. 1989; Khorassani et al. 2011). Most of the other crops are colonized by arbuscular mycorrhizal fungi and access only Pi from the soil solution (Morel and Plenchette 1994).

After harvest, plant residues decompose under the influence of the soil fauna and of the microbial biomass and release nutrients to the soil solution. Experiments conducted with crop residues labeled with radioactive P have shown that P added as crop residues has a smaller or equal availability to the following crop than a fresh addition of water soluble P fertilizer (McLaughlin and Alston 1986; Thibaud et al. 1988; Nachimuthu et al. 2009). Similarly, the P use efficiency of "processed" plant products, such as manure or compost, measured with isotopic methods, has been reported to be smaller than that of water soluble P (Sinaj et al. 2002; Oberson et al. 2010). This smaller efficiency has been explained by the presence of scarcely soluble Pi and of organic P forms in manure and compost (Sinaj et al. 2002; Oberson et al. 2010).

The Role of Microorganisms

As for plants, the most important source of P for soil organisms is Pi that is taken up from the soil solution. Microorganisms, moreover, are also able to take up small organic P compounds (Dyhrman et al. 2007). The microbial P content in the upper soil horizon varies widely, e.g. between 2.4 mg P kg^{-1} soil under cropping and 77 mg P kg^{-1} soil under grassland (Oberson and Joner 2005).

Microorganisms can also adapt to low P availability by activating high affinity Pi transporters and releasing organic acids and protons, siderophores and phosphatase enzymes into the external environment (Jones and Oburger 2011; Richardson et al. 2011). Because soil microbes are very efficient to gather P, their growth is rarely P-limited. For instance, Göransson et al. (2011) studying the soils of a glacier forefield in the Swiss Alps, and Ehlers et al. (2010) working on an oxisol from Kenya could not observe any P limitation in the growth of soil microorganisms whereas both soils contained very little available Pi. Similarly, Bünemann et al. (2012) demonstrated that microorganisms were able to immobilize Pi very rapidly in a soil that was rich in organic matter and that contained very little available Pi.

The turnover time of microbial P in soil (time needed to totally renew the P content of the microbial biomass) has been reported to vary between 40 days and 2.5 years (Oberson and Joner 2005). Achat et al. (2010) studying a low P forest soil suggest that 80 % of the microbial P can be ascribed to a pool with a turnover time as short as 9 days while the rest of the microbial P would have a turn over time of 80 days or more. This is important information as it shows how fast the P held in the microbial biomass can be released to the soil solution. Besides soil Pi availability, carbon (C) and nitrogen (N) contents also affect microbial P turnover. The grazing of microorganisms by soil microfauna, such as nematodes, amoeba and protozoa can result in the release of Pi to the solution and to an increased uptake of Pi by plants (Irshad et al. 2011). Finally, strong changes in osmotic pressure in the environment surrounding microorganisms caused for instance by repeated rapid drying/rewetting cycles lead to microbial cell lysis and to P release to the solution (Blackwell et al. 2010).

Organic P Forms

The above discussed biological reactions can release or consume organic P compounds. These compounds are classified in P monoesters (such as inositol phosphates or sugar phosphates), P diesters (such as RNA, DNA, phospholipids), and phosphonates. Phosphonates are usually present in small concentrations in soil. The real concentration of different forms of monoesters and diesters is still a matter of debate as the method we use to analyze them (liquid state ^{31}P Nuclear Magnetic Resonance following a soil extraction with NaOH-EDTA) can produce a number of artifacts (Vestergren et al. 2012). Condensed forms of P (pyrophosphate, polyphosphate) are also produced by microorganisms, but they are not present in large concentrations in soils. Finally, similar to Pi, also organic P forms can be stabilized on soil particles. Myoinositol hexakisphosphate, in particular, can be rapidly and strongly sorbed on the surface of calcite or iron oxides (Celi and Barberis 2005). This sorption can decrease the accessibility of the organic P substrate to microorganisms and to enzymes.

Importance of the "Organic Pathway" for the P Cycle in Natural Ecosystems

Studying P dynamics in ecosystems that are little affected by human activities can give a good idea of the potential importance of the "organic pathway" as such systems obtain their P from weathering of parent material or from dust and then recycle P through plant residues and the soil food web (Peltzer et al. 2010). More specifically, studying the dynamics of P in soil chronosequences, i.e. on soils developed on the same parent material since different times, can provide information on this pathway. This discussion will show whether mimicking the functioning of natural ecosystems to design sustainable cropping systems as suggested by Doré et al. (2011) and Malézieux (2012) is relevant for improving P use by plants and decreasing P losses.

Importance of the "Organic Pathway" for P Use

Since the seminal publication of Walker and Syers (1976) a lot of research has been done on the relationships between soil phosphorus dynamics and soil development. According to our current paradigm, P is mainly present as apatite in the early stage of soil development, the stock of apatite then decreases while the stock of P present in organic forms increases, and finally at late soil development stages the stock of organic P forms decreases and the fraction of Pi bound to aluminum and iron oxides increases. At the beginning of pedogenesis, plant growth appears to be mainly

N-limited; while as the soil gets older the vegetation becomes more and more P-limited. After a certain time, which depends on the climate and the local soil conditions, soil P becomes so limiting that the ecosystem enters in retrogression (Peltzer et al. 2010).

Studies conducted in the Franz Josef Glacier chronosequence in New Zealand on soils developed since 5–120,000 years on schist and greywacke show that these changes in P forms and stocks are paralleled by changes in plant species, soil microbial communities and soil microfauna. At the younger sites, evergreen angiosperms are dominant while conifers (*Podocarpaceae*) appear in the 500 year-old site and become dominant at the older sites (Richardson et al. 2004). At the younger stages, bacteria dominate the microbial biomass, while at older stages fungi become dominant (Allison et al. 2007). Finally, microbial-feeding nematodes and omnivorous nematodes increase from the younger site to the 5000 year-old site, while at the older stages fungal-feeding nematodes are dominant (Doblas-Miranda et al. 2008). In parallel, Allison et al. (2007) measured an increase in phosphatase activities with increasing soil age. Turner et al. (2007) assessed organic P forms in this chronosequence by liquid state ^{31}P NMR following a NaOH-EDTA extraction and observed an increase of all organic P forms with soil age until 500 years and then a decrease of these forms at the older sites. Interestingly, the proportion of organic P present as extractable scylloinositol hexakisphosphate decreased after 5000 years, although this form of organic P was supposed to be very stable. Furthermore, with increasing soil age these authors observed an increase in the proportion of organic phosphorus present as DNA and in the proportion of extractable inorganic phosphorus present as pyrophosphate, although these forms were supposed to be rapidly mineralizable.

The importance of soil microorganisms for the soil P available pool has been re-evaluated recently by Tamburini et al. (2012) in another soil chronosequence which is located in the forefield of the Damma glacier in the Central Swiss Alps. This chronosequence focuses on early soil development on a granitic parent material (Bernasconi et al. 2011). Tamburini et al. (2012) analysed the natural abundance of $\delta^{18}O$ associated to phosphate, noted $\delta^{18}O_P$, in soil and plant P pools. Under ambient temperature and pressure, the $\delta^{18}O_P$ is mostly determined by the role of phosphatase enzymes that transfer O atoms from water molecules to phosphate and to a lesser extent by biological uptake (Tamburini et al. 2012). In their study, Tamburini et al. (2012) showed that the metabolic P (i.e. the P pool that is rapidly turned-over in leaf cells) of plants sampled on the forefield had a mean $\delta^{18}O_P$ of +22 ‰ while the parent material had a mean $\delta^{18}O_P$ of +7 ‰ which is to be expected for apatite of igneous origin. Soil available Pi (resin extractable Pi), however, showed an intermediate enrichment ($\delta^{18}O_P$ between +11 and +19 ‰), which was explained by a complete equilibration between O in phosphate and O in water within soil organisms at the temperature observed in the forefield. These results demonstrate that the entire pool of available Pi has been going through the soil microbial biomass at some point in time. Moreover, these results suggest that at the youngest site (12 years old) the P in the microbial biomass has a very short turn over time, probably as short as a few weeks.

Altogether, these results suggest that in these natural systems a complex set of processes develops with time involving plants, soil fauna and microorganism to maximize P use at ecosystem level (Peltzer et al. 2010). Here the "organic pathway" is definitively responsible for a high P use by organisms, and a better understanding of these mechanisms might help to increase P use efficiency in agricultural systems.

Importance of the "Organic Pathway" for P Losses

Is the "organic pathway" a guarantee that no P is lost from soils in natural systems? Unfortunately, the answer to this question is no. Pedogenetic P losses from soils actually lead to ecosystem retrogression (Peltzer et al. 2010). Walker and Syers (1976) had shown a strong decrease in soil total P with time in all the chronosequences they studied. In the Franz Josef Glacier chronosequence Walker and Syers (1976) observed a decrease in soil total P from 1292 g P m^{-2} in the initial soil to 76 g m^{-2} after 22,000 years (yielding an annual loss rate of 0.05 g P m^{-2} $year^{-1}$, while in another chronosequence (Manawatu) soil total P decreased from 509 g P m^{-2} in the initial soil to 300 g P m^{-2} after 10,000 years (0.02 g P m^{-2} $year^{-1}$). Using a pedogenic index, St. Arnaud et al. (1988) and Letkeman et al. (1996) showed that even in the relatively cold and moisture-limited environment of the Prairies in Western Canada, soil profiles could lose between 10 and 40 % of their total P during their relatively short pedogenesis (13,000 years). The maximum rate of P loss observed by Letkeman et al. (1996) was 0.006 g P m^{-2} $year^{-1}$. Letkeman et al. (1996) suggest that the redistribution of P within the profile and the P losses from the solum are strongly affected by the local soil environment (i.e. whether the profile is situated on a knoll or in a depression) and not only by the climatic zone in which the soil has developed. Furthermore, their results suggest that P is redistributed within the profile as Pi. In contrast, Schoenau and Bettany (1987) explained P losses from soil profiles from the same region by the leaching of low molecular weight organic P. More recently, Porder and Chadwick (2009) showed that P losses from soil profiles developed on basaltic lava of ages ranging from 10,000 to 370,000 years across a rainfall gradient ranging from 500 to 2000 mm $year^{-1}$ were highest at highest rainfall, and that they were related to Fe losses.

Importance of the "Organic Pathway" for P Availability and Losses in Western and Northern European Agricultural Systems

This part of the chapter first provides background information on assessing P use efficiency for crops. Then it discusses whether P going through the "organic pathway" can be used at 100 %. Finally it assesses whether the "organic pathway" allows retaining all the P in the soil profile.

Estimating the P Use Efficiency in Agricultural Systems

There is no absolute definition for P use efficiency for crops over the long-term (Syers et al. 2008; Chien et al. 2012). Chien et al. (2012) propose to calculate the apparent per cent recovery of fertilizer P (ARecov%) using Eq. 1.

$$\text{ARecov\%} = 100 \times \left(\text{Puptake}_{+\text{P}} - \text{Puptake}_{\text{OP}}\right)/(\text{Pfertilizer}) \qquad (1)$$

where $\text{Puptake}_{+\text{P}}$ is the amount of P taken up by a crop in the presence of a P fertilizer (kg P ha^{-1}), $\text{Puptake}_{\text{OP}}$ (kg P ha^{-1}) is the amount of P taken up by the same crop in the absence of P fertilization, and Pfertilizer (kg P ha^{-1}) is the total amount of P added to the crop as organic and/or mineral fertilizers. Equation 1 can be used to calculate an apparent recovery of fertilizer P for a given crop, and it can be used to calculate a cumulated apparent recovery (cumARecov%) for several rotations if we consider the cumulated amounts of P added to and removed from the system. Although Eq. 1 can be used in field trials in which different types of P applications are compared with a treatment receiving no P, there is a limit to it. P uptake by a plant from a fertilized soil ($\text{Puptake}_{+\text{P}}$) is the sum of P taken up from the fertilizer (Pdff, in kg P ha^{-1}) and of P taken up from the soil (Pdfs, in kg P ha^{-1}). Equation 1 can be rewritten as follows:

$$\text{ARecov\%} = 100 \times ((\text{Pdff} + \text{Pdfs}_{+\text{P}}) - \text{Pdfs}_{\text{OP}})/(\text{Pfertilizer}) \qquad (2)$$

With $\text{Pdfs}_{+\text{P}}$ = the amount of P derived from the soil and taken up by the crop in the fertilized treatment, and Pdfs_{OP} = amount of P derived from the soil and taken up by the crop in the non-fertilized treatment. Equation 1 implicitly assumes that $\text{Puptake}_{+\text{P}} - \text{Puptake}_{\text{OP}}$ is equal to Pdff or, in other words, that $\text{Pdfs}_{+\text{P}}$ and Pdfs_{OP} are identical. Therefore, ARecov% will be the true recovery of P added with fertilizer by the crop only if $\text{Pdfs}_{+\text{P}}$ and Pdfs_{OP}. are equal. However, usually field experiments do not allow quantifying $\text{Pdfs}_{+\text{P}}$ and Pdfs_{OP}. This is why ARecov% cannot show a real recovery rate but only an apparent one.

Only radioisotope labelling of the soil or the fertilizer with ^{32}P or ^{33}P can allow distinguishing the P derived from the soil from the P derived from a fertilizer in a plant that has been growing in the presence of P fertilizer (Frossard et al. 2011). Studies conducted with P radioisotopes in the presence of a freshly added organic or mineral fertilizer show that $\text{Pdfs}_{+\text{P}}$ and Pdfs_{OP} are often different from each other. For instance, white clover amended with compost as the only source of P in an acid soil had a higher Pdfs than when it had not received any P (so in this case $\text{Pdfs}_{+\text{P}}$ was higher than Pdfs_{OP}; Sinaj et al. 2002). This was explained by a positive influence of the compost addition on root system development followed by a better use of soil P in this treatment. On the contrary, Morel and Fardeau (1989) showed that $\text{Pfds}_{+\text{P}}$ decreased when English ryegrass was growing on high P soil with increasing addition of diammonium phosphate (so in this case $\text{Pdfs}_{+\text{P}}$ was smaller than Pdfs_{OP}). Therefore interactions between P inputs and P uptake by the crop from the soil exist and cannot be ignored. The main problem with radioisotope approaches is the short half-life of ^{32}P and ^{33}P, which prevent long-term studies. Nevertheless, knowing on the one side the cumARecov% of a given treatment over decades, and on the other side the short-term efficiency of a given type of fertilizer measured with isotopes, we should be able to draw relevant conclusions on the use of fertilizer and soil P by crops over the long-term.

Importance of the "Organic Pathway" in Determining P Use Efficiency

We analyze here one example each of cropping and grassland systems for which we have sufficient data.

Cropping systems Let us consider the DOK field experiment in which two conventional cropping systems (CONMIN and CONFYM) are being compared with two organic cropping systems (BIOORG and BIODYN) and with a non-fertilized treatment (NOFERT). This experiment is located near Basel (Switzerland) on a deep luvisol and has been running since 1978. This trial has been described many times in the literature (see e.g. Mäder et al. 2006). The main characteristics of the treatments are given in Table 1. A lot of results have been gathered on P dynamics in this trial

Table 1 Fertilization and plant protection in the organic and conventional cropping systems and in the unfertilized control of the DOK long-term field experiment, with average P input and P balances for the period 1978–2006

	Control	Organic systems		Conventional systems	
	NOFERT	BIODYN	BIOORG	CONFYM	CONMIN[a]
	Unfertilized control	Bio-dynamic	Bio-organic	Conventional with manure	Conventional, exclusively mineral fertilizers
Fertilization: type and level	None	Composted farmyard manure (FYM), aerated slurry from 1.2/1.4 LU ha^{-1} year^{-1}[b], preparations[c]	Slightly rotted manure and aerated slurry from 1.2/1.4 LU ha^{-1} year^{-1}[b, d]	Manure, slurry and complement of mineral fertilizer according to official guidelines	Mineral fertilizer according to official guidelines
Plant protection					
Weed control	Mechanical			Mechanical and herbicides	
Disease control	Indirect methods		Indirect methods, until 1991 copper	Fungicides (thresholds)	
Insect control	Plant extracts, bio-control			Insecticides (thresholds)	
Special treatments	Bio-dynamic preparations		None	Plant growth regulators	

[a]CONMIN was unfertilized in the 1st crop rotation period from 1978 to 1984. [b]Increase from 1.2 to 1.4 livestock units (LU) ha^{-1} year^{-1} at the beginning of the 3rd crop rotation period (1992–1998). [c]Biodynamic composting additive. [d]From 1978 till 1991 about 14 % of P input was applied in form of rock phosphate

(see e.g. Oberson et al. 2011; Keller et al. 2012). Selected characteristics of soils sampled under different cropping systems are shown in Table 2. These data show that the BIODYN treatment has the highest total C, N, and microbial C, N and P contents, suggesting that the soil in this treatment has the highest biological activity. The same conclusion has been drawn from other studies conducted in the DOK field experiment (Mäder et al. 2006; Fliessbach et al. 2007). It is therefore sound to assume that more P is being transferred through the "organic pathway" in the BIODYN treatment than in the other treatments.

The cumulated apparent recovery of fertilizer P (cumARecov%) calculated for the different treatments is shown in Table 3 for the entire duration of the trial. The small cumARecov% observed at the end of rotation 1 in all treatments (Table 3) can be explained by a high P uptake by crops in the non-fertilized treatment at the beginning of the trial. The amount of P taken up by crops in the non-fertilized treatment then decreased and stabilized explaining the subsequent changes in cumARecov% with time. None of the treatments allows for a 100 % apparent recovery of the added P fertilizers (cumARecov%, Table 2). The maximum cumARecov% is 49 % and is observed in the BIODYN treatment at the end of the 4th rotation, while the other treatments show cumARecov% ranging between 40 and 42 %. Will it be possible to reach a cumulative recovery rate of 100 % in the future? We argue that it will be not possible as this parameter is negatively correlated with the amount of soil available Pi (Fig. 2a, $p = 0.0002$ when all points are considered in the regression). The same type of relation is observed between cumARecov% and total inorganic P (Fig. 2b, $p = 0.001$ when all points are considered in the regression). On the contrary, a positive relation is observed between total organic P and cumARecov% (Fig. 2c, $p = 0.03$ when all points are considered in the regression).

How can the higher cumARecov% observed in the treatment BIODYN at the end of the 3rd and of the 4th rotation be explained? Is this really due to an increased transfer of fertilizer P through the "organic pathway"? Or is it an artifact? The first explanation for this high cumARecov% in BIODYN is the low rate of P inputs. It is indeed known that P use efficiency of freshly added fertilizer decreases when soil Pi availability increases (Gallet et al. 2003; Oberson et al. 2010). It is possible that this higher cumARecov% is also due to an increased transfer of P through the "organic pathway" as the soil of the BIODYN treatment presents the highest values in organic P, in microbial P, N and C and the highest rate of P mineralization rates (Table 2). However, this high cumARecov% could also be an artifact, concealing a stronger use of P derived from the soil by the crops growing in the BIODYN treatment than in the other treatments (i.e. this could be a case where $Pdfs_{+P} > Pdfs_{OP}$). In a pot experiment with Italian ryegrass, Oberson et al. (2010) indeed showed that the residual effect of P fertilizers added in the BIODYN treatment was lower than the residual effect of P fertilizers added in the BIOORG, CONMIN and CONFYM treatments. In the absence of any fresh P addition, the efficiency of residual P fertilizers reached 8.7 % in BIODYN, 12.2 % in BIOORG, 13.4 % in CONMIN, and 15 % in CONFYM (Oberson et al. 2010). The similar residual P efficiency in BIOORG and CONMIN would suggest that all the P added as manure and slurry in BIOORG has been finally released as Pi and used by the

Table 2 Total soil C, N, inorganic P, organic P, soil available Pi (amount of P isotopically exchangeable within 1 min, E1 min), soil microbial C, P and N measured in different treatments of the DOK field experiment and daily rate of organic P mineralization

		Data from Keller et al. (2012)								Data from Oehl et al. (2004)
Treatments	Sampling year	Total C (g kg^{-1})	Total N (g kg^{-1})	Total Pi (mg kg^{-1})	Total Po (mg kg^{-1})	Available Pi (E1 min) (mg kg^{-1})	Microbial biomass			Net P mineralization rate (mg P kg^{-1} day^{-1})
							P (mg kg^{-1})	C (mg kg^{-1})	N (mg kg^{-1})	
Original	1977			386	412	12.1				nd
BIODYN	2007	16.7	1.75	260	454	3.1	19.2	160	27	2.0
BIOORG	2007	14.1	1.44	279	419	4.0	14.2	118	21	1.4
CONFYM	2007	13.6	1.42	346	403	8.0	14.9	128	19	nd
CONMIN	2007	14.2	1.50	317	383	5.9	12.1	118	18	1.3
NOFERT	2007	12.1	1.31	222	351	1.0	7.2	88	14	nd

Table 3 Average P inputs and outputs, monitored for each rotation period, as well as the cumulated apparent recovery of P fertilizers (cumARecov%) in the different treatments of the DOK field

Rotation	1st rotation (1977–1985)			2nd rotation (1986–1991)			3rd rotation (1992–1998)			4th rotation (1999–2005)		
Treatments	Input (kg P ha^{-1} $year^{-1}$)	Output (kg P ha^{-1} $year^{-1}$)	cumARecov (%)	Input (kg P ha^{-1} $year^{-1}$)	Output (kg P ha^{-1} $year^{-1}$)	cumARecov (%)	Input (kg P ha^{-1} $year^{-1}$)	Output (kg P ha^{-1} $year^{-1}$)	cumARecov (%)	Input (kg P ha^{-1} $year^{-1}$)	Output (kg P ha^{-1} $year^{-1}$)	cumARecov (%)
BIODYN	28.5	30.5	15.1	27.5	32.6	28.9	16.6	33.0	46.0	20.4	30.3	49.4
BIOORG	29.3	31.2	17.1	25.2	33.7	33.0	25.0	31.6	42.5	24.8	26.3	40.6
CONFYM	51.1	35.9	19.0	43.9	39.2	29.7	34.0	42.6	42.6	40.6	34.0	42.0
CONMIN	0.0	27.5		46.1	34.1	31.9	36.3	35.7	42.0	36.7	33.2	42.0
NOFERT	0.0	26.2		0.0	20.7		0.0	15.8		0.0	17.8	

Note that the CONMIN treatment did not receive any fertilization during the first rotation

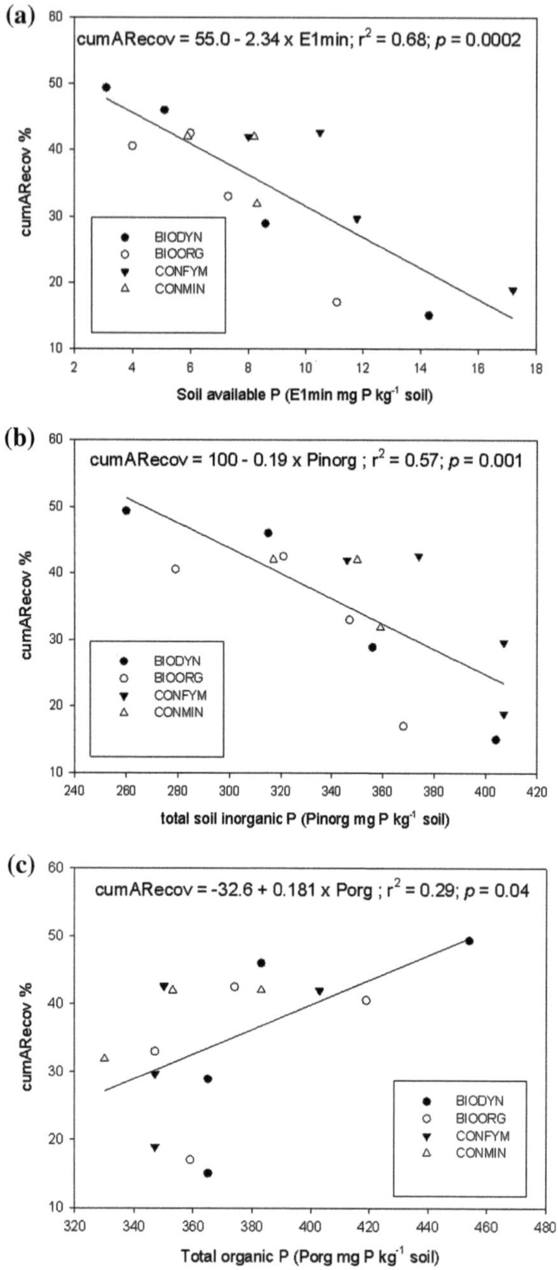

Fig. 2 **a** Relationship between the cumulated apparent recovery of added fertilizer calculated after the 1st, 2nd, 3rd and 4th rotations (cumARecov%) and soil available Pi measured as the amount of P isotopically exchangeable within 1 min (E1 min, mg P kg^{-1} soil); **b** relationship between cumARecov% and total soil inorganic P; **c** relationship between cumARecov% and total soil organic P in the four fertilized treatments of the DOK field trial (Data from Oehl et al. (2002) and from Astrid Oberson unpublished)

plants to the same degree as the P added as water soluble Pi in the CONMIN treatment. Indeed, Keller et al. (2012) comparing soil samples taken in the DOK field experiment in 1977 to those taken in 2007 did not observe any significant change in organic P in the organically fertilized treatments while a strong decrease in total inorganic P was observed. They concluded that in the long term the organic forms of P added with manure and slurry had been totally mineralized. The low residual P fertilizers effect of P added in the BIODYN treatment can therefore not be accounted by a lower fertilizing value of organic P forms and must be searched elsewhere. Oberson et al. (2010) proposed the following hypotheses to explain the low residual P fertilizer effect observed in BIODYN: (i) the production of scarcely soluble calcium phosphates during manure composting, (ii) the maintenance of soil pH at 6.5 hindering the dissolution of these calcium phosphates, and (iii) an increased competition between the soil microbial biomass and the plant for P. Altogether, this would in the long term lead to an increased P uptake from the soil by the crops and to a relative enrichment with scarcely soluble forms of P in the BIODYN treatment. The low rate of P inputs and the high carbon content in the BIODYN treatment would also be in line with an increased P uptake from the soil as low Pi availability favours root colonization by arbuscular mycorrhizal fungi (Hildermann et al. 2010) while organic matter improves soil structure (Siegrist et al. 1998). A similar effect of organic matter improving soil structure and the use of soil P has been reported by Syers et al. (2008). As improving soil P use is a globally pursued goal (Richardson et al. 2011; Simpson et al. 2011; Sattari et al. 2012), analyzing the responsible mechanisms in the BIODYN treatment is relevant. However, after 30 years, P availability is becoming limiting for plant productivity in the BIODYN treatment, and thus this management option has its limits.

Permanent grasslands We will consider two field trials conducted in 2001/2002 in the vicinity of Zurich (Switzerland) on a regosol/phaenozem complex at site I and on a cambisol at site II. These experiments, which were conducted to evaluate methods to limit P losses from permanent grasslands are described in Schärer et al. (2007). We consider here 3 of the 4 treatments studied by Schärer et al. (2007): control, the grassland was managed as it used to be before the start of the study, but nutrients were brought in a water soluble form at the rate of 40 kg P ha^{-1} $year^{-1}$ at site I and at the rate of 22 kg P ha^{-1} $year^{-1}$ at site II; zero P, the grassland was managed as before but without P addition; tillage, the grassland was tilled in May 2001 at 25 cm depth and re-sown with a mixture of grasses and legumes. In both years, the forage yield varied between 11 and 15 t DM ha^{-1} at site I and between 6 and 12 t DM ha^{-1} at site II. Data shown in Table 4 demonstrate that available Pi and microbial P were higher in the 0–4 cm than in the 4–20 cm layer. The amount of available Pi was not strongly affected by the treatments, except by tillage, which reduced P availability in the 0–4 cm. The amount of microbial P was extremely high in the 0–4 cm horizon (between 77 and 110 mg P kg^{-1} soil) at both sites in the control and zero P treatments while it decreased strongly after ploughing. In the 4–20 cm horizon microbial P was still high but unaffected by the different treatments (between 40 and 50 mg P kg^{-1} soil). If we consider for the 0–4 cm horizon an apparent density of 1 T m^{-3} and for the 4–20 cm an apparent density of 1.2 T m^{-3}, we come to the conclusion that in summer 2001 the upper horizon of site I contained 86–136 kg P ha^{-1} in microbial form, with similar numbers (93–128 kg P ha^{-1}) at site II. Similar high microbial P contents were reported in other fertilized permanent grasslands (see e.g. Liebisch 2011; Bünemann et al. 2012). Assuming further a turnover rate of 3 months for microbial P and a growing season of 6 months, we come to the conclusion that the fluxes of P through the microbial biomass are much larger than the amount of P removed by cuts (Table 4). It is plausible that the added P fertilizer gets recycled through the microbial biomass before being taken up by the plants, but this should be confirmed experimentally. Altogether these data suggest that the "organic pathway" is important for plant nutrition in grassland systems. This "organic pathway" is even more significant when P is applied as manure or as slurry, which is the rule rather than the exception as production systems that include grasslands also include animal keeping. As in cropping systems however the fact that a high flux of P is going through the "organic pathway" is not a guarantee that P fertilizer is used at 100 %. P removal data shown in Table 4 and P inputs mentioned above allow calculating the cumARecov% for the control treatment which reaches over the 2 years 5.5 % in site I and 9.5 % in site II when compared to the zero P treatment. As for the DOK example, the relatively low cumARecov% (well below 100 %) is related to the high P uptake from the non-fertilized treatment at both sites.

Importance of the "Organic Pathway" for P Losses

We have seen in a previous section that soils under natural systems can lose a substantial amount of P during their pedogenesis. But can the "organic pathway" hinder these losses in agricultural systems? Letkeman et al. (1996) studied a cultivated site and showed that it had lost more P than a similar site under native vegetation. The increases in losses were larger than the estimated Premoval by the crops. So cultivation led to increased rates of P losses.

P loss from soils is a function of soil properties (including Pi availability), water dynamics and crop management, while P delivery to water bodies is a function of topography and connectivity between P sources and sinks (Haygarth et al. 2005). Measures that lead to a decrease in soil Pi availability can decrease Pi losses (Quinton et al. 2003). In some cases, when Pi availability becomes very low, then crops can uplift P from deeper horizons to the surface (Oehl et al. 2002). In an experiment conducted in pots with a low organic matter sandy soil and in the presence of high rainfall intensity (1 mm min^{-1}), inoculation with arbuscular mycorrhizal fungi was shown to reduce Pi leaching (van der Heijden 2010). But whether arbuscular mycorrhizal fungi could actually be managed in agricultural systems so as to decrease P losses seems doubtful. Blackwell et al. (2010) demonstrated that freezing/thawing and drying/wetting cycles led to the release of microbial P to the solution but they also concluded that this would have a limited impact on P losses from soils.

As P losses and delivery are very heterogeneous in space and time, it is actually very difficult to control P losses from agro-ecosystems. Whereas the "organic pathway" may help to maintain P in organic forms, this is not a guarantee that P will not be lost from the soil. Here the best strategy to minimize losses is to decrease P

Table 4 Amounts of P that is isotopically exchangeable within 1 min (E1min), of microbial P and of P exported by forage in 2 permanent grassland soils (site I and site II) that have been submitted to different treatments: business as usual (control; P added according to plant needs), business as usual but no P added (zero P), pasture tilled and re-sown without P addition (tillage)

		Isotopically exchangeable P E(1 min) (mg P kg^{-1} soil)		Microbial P* P_{mic} (mg P kg^{-1} soil)			P removed kg P ha^{-1} $year^{-1}$	
Depth/site	Treatment	July 2001	Sept 2001	July 2001	Sept 2001	Oct 2002	2001	2002
Site I 0–4 cm	Control	7.3 ± 1.28	6.9 ± 2.04	106.6 ± 22.4	77.4 ± 11.2	105.9 ± 12.8	54.0 ± 7.1	41.8 ± 3.6
	Zero P	10.5 ± 2.41	7.8 ± 1.72	104.2 ± 21.1	88.6 ± 19.4	110.0 ± 5.0	54.2 ± 10.2	37.2 ± 3.4
	Tillage	4.1 ± 1.28	3.2 ± 1.15	19.8 ± 2.35	30.4 ± 8.79	48.1 ± 13.0	38.0 ± 2.9	45.3 ± 4.5
Site I 4–20 cm	Control	2.6 ± 0.53	1.8 ± 1.04	42.8 ± 17.3	40.6 ± 7.11	nd		
	Zero P	3.7 ± 0.88	2.6 ± 1.31	49.1 ± 15.4	38.1 ± 8.47	nd		
	Tillage	3.2 ± 0.38	3.8 ± 1.59	40.9 ± 8.67	44.0 ± 10.0	nd		
Site II 0–4 cm	Control	7.3 ± 1.53	7.5 ± 1.03	103.0 ± 17.9	81.0 ± 17.8	86.9 ± 11.6	34.8 ± 1.4	50.6 ± 3.2
	Zero P	7.6 ± 2.95	6.0 ± 1.64	103.4 ± 8.42	93.8 ± 12.3	79.8 ± 6.6	35.0 ± 3.2	46.2 ± 2.2
	Tillage	6.3 ± 1.68	4.2 ± 0.41	62.4 ± 26.9	38.9 ± 6.67	37.9 ± 8.0	26.6 ± 3.3	49.3 ± 3.7
Site II 4–20 cm	Control	4.0 ± 0.82	3.8 ± 1.03	45.2 ± 20.5	41.2 ± 3.20	nd		
	Zero P	3.7 ± 0.81	3.1 ± 0.71	43.0 ± 8.13	38.8 ± 5.83	nd		
	Tillage	5.0 ± 0.97	4.1 ± 0.41	49.6 ± 0.72	40.6 ± 4.58	nd		

Mean and standard deviations are shown

*Microbial P has been extracted by a resin in the presence of hexanol. The results are given without the use of a kp factor. *nd* not determined

inputs to the absolute minimum that is required to achieve expected yields and if necessary to implement additional methods as described by the research groups who worked in the COST action 869 (http://www.cost869.alterra.nl/).

References

Achat DL, Morel C, Bakker MR, Augusto L, Pellerin S, Gallet-Budynek A, Gonzalez M (2010) Assessing turnover of microbial biomass phosphorus: combination of an isotopic dilution method with a mass balance model. Soil Biol Biochem 42:2231–2240

Allison VJ, Condron LM, Peltzer DA, Richardson SJ, Turner BL (2007) Changes in enzyme activities and soil microbial community composition along carbon and nutrient gradients at the Franz Josef chronosequence, New Zealand. Soil Biol Biochem 39:1770–1781

Benitez-Nelson CR (2000) The biogeochemical cycling of phosphorus in marine systems. Earth Sci Rev 51:109–135

Bernasconi SM, Bauder A, Bourdon B, Brunner I, Bünemann E, Christl I, Derungs N, Edwards P, Farinotti D, Frey B, Frossard E, Furrer G, Gierga M, Göransson H, Gülland K, Hagedorn F, Hajdas I, Hindshaw R, Ivy-Ochs S, Jansa J, Jonas T, Kiczkac M, Kretzschmar R, Lemarchand E, Luster J, Magnusson J, Mitchell EAD, Olde Venterink H, Plötze M, Reynolds B, Smittenberg R, Stählin M, Tamburini F, Tipper ET, Wacker L, Welc M, Wiederhold JG, Zeyer J, Zimmermann S, Zumsteg A (2011) Chemical and biological gradients along the Damma Glacier soil chronosequence (Switzerland). Vadose Zone J 10:867–883

Blackwell MSA, Brookes RC, de la Fuente-Martinez N, Gordon H, Murray PJ, Snars KE, Williams JK, Bol R, Haygarth PM (2010) Phosphorus solubilization and potential transfer to surface waters from the soil microbial biomass following drying-wetting and freezing-thawing. Adv Agron 106:1–35

Bünemann EK, Oberson A, Frossard E (2011) Phosphorus in action. Soil Biol, vol 26. Springer, Berlin

Bünemann EK, Oberson A, Liebisch F, Keller F, Annaheim KE, Huguenin-Elie O, Frossard E (2012) Rapid microbial phosphorus immobilization dominates gross phosphorus fluxes in a grassland soil with low inorganic phosphorus availability. Soil Biol Biochem 51:84–95

Celi L, Barberis E (2005) Abiotic stabilization of organic phosphorus in the environment. In: Turner BL, Frossard E, Baldwin DS (eds) Organic phosphorus in the environment. CABI, Wallingford, pp 113–132

Ceulemans T, Merckx R, Hens M, Honnay O (2011) A trait-based analysis of the role of phosphorus vs. nitrogen enrichment in plant species loss across North-west European grasslands. J Appl Ecol 48:1155–1163

Chien SH, Sikora FJ, Gilkes RJ, McLaughlin MJ (2012) Comparing of the difference and balance methods to calculate percent recovery of fertilizer phosphorus applied to soils: a critical discussion. Nutr Cycl Agroecosys 92:1–8

Doblas-Miranda E, Wardle DA, Peltzer DA, Yeates GW (2008) Changes in the community structure and diversity of soil invertebrates across the Franz Josef Glacier chronosequence. Soil Biol Biochem 40:1069–1081

Doré T, Makowski D, Malézieux E, Munier-Jolain N, Tchamitchian M, Tittonell P (2011) Facing up to the paradigm of ecological intensification in agronomy: revisiting methods, concepts and knowledge. Eur J Agron 34:197–210

Dyhrman ST, Ammerman JW, Van Mooy BAS (2007) Microbes and the marine phosphorus cycle section IV. Processes, chapter 9. Microbes and major elemental cycles. Oceanography 20: 110–116

Ehlers K, Bakken LR, Frostegård Å, Frossard E, Bünemann EK (2010) Phosphorus limitation in a Ferralsol: impact on microbial activity and cell internal P-pools. Soil Biol Biochem 42: 558–566

FAO (2003) http://www.fao.org/docrep/meeting/006/y8704e.htm, checked 14 April 2012
Filippelli GM (2011) Phosphate rock formation and marine phosphorus geochemistry: the deep time perspective. Chemosphere 84:759–766
Fliessbach A, Oberholzer HR, Gunst L, Mader P (2007) Soil organic matter and biological soil quality indicators after 21 years of organic and conventional farming. Agric Ecosys Environ 118:273–284
Follain S, Schvartz C, Denoroy P, Villette C, Saby NPA, Arrouays D, Lemercier B, Walter C (2009) A method for assessing available phosphorus content in arable topsoils over large spatial scales. Agron Sustain Dev 29:371–379
Follmi KB (1996) The phosphorus cycle, phosphogenesis and marine phosphate-rich deposits. Earth Sci Rev 40:55–124
Frossard E, Bünemann E, Jansa J, Oberson A, Feller C (2009) Concepts and practices of nutrient management in agro-ecosystems: can we draw lessons from history to design future sustainable agricultural production systems? Die Bodenkultur 60:43–60
Frossard E, Achat DL, Bernasconi SM, Bünemann EK, Fardeau JC, Jansa J, Morel C, Rabeharisoa L, Randriamanantsoa L, Sinaj S, Tamburini F, Oberson A (2011) The use of tracers to investigate phosphate cycling in soil/plant systems. In: Bünemann EK, Oberson A, Frossard E (eds) Phosphorus in action. Soil Biol, vol 26. Springer, Berlin, pp 59–91
Frossard E, Bünemann EK, Oberson A, Jansa J, Kertesz MA (2012) Phosphorus and sulfur in soil. In: Huang PM, Li Y, Summer ME (eds) Part IV soil biology and biochemistry of the handbook of soil sciences properties and processes, vol 26, second edn. CRC Press, Boca Raton, pp 2–15
Gallet A, Flisch R, Ryser J-P, Nösberger J, Frossard E, Sinaj S (2003) Uptake of residual phosphate and freshly applied diammonium phosphate by *Lolium perenne* and *Trifolium repens*. J Plant Nutr Soil Sci 166:557–567
George TS, Fransson AM, Hammond JP, White PJ (2011) Phosphorus nutrition: rhizosphere processes, plant response and adaptation. In: Bünemann EK, Oberson A, Frossard E (eds) Phosphorus in action. Soil Biol, vol 26. Springer, Berlin, pp 245–274
Göransson H, Venterink HO, Bååth E (2011) Soil bacterial growth and nutrient limitation along a chronosequence from a glacier forefield. Soil Biol Biochem 43:1333–1340
Granger SJ, Bol R, Anthony S, Owens PN, White SM, Haygarth PM (2010) Towards a holistic classification of diffuse agricultural water pollution from intensively managed grasslands on heavy soils. Adv Agron 105:83–115
Haygarth PM, Condron LM, Heathwaite AL, Turner BL, Harris GP (2005) The phosphorus transfer continuum: Linking source to impact with an interdisciplinary and multi-scaled approach. Sci Total Environ 344:5–14
Hildermann I, Messmer M, Dubois D, Boller T, Wiemken A, Mader P (2010) Nutrient use efficiency and arbuscular mycorrhizal root colonisation of winter wheat cultivars in different farming systems of the DOK long-term trial. J Sci Food Agric 90:2027–2038
Hoffland E, Findenegg GR, Nelemans JA (1989) Solubilization of rock phosphate by rape. II. Local root exudation of organic acids as a response to P starvation. Plant Soil 113:161–165
Irshad U, Villenave C, Brauman A, Plassard C (2011) Grazing by nematodes on rhizosphere bacteria enhances nitrate and phosphorus availability to *Pinus pinaster* seedlings. Soil Biol Biochem 43:2121–2126
Jansa J, Finlay R, Wallander H, Smith FA, Smith SE (2011) Role of mycorrhizal symbioses in phosphorus cycling. In: Bünemann EK, Oberson A, Frossard E (eds) Phosphorus in action. Soil Biol, vol 26. Springer, Berlin, pp 137–168
Jones DL, Oburger E (2011) Solubilization of phosphorus by soil microorganisms. In: Bünemann EK, Oberson A, Frossard E (eds) Phosphorus in action. Soil Biol, vol 26. Springer, Berlin, pp 169–198
Keller M, Oberson A, Annaheim KA, Tamburini F, Mäder P, Mayer J, Frossard E, Bünemann EK (2012) Phosphorus forms and enzymatic hydrolysability of organic phosphorus in soils after 30 years of organic and conventional farming. J Plant Nutr Soil Sci 175:385–393

Khorassani R, Hettwer U, Ratzinger A, Steingrobe B, Karlovsky P, Claassen N (2011) Citramalic acid and salicylic acid in sugar beet root exudates solubilize soil phosphorus. BMC Plant Biol 11 (Article Number: 121). doi:10.1186/1471-2229-11-121

Lemercier B, Gaudin L, Walter C, Aurousseau P, Arrouays D, Schvartz C, Saby NPA, Follain S, Abrassart J (2008) Soil phosphorus monitoring at the regional level by means of a soil test database. Soil Use Manage 24:131–138

Letkeman LP, Tiessen H, Campbell CA (1996) Phosphorus transformations and redistribution during pedogenesis of western Canadian soils. Geoderma 71:201–218

Liebisch F (2011) Plant and soil indicators to assess the phosphorus nutrition status of agricultural grasslands. PhD thesis, Dissertation. ETH Zurich, Switzerland. URL http://e-collection.library.ethz.ch/view/eth:2757, 144 p

Mäder P, Fliessbach A, Dubois D, Gunst L, Jossi W, Widmer F, Oberson A, Frossard E, Oehl F, Wiemken A, Gattinger A, Niggli U (2006) The DOK experiment (Switzerland). In: Raupp J, Pekrun C, Oltmanns M, Köpke U (eds) Long term field experiments in organic farming. Verlag Dr. Köster, Berlin, pp 41–58

Malézieux E (2012) Designing cropping systems from nature. Agron Sustain Dev 32:15–29

McLaughlin MJ, Alston AM (1986) The relative contribution of plant residues and fertiliser to the phosphorus nutrition of wheat in a pasture/cereal system. Aust J Soil Res 24:517–526

Morel C, Fardeau JC (1989) Native soil and fresh fertilizer phosphorus uptake as affected by rate of application and P fertilizers. Plant Soil 115:123–128

Morel C, Plenchette C (1994) Is the isotopically exchangeable phosphate of a loamy soil the plant available P? Plant Soil 158:287–297

Nachimuthu G, Guppy C, Kristiansen P, Lockwood P (2009) Isotopic tracing of phosphorus uptake in corn from P-33 labelled legume residues and P-32 labelled fertilisers applied to a sandy loam soil. Plant Soil 314:303–310

Oberson A, Joner EJ (2005) Microbial turnover of phosphorus in soil. In: Turner BL, Frossard E, Baldwin DS (eds) Organic phosphorus in the environment. CABI, Wallingford, pp 133–164

Oberson A, Tagmann HU, Langmeier M, Dubois D, Mäder P, Frossard E (2010) Fresh and residual phosphorus uptake by ryegrass from soils having different fertilization histories. Plant Soil 334:391–407

Oberson A, Bünemann E, Frossard E, Pypers P (2011) Management impacts on biological P cycling in cropped soils. In: Bünemann EK, Oberson A, Frossard E (eds) Phosphorus in action. Soil Biol, vol 26. Springer, Berlin, pp 431–458

Oehl F, Oberson A, Tagmann HU, Besson JM, Dubois D, Mäder P, Roth HR, Frossard E (2002) Phosphorus budget and phosphorus availability in soils under organic and conventional farming. Nutr Cycl Agroecosys 62:25–35

Oehl F, Frossard E, Fliessbach A, Dubois D, Oberson A (2004) Basal organic phosphorus mineralization in soils under different farming systems. Soil Biol Biochem 36:667–675

Paul EA (2007) Soil microbiology, ecology and biochemistry, 3rd edn. Academic Press, Burlington

Peltzer DA, Wardle DA, Allison VJ, Baisden WT, Bardgett RD, Chadwick OA, Condron LM, Parfitt RL, Porder S, Richardson SJ, Turner BL, Vitousek PM, Walker J, Walker LR (2010) Understanding ecosystem retrogression. Ecol Monogr 80:509–529

Porder S, Chadwick OA (2009) Climate and soil-age constraints on nutrient uplift and retention by plants. Ecology 90:623–636

Quinton JN, Strauss P, Miller N, Azazoglu E, Yli-Halla M, Uusitalo R (2003) The potential for soil phosphorus tests to predict phosphorus losses in overland flow. J Plant Nutr Soil Sci 166:432–437

Richardson SJ, Peltzer DA, Allen RB, McGlone MS, Parfitt RL (2004) Rapid development of phosphorus limitation in temperate rainforest along the Franz Josef soil chronosequence. Oecologia 139:267–276

Richardson AE, Lynch JP, Ryan PR, Delhaize E, Smith FA, Smith SE, Harvey PR, Ryan MH, Veneklaas EJ, Lambers H, Oberson A, Culvenor RA, Simpson RJ (2011) Plant and microbial strategies to improve the phosphorus efficiency of agriculture. Plant Soil 349:121–156

Sanchez PA, Shepherd KD, Soule MJ, Place FM, Buresh RJ, Izac A-MN (1997) Soil fertility replenishment in Africa: an investment in natural resource capital. In: Buresh RJ, Sanchez PA, Calhoun F (eds) Replenishing soil fertility in Africa, SSSA special publication number 51. Soil Science Society of America, American Society of Agronomy, Madison, Wisconsin, USA, pp 1–47

Sattari SZ, Bouwman AF, Giller KE, van Ittersum MK (2012) Residual soil phosphorus as the missing piece in the global phosphorus crisis puzzle. Proc Nat Acad Sci USA 109:6348–6353

Schärer M, Stamm C, Vollmer T, Frossard E, Oberson A, Flühler H, Sinaj S (2007) Reducing phosphorus losses from over-fertilized grassland soils proves difficult in the short term. Soil Use Manage 23:154–164

Schoenau JJ, Bettany JR (1987) Organic matter leaching as a component of carbon, nitrogen, phosphorus and sulphur cycles in a forest, grassland and gleyed soil. Soil Sci Soc Am J 51:646–651

Schröder JJ, Cordell D, Smit AL, Rosemarin A (2010) Sustainable use of phosphorus. EU Tender ENV.B.1/ETU/2009/0025 Plant Research International, Wageningen UR, report 357, 123 pages + annexes

Siegrist S, Schaub D, Pfiffner L, Mäder P (1998) Does organic agriculture reduce soil erodibility? The results of a long-term field study on loess in Switzerland. Agric Ecosys Environ 69: 253–264

Simpson RJ, Oberson A, Culvenor RA, Ryan MH, Veneklaas EJ, Lambers H, Lynch JP, Ryan PR, Delhaize E, Smith FA, Smith SE, Harvey PR, Richardson AE (2011) Strategies and agronomic interventions to improve the phosphorus-use efficiency of farming systems. Plant Soil 349:89–120

Sims T, Sharpley AN (eds) (2005) Phosphorus: agriculture and the environment. Agronomy monograph 46. ASA/CSSA/SSSA, Madison, WI

Sinaj S, Traore O, Frossard E (2002) Effect of compost and soil properties on the availability of composts phosphate for white clover (*Trifolium repens* L.). Nutr Cycl Agroecosys 62:89–102

Sinaj S, Richner W, Flisch R, Charles R (2009) Données de base pour la fumure des grandes cultures et des herbages. Rev Suisse d'agriculture 41:1–98

Spiess E (2011) Nitrogen, phosphorus and potassium balances and cycles of Swiss agriculture from 1975 to 2008. Nutr Cycl Agroecosys 91:351–365

St Arnaud RJ, Stewart JWB, Frossard E (1988) Application of the "pedogenic index" to soil fertility studies in Saskatchewan. Geoderma 43:21–32

Syers JK, Johnston AE, Curtin D (2008) Efficiency of soil and fertilizer phosphorus use. Reconciling changing concepts of soil phosphorus behaviour with agronomic information. FAO fertilizer and plant nutrition bulletin 18, Food and Agriculture Organization of the United Nations Rome

Tamburini F, Pfahler V, Bünemann EK, Guelland K, Bernasconi SM, Frossard E (2012) Oxygen isotopes unravel the role of microorganisms in phosphate cycling in soils. Environ Sci Technol 46:5956–5962

Thibaud MC, Morel C, Fardeau JC (1988) Contribution of phosphorus issued from crop residues to plant nutrition. Soil Sci Plant Nutr (Japan) 34:481–491

Turner BL, Condron LM, Richardson SJ, Peltzer DA, Allison VJ (2007) Soil organic phosphorus transformations during pedogenesis. Ecosystems 10:1166–1181

van der Heijden MAG (2010) Mycorrhizal fungi reduce nutrient loss from model grassland ecosystems. Ecology 91:1163–1171

Vestergren J, Vincent AG, Jansson M, Persson P, Ilstedt U, Gröbner G, Giesler R, Schleucher J (2012) High-resolution characterization of organic phosphorus in soil extracts using 2D $^{1}H–^{31}P$ NMR correlation spectroscopy. Environ Sci Technol 46:3950–3956

Walker TW, Syers JK (1976) Fate of phosphorus during pedogenesis. Geoderma 15:1–19

Wassen MJ, Olde Venterink H, Lapshina ED, Tanneberger F (2005) Endangered plants persist under phosphorus limitation. Nature 437:547–550

Determination of Plant Available P in Soil

Markku Yli-Halla, Judith Schick, Sylvia Kratz and Ewald Schnug

Abstract Fertilization should be based on a proper diagnosis of the plant nutritional status. Without, there is always the risk of under fertilization causing economic losses by not exploiting the yield potential of the site and its crop. On the opposite, over fertilization not only hampers fertilizer economy through inefficient nutrient rates, but also causes serious environmental impacts on neighboring ecosystems. There are four basic diagnostic methods each of which has its advantages and disadvantages: *ceteris paribus* fertilizer trials are complicated and time consuming delivering results far too late for fertilization planning. Visual assessment is fast but requires experts eyes and works only for a few nutrients and only when severe nutrient deficiency occurs. Plant analysis is accurate, however, bound to well defined growth stages, but shows only the actual stage of supply with little information about available reserves in the soil or growth media and is also too late with its results for practical fertilization. Last but not least: analyzing the soil can be independent of crop development, well in advance to contribute for the fertilizer design for the actual crop and offers insight into the reserves in the substrate. However, its value for fertilization planning is strongly depending on the quality of the calibration of the results. Of all essential plant nutrients phosphorus is the one for which the most and intensive research work on assessing soils has been conducted in the past. This chapter introduces the basic conceptions of soil analysis for plant available P, provides an overview on available methods and discusses their advantages and disadvantages.

Keywords Soil analysis · Extraction of phosphorus · Comparability of soil test results

M. Yli-Halla (✉)
Department of Food and Environmental Sciences, University of Helsinki,
P.O. Box 27, 00014 Helsinki, Finland
e-mail: markku.yli-halla@helsinki.fi

J. Schick · S. Kratz · E. Schnug
Federal Research Centre for Cultivated Crops, Institute for Crop and Soil Science,
Julius Kühn-Institut (JKI), Brunswick, Germany

E. Schnug and L.J. De Kok (eds.), *Phosphorus in Agriculture: 100 % Zero*,
DOI 10.1007/978-94-017-7612-7_5

Introduction

The P-demand of agricultural crops is usually covered by the application of mineral P-fertilisers or organic P-sources. Regarding the finiteness of rock phosphates, the basic raw material for the production of mineral fertilisers, sustainable handling of P in agriculture is compulsory to save this resource. Nevertheless, many agricultural areas currently suffer from P-surpluses resulting from inaccurate fertiliser applications which exceed the actual demand of the crop. This, in turn, leads to the transfer of P from agricultural fields into water bodies, enhancing their eutrophication. It becomes apparent that it is urgently required to adjust P-fertiliser applications to the actual P supply of the particular soil, and matching the real P-demand of the crop. Furthermore, it is necessary to identify spots, which are source areas of P transport and to prevent or at least reduce further P-losses. In order to determine the actual P-status of a particular soil, different soil P tests basing on different mechanisms can be applied. The following chapter presents an overview of various soil tests and discusses their mechanisms, suitability and shortcomings for the determination of plant-available P.

Why Is Plant Available P Determined?

In the first place, the determination of plant available P in agronomic soil testing aims at estimating the ability of a soil to supply plants with P, or the abstract concept of “soil P status”. The soil P status is supposed to be reflected by the operationally defined “soil test P” (STP). It is usually concluded that the lower the amount of plant available P in the soil, the poorer is the P supply to a crop, and the more additional P is needed in the respective field. The adequate level of plant available P in soil and P fertilisation requirement differ from crop to crop and from one soil to another. Usually, a lower STP is sufficient in fine-textured soils than in coarse or organic soils.

In addition to their use in the field of plant nutrition, traditional soil test methods have also been evaluated for being used as “environmental soil tests” (e.g. Sims 1993; Daly and Casey 2005; Carmo Horta and do Torrent 2007). It was shown that high STP contributes to a higher dissolved P concentration in runoff water (Sharpley 1995; Yli-Halla et al. 1995; Pote et al. 1996), and higher P release during the erosion process and in the eventual water body. As an example, the so-called “change-point” (60 mg Olsen P kg^{-1}) was elaborated to assess the probability of P leaching from agricultural soils (Hesketh and Brookes 2000). In drainage water from soils with a concentration of <60 mg Olsen P kg^{-1}, very little P (<0.15 mg P l^{-1}) was found, while a linear increase of the P concentrations in drainage water from soils above 60 mg Olsen P kg^{-1} was observed (Hesketh and Brookes 2000). Thus, the exact knowledge of the amount of available P in a particular soil can be used to identify spots, which are prone to P leaching. Since information on STP is commonly

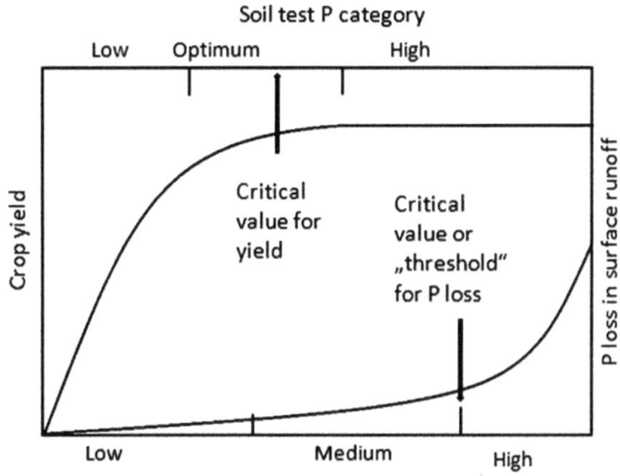

Fig. 1 Influence of soil test phosphorus (STP) on crop yield and P loss in runoff (Maguire et al. 2005)

available from agricultural soils, it is also often used as an input variable in simulation models such as EPIC (Jones et al. 1984; Williams and Sharpley 1989), CREAMS (Knisel 1980; Sharpley et al. 1982) and ICECREAM (Tattari et al. 2001; Yli-Halla et al. 2005), which are applied in the estimation of P loading to waters.

Besides the mere identification of agricultural fields or single areas which are vulnerable to P losses, soil P testing can be applied to develop a balanced fertilisation regime that is adjusted to the actual P supply in the soil and the demand of the particular crop. This, in turn, helps to reduce the amount of available P in soils and thus run-off of nutrients from agricultural fields. To assess accurate, field related P-requirements, it is beneficial to monitor the quantity of available P over time, which allows to develop response curves and finally make appropriate fertiliser recommendations (see Fig. 1) (Fixen and Grove 1990; Maguire et al. 2005).

What Does the Term "Plant Available P" Stand for?

To understand the mechanism of the different soil P-tests, the occurrence of different P-forms and their behaviour in the soil have to be known. Dissolved orthophosphate-P ($H_2PO_4^-$ or HPO_4^{2-}) in the soil solution is the immediate source of P to plants. According to Sibbesen (1983), the process of P-uptake via the plant root depends on the following factors: (a) the P concentration in the soil solution, (b) the soil as a medium for P-diffusion and (c) the dissolution of P in the solid phase as a result of the parameters "decreasing P-concentration in the solution", "time", "changing activity of HCO^{3-} and H^+", "changing activity of exuded organic anions" and "changing activity of phosphate precipitation cations".

The adsorption of P from the surrounding soil solution by the plant root triggers the diffusion of P from the soil solution to the plant root and dissolution or desorption of P from the solid phase. The plant root itself influences soil P since it enriches the rhizosphere with HCO_3^- or H^+. In addition, plant roots also release CO_2 and other organic compounds (Sibbesen 1983). Soil solution P is, however, seldom determined because this fraction is usually only a very small part of the share of soil P that is available to plants and the extraction of a soil solution in amounts sufficient for analysis is laborious. Even though the dissolved P concentration of the soil solution has been estimated by the extraction with water (e.g. Torrent and Delgado 2001), soil solution P concentration is not used in the estimation of soil P status. Neither has the total P concentration of soil much value in the estimation of plant available P. In the determination of total P, soil minerals and organic matter are completely dissolved with strong chemicals such as perchloric acid and hydrogen fluoride, or solubilised by alkali carbonate fusion. A slightly less drastic dissolution is obtained with aqua regia (concentrated nitric acid and hydrochloric acid, 1:3), and the result can be called "semi-total P". In most soils, the bulk of soil P is contained in the "stabile P pool" which is strongly retained in mineral structures or in stable organic matter, not contributing to the P concentration in the soil solution within the time frame that is relevant from the agricultural point of view.

In contrast, the pool of "labile P" which forms the stock of plant-available P is usually only a tiny part of total P and consists of (i) P retained on soil particles by surface adsorption, (ii) P contained in soil as relatively easily dissolved calcium phosphates and (iii) P incorporated in fresh organic matter that is mineralized rapidly. The quantity of available P varies considerably, which makes its prediction in a particular soil a very difficult task. Particularly, phosphate adsorbed to aluminium and iron (hydr)oxides and occurring as fresh calcium phosphate precipitates control the solution P concentration and P is released from these sources as a response of P removal from solution by plant uptake. Therefore it is this reserve ("capacity"), or fraction of it, that usually is determined when the P supplying power of soil is estimated.

Determination of Plant Available P

"Plant available P" is a theoretical concept and consists of those P reserves that can in principle be utilized by plants during their life cycle. It is usually operationally defined as the concentration of P extracted from soil with various solutions. In agricultural soil testing, aiming at analysing plenty of samples at low cost, simple and rapid extraction methods are preferred. The purpose of the extraction is to transfer P from the solid phase to the solution. Due to the various chemical P-forms that contribute to the size of the plant-available P pool, the prediction of available P in a particular soil is challenging. However, a number of different soil P tests, which show a satisfactory correlation with the actual P uptake by plants have been developed (Fixen and Grove 1990). Usually, the extracts base on diluted strong or weak acids, sometimes combined with complexing agents aiding to dissolve the nutrients. In some cases, buffered alkaline solutions are used, as well. The basic principle of soil P tests is the mixing of a defined amount of soil with the extractant which is then shaken for a defined period of time. The actual determination of P in solution is usually carried out by measuring phosphate with colorimetric methods or with inductively coupled plasma emission spectroscopy (ICP). The latter generally results in higher P-amounts (Maguire et al. 2005) since colorimetry only detects dissolved inorganic P (in orthophosphate form) while ICP measures the total amount of P (organic and inorganic) dissolved in an extract.

All soil P tests aim at estimating the amount of P in the soil that can be taken up by the plant, thus, most soil P-tests base on the concept of soil extraction with an extractant that dissolves a proportion of P from the solid phase. According to Sibbesen (1983), one of the following four different approaches is generally applied: (a) The solution-to-soil ratio is increased to decrease the solution P concentration. This disturbs the equilibrium between solid phase P and solution P. If the extractant is water, then the activity of P precipitating cations in the solution decreases, too. (b) The pH-value is changed by using dilute acids, pH-buffers or dilute bases. (c) Anions are introduced that complex and/or precipitate phosphate-precipitating cations such as calcium, aluminium and ferric cations. (d) Anions are introduced which desorb phosphate and limit the re-adsorption of already dissolved phosphate. Most standard extraction methods follow these principles and introduce anions that complex and/or precipitate cations (calcium, aluminium and ferric cations) that would otherwise be precipitated with phosphate. However, if the method is not based on complexing actions, the pH of the solution influences the amount of P that is extracted. It is worth mentioning that in this case the pH does not only influence the dissolution but also the re-precipitation of the dissolved P (Gassner 2003). Generally, acidic extractants are regarded as more suitable for acidic soils and bicarbonate containing reagents more suitable for calcareous soils (Fixen and Grove 1990).

While plant roots remove P from the soil solution, a chemical extractant induces the accumulation of the dissolved P in the liquid phase. Usually, a certain amount of dissolved P is re-precipitated and/or re-adsorbed because chemical reagents break the bonds between phosphate and soil constituents and favour the formation of other bonds. Thus, the amount of P extracted by an extractant is the combination of two opposing reactions, underlining the empirical nature of these soil P-tests (Sibbesen 1983). P tests that are based on "mild" extractants keep the soil chemically intact while methods which use strong acids or strongly complexing agents cause significant chemical changes in the soil which, compared to the chemical effects of the plant root, are more drastic. Hence, the use of acidic P-tests often leads to the overestimation of the P-status of neutral and particularly calcareous soils since sparingly soluble calcium phosphates are dissolved, which would not have been dissolved by the activity of plant roots (Sibbesen 1983).

In the past, a multitude of methods have been developed to determine the amount of extractable P, which reflects plant availability best and allows to assess potential environmental impacts. There is no universally agreed method for the extraction of plant available P, and more methods are probably available for P than for any other plant nutrient. Table 1 lists extraction methods commonly used in soil P testing and reflects the fact that many of the methods in use were developed at least 50 years ago. Usually, the choice of a method as the standard soil test is the result of the suitability of the extractant for the majority of soils in that particular country. The extraction methods differ concerning parameters such as pH, extraction time, solution-to-soil ratio, temperature and concentration of active agents.

Table 1 Methods used for the extraction of plant available P from soil

Method	Extractant	Countries used	References
Bray 1	0.025 M HCl + 0.03 M NH_4F, pH 2.6; 7:1, 1 min	USA	Bray and Kurtz (1945)
Mehlich I	0.0125 M H_2SO_4 + 0.05 M HCl, pH 1.25; 4:1, 5 min	USA	Mehlich (1953)
Mehlich III (M III)	0.2 M CH_3COOH, 0.25 M NH_4NO_3, 0.015 M NH_4F, 0.013 M HNO_3, 0.001 M EDTA, pH 2.5; 10:1, 5 min	USA, Estonia	Mehlich (1984)
Olsen	0.5 M $NaHCO_3$, pH 8.5, 20:1, 30 min	UK, Denmark, Israel, USA	Olsen et al. (1954)
Colwell	0.5 M $NaHCO_3$, pH 8.5, 100:1, 16 h	Australia	Colwell (1963)
Morgan	0.52 M CH_3COOH, 0,.2 M CH_3COONa, pH 4.8, 5:1, 15 min	Ireland, USA	Morgan (1941)
Truog	0.001 M H_2SO_4, 0.3 % $(NH_4)SO_4$, pH 3.0, 100:1, 30 min		Truog (1930)
AB-DTPA	1 M NH_4HCO_3 + 0.005 M DTPA, pH 7.6; 2:1, 15 min		Soltanpour and Schwab (1977)
Ammonium acetate (AAAc)	0.5 M CH_3COOH, 0.5 M CH_3COONH_4, pH 4.65; 10:1, 1 h	Finland	Vuorinen and Mäkitie (1955)
Ammonium acetate—EDTA (AAAc + EDTA)	0.5 M CH_3COOH, 0.5 M CH_3COONH_4, 0.02 M Na_2-EDTA, pH 4.65	Switzerland	Lakanen and Erviö (1971), Demaria et al. (2005)
Ammonium lactate (AL)	0.1 M NH_4-lactate, 0.4 M CH_3COOH, pH 2.75; 20:1, 30 min	Lithuania, Norway, Sweden	Egnér (1954)
Double lactate (DL)	0.02 M Ca-lactate, 0.02 M HCl, pH 3.5–3.7, 50:1, 1.5 h	Germany, Latvia, Poland	Riehm (1942)

(continued)

Table 1 (continued)

Method	Extractant	Countries used	References
Calcium lactate acetate (CAL)	0.1 M Ca-lactate + 0.1 M $(CH_3COO)_2Ca$ + 0.3 M CH_3COOH, pH 4.1, 20:1, 2 h	Germany	Schüller (1969)
$CaCl_2$	0.01 M $CaCl_2$, 10:1, 2 h	The Netherlands	Houba et al. (1998)
Water (W)	Deionized H_2O, 60:1, 22 h	The Netherlands	Sissingh (1971)
CO_2 saturated water (CO_2)	CO_2 saturated nanopure H_2O; pH 3.5–4, p_{CO2} 6 bar; 2.5:1, 1 h	Switzerland	Dirks and Scheffer (1930); Demaria et al. (2005)
Desorption techniques (P sink methods)			
P_i	$Fe(OH)_3$ impregnated filter paper in 0.005 M $CaCl_2$; 40:1, 24 h		van der Zee et al. (1987), Chardon et al. (1996)
Anion exchange resin (AER)	Anion exchange resin + 1 H_2O; 1.0 g resin; 100:1; n.a. (Amer et al. 1955); 2.2 g resin; 25:1; 24 h (Sibbesen 1977)		Amer et al. (1955), Sibbesen (1977)
Anion exchange membrane (AEM)	Anion exchange membrane + H_2O; 100:1; 16 h (Saunders 1964) 25:1; 72 h (Fernandes and Warren 1996)		Saunders (1964), Fernandes and Warren (1996)
Diffusive gradients in thin films (DGF)	Diffusive gradients in thin films		Zhang et al. (1998)
Electro-ultrafiltration (EUF)	Deionized H_2O 10:1, 30 + 5 min (2 fractions)	Germany	Nemeth (1980, 1982), VDLUFA (2012) methods Vol. 1, A 6.4.2 and A 6.1.7.3
Isotope techniques			
IEP			McAuliffe et al. (1947); cited after Amer et al. (1955), Cabeza et al. (2011), Frossard et al. (2011)

Therefore, the amount of extracted P can vary considerably since the extractants differ in extraction power. However, the results produced by different extractants tend to be correlated (Sibbesen 1983; Maguire et al. 2005). Besides these conventional chemical extraction methods, several methods have been developed which follow a different approach to determine the amount of plant-available P in the soil. So-called desorption techniques use strong sinks for P to remove it from the solution, such as an anion exchange resin or membrane, ferrihydrite oxide gels (diffusive gradients in thin films) or iron oxide impregnated filter paper (P_i). Isotope techniques estimate plant-available P by means of ^{32}P or ^{33}P-labelled orthophosphate, which bases on the equilibration of ^{32}P or ^{33}P with soil P.

Characteristics of Various Extraction Methods

Criteria for a Method to Extract Plant Available P from Soil

A suitable method for assessing the soil P status extracts P from the same pool that is utilised by plants. The results should correlate well with plant P uptake (Sibbesen 1983) and successfully predict the yield increase obtained with P fertilization. The method used to assess the available P amounts in soils should allow to place the soil sample in a predetermined group or category, for example using the terms low, medium and high. This helps to make fertiliser recommendations (Fixen and Grove 1990). Several important criteria have been formulated for a good extraction method. According to Thomas and Peaslee (1973) the method should ensure that a reproducible and consistent proportion of the soil's plant-available P is removed, and the extent and nature of reactions between soils and added P are reflected. According to Kamprath and Watson (1980) the method should extract a proportionate fraction of plant-available P from soils differing in other properties, be applicable with reasonable accuracy and speed and ensure that the determined P value is correlated with crop growth response or P uptake. The method has to be robust, rather simple and quick to perform in routine soil testing preferably with commonly available laboratory device. The concentration of P in the extract has to be high enough to be easily measurable and not significantly affected by slight contaminations. These practical criteria were particularly important earlier in all laboratories and are still highly valid at least in developing countries. The simplicity of a method often implies that the same extraction can be used for the determination of several nutrients, since soil testing usually includes also other nutrients than P only. An ideal "universal extraction method" would give relevant results for all nutrients. Unfortunately, such extraction methods do not yet exist, even though many of the methods mentioned in Table 1 are used for several elements.

The presented methods extract P from soil with different mechanisms and efficiencies, and they attack different soil constituents. P in acidic soils is retained predominantly by surface adsorption to aluminium and iron hydroxides and in calcareous soils as calcium phosphate precipitates. Therefore a particular chemical extractant is not optimal to all soils. Deionized water, dilute neutral salt solutions and the P sink methods probably give relevant results for a wider range soil pH values than the extraction methods utilizing different (acid or alkaline) chemicals for P dissolution.

Methods Involving Strong Acids

Many of the traditional soil testing methods still surviving, such as Bray 1, Mehlich I, Mehlich III and Truog methods, base on strong acids and often very short extraction times and their pH is very low (≤3.0). These methods are recommended to be used in acidic soils, and they likely dissolve P bound by Fe and Al as well as fresh Ca phosphate precipitates originating from fertilisers. In calcareous soils these methods yield results that do not properly reflect the plant available P content of the soil. The Bray 1 method, traditionally performed at a low solution-to-soil ratio (originally 7:1, now often 10:1), gives low results for calcareous soils, because the acidity of the extractant is neutralised by soil carbonate, while fluoride (F^-), complexing Al and thus promoting P desorption in acidic soils, can be precipitated as CaF_2 (summarised by Olsen and Sommers 1982). All these strongly acid extractants also have a potential to dissolve non-labile Ca phosphates, thus overestimating the plant available P in soils where rock phosphates have recently been applied as fertilisers or in geologically young soils that contain calcium phosphates as primary minerals. The Mehlich III method, probably the most complex in composition among all soil test methods, has been launched as a universal extractant for P, K, Ca, Mg, Cu, Mn and Zn. In contrast to the Mehlich I method, the Mehlich III extract is considered to be also suitable for neutral and alkaline soils because its acidity is neutralised less by soil carbonates than Mehlich I (Tran and Simard 1993; cited after Sims 2000).

Methods Involving Bicarbonate

Extractants containing bicarbonate are most commonly used in calcareous, neutral and alkaline soils. The Olsen and Colwell procedures have a similar pH of 8.5 but they differ in solution-to-soil ratio (Colwell 100:1; Olsen 20:1) and extraction time (Olsen 30 min; Colwell 16 h). The Olsen-P test was originally developed to assess the amount of plant available P in calcareous soils with a $CaCO_3$ content >2 % by using a weak $NaHCO_3$-solution (Watson and Mullen 2007). Basically, the method aims at reducing the concentration of soluble Ca^{2+}, Al^{3+} and Fe^{3+} in the soil solution by precipitating Ca^{2+} as $CaCO_3$ or forming Al- and Fe-oxyhydroxides with HCO_3^-, CO_3^{3-}, and OH^-, which are components of the extracting agent. Since these processes lead to higher negative surface charges and/or decreasing numbers of sorption sites on Fe- and Al-oxide surfaces at high pH-levels, the desorption of available P into solution and thus the P solubility is enhanced (Sims et al. 1998; Sims 2000). Colwell (1963) noted that the Olsen method had its shortcomings for soils with a low amount of extractable P. Thus, he increased extraction time and solution-to-soil ratio, which usually leads to higher amounts of extracted P compared to the Olsen method. The more recently launched Ammonium-bicarbonate (AB)-DTPA-method (pH 7.6) is a universal extractant for P, K, Ca, Mg, Cu, Fe,

Mn, dissolving micronutrients by chelation and exchanging cations with NH_4^+ ions (Soltanpour and Schwab 1977). In the extraction with bicarbonate in calcareous soils, $CaCO_3$ is precipitated, which promotes dissolution of labile calcium phosphates and increases the P concentration in the extract. Owing to their high pH, these solutions result in the desorption of P from Al and Fe hydroxides with the ligand exchange by OH^- and promote the solubility of Al and Fe phosphates. A practical problem to be taken into account when using bicarbonate-containing solutions in acidic soils is the evolution of CO_2 gas, which increases the pressure in the closed extraction vessel during the extraction.

Development of Methods Involving Lactate

The use of lactate as an extractant bases on its very strong solubilisation and complexing properties for Ca, Al and Fe which makes it capable to chelate these ions (Kamprath and Watson 1980; Vanderdeelen 1994). Thus, lactate is suitable to desorb P from (hydrous) Fe-/Al-oxides in soils (as well as from soluble Ca-phosphate compounds if Ca-free lactate is used) and to prevent its re-adsorption (Vanderdeelen 1994; Eriksson 2009). In 1938, Egnér et al. developed the common "single" lactate method which had its weakness when used to extract P from calcareous soils because the acid lactate-solution was neutralised at $CaCO_3$ concentrations >2 % in the soil. Furthermore, a decrease of P solubility with increasing pH was observed, indicating that the amount of soluble P would be underestimated in calcareous soils. To solve these shortcomings, Riehm (1942) modified the method by doubling the lactate concentration in the extract (double lactate method). This lead to an increase of the acidity and buffering capacity of the solution (Egnér et al. 1960). The application of this method led to results which corresponded well to those obtained by the single lactate method, however, the method was still not suitable for calcareous soils and reached its limits at $CaCO_3$ concentrations >4 % (Riehm 1958; Egnér et al. 1960). This can be attributed to the insufficient buffering capacity of the solution for soils with a higher $CaCO_3$-content or rather, a higher reactivity (Werner 1974). Thus, the DL-method was again modified to allow the assessment of available P from calcareous soils as well. The newly developed method was described as ammonium lactate method (P-AL) by Egnér (1954), Riehm (1958) and Egnér et al. (1960). While in the DL-method, calcium lactate acts as complexing ion in the extracting agent, ammonium lactate is used for the AL-method, thereby also allowing for the determination of Ca. In addition, the acidity of the extract is increased by the addition of acetic acid to a pH value of 3.6 to allow the assessment of available P in soils containing up to 20 % $CaCO_3$ (Riehm 1958). As the addition of acetic acid and the higher amount of lactate increases the total concentration of chelating ions in the AL-extract, the P solubility is less related to the pH-value of the soil (Egnér et al. 1960). However, investigations revealed that both, the DL- and the AL-method, still had a shortcoming when assessing the P availability in soils containing high amounts of apatitic structures. Due to the acidic nature of both methods, P from apatite, highly insoluble in the actual soil pH, is also dissolved and assessed as plant available, which may lead to an overestimation of the STP in these soils. On the other hand, the DL-extract will not dissolve plant available amorphous Al- and Fe-phosphates (Schüller 1969; Werner 1974). Therefore, Schüller (1969) further modified the DL-method into the calcium acetate lactate (P-CAL) method to allow an accurate assessment of the P supply in calcareous soils and soils fertilised with rock phosphates. He increased the Ca-concentration, and thus the pH of the CAL-extracting agent to 4.1, which lead to a higher buffering capacity (Werner 1974). It turned out that CAL allowed a better distinction between the various Ca-phosphates of different plant availability than the DL- and AL-methods did. As shown by vegetation trials, mono-calcium (MCP) phosphate and ammonium phosphates are completely and instantly plant available, while di-calcium phosphate (DCP) is also available but needs more time to become effective. Silica containing phosphates also show a good plant availability, while the agronomic effectiveness of basic calcium phosphates/apatites is comparably low. The application of the CAL-extract reflects these differences, as long as the content of free, reactive $CaCO_3$ remains below 15 %. A $CaCO_3$ content above 15 % will lead to the neutralisation of some of the CAL-solution by the carbonate and the extractant will not dissolve the complete amount of potentially plant available Ca-phosphates. The same limit applies to the dissolution of amorphous labile Fe- and Al-phosphates by CAL, which are available to plants. Thus, an underestimation of available P may occur in soils with a free, reactive $CaCO_3$ content above 15 % (Werner 1974).

Acetate Solutions as Extractants of Soil P

Acetic acid and sodium acetate are the main components of the extract according to Morgan (1941), which is used as the standard extractant for P, K, Ca, Mg, Cu, Mn and Zn in a few north eastern and north western states of the USA and in Ireland. The pH of this extract is buffered at 4.8 and it usually extracts much less P than the methods mentioned above but is considered suitable for soils of a rather wide pH range. The method has been modified a few times regarding the concentrations of its components. In one modified Morgan test, sodium acetate was replaced with ammonium acetate, keeping the original pH (McIntosh 1969). Already earlier, another modification of the Morgan method had been adopted as the Finnish routine method for estimating the available amount of all macronutrients except N in soils (Vuorinen and Mäkitie 1955; Saarela 2002). The method was modified with an equal molarity of acetic acid and ammonium acetate, but with a slightly lower pH (4.65). A comparison of both methods showed that the AAAc-method extracts significantly higher P-amounts than the Morgan method (Vuorinen and Mäkitie 1955; Saarela 2002). An overestimation of plant available P by AAAc was observed in heavily limed soils by Jaakkola et al. (1997). They attributed it to the dissolution of non-labile calcium phosphates which occurred because the pH of the

AAAc extract is buffered at the pH-value of 4.65, which is much lower than the originally neutral pH of these soils.

The AAAc-method was further modified by the addition of 0.02 M Na_2-EDTA. This modification was originally meant for extraction of metallic micronutrients (Lakanen and Erviö 1971). The acid ammonium acetate solubilises P forms by decreasing the pH of the soil solution, or rather, by adjusting it to the conditions found at the contact zone of plant roots in buffered agricultural soils (Schnug et al. 1999; Gallet 2001). Similar to its function in the M3-extract, EDTA is added to complex phosphate binding cations such as Ca, Al and Fe, as well as a number of trace metals such as Cu, in order to prevent their re-precipitation with phosphates, which might otherwise take place during the extraction process as a secondary reaction (Cottenie et al. 1979; Gallet 2001). During the activities of the FAO European Cooperative Research Network on Trace Elements, established in 1977, this AAAc-EDTA method was endorsed and commonly it is referred to as a method by Cottenie (e.g. Cottenie et al.1979). It extracts a larger amount of soil P (Lakanen and Erviö 1971) or added ^{32}P (Paasikallio and Häkkinen 1977) than AAAc alone and is suitable for simultaneous extraction of both plant available micro- and macronutrients. However, Saarela and Sippola (1990) considered it unsuitable for testing P in Finnish soils because they assumed that the extract was too destructive to hydrous Fe- and Al-oxides, the main sorbing sites in acidic soils. Thus, the amount of P extracted by AAAc-EDTA might overestimate the amount of available P, especially in acidic soils. Even though it has not been successful in predicting the response to P fertilisation in Finland, the method has been successfully adopted as standard extraction method for P in Switzerland and Belgium.

Non-destructive Methods for the Estimation of Plant-Available P

Deionized water is a non-aggressive extractant, which can be regarded as a more "natural" way of extracting P from soil than the solutions containing compounds that actually dissolve soil constituents. After comparing isotopically exchangeable soil phosphates which were determined in pot trials and aqueous soil suspensions it was concluded that water extracts those P forms which are mobile under pot trial conditions. The correlation of P_w-values with crop response, also in combination with P_{AL}-values, was investigated in a pot trial with different soils which varied widely in terms of soil condition and type. The P_W-values showed a closer relationship to crop response than the common P_{AL}-values. P_{AL} revealed its shortcomings especially for soils other than loess (van der Paauw et al. 1971). The extraction of some P forms with a low availability occurs with many other extractants when they are more acid or alkaline than the soil solution (Self-Davis et al. 2000). Therefore, the water extraction is a valid alternative to these extractants, since it should maintain the soil pH within one unit of its original value. In addition, the water extract is interesting from an environmental perspective, because it can be expected to simulate the release of P to run-off or leaching water more accurately than any of the stronger chemical extractants used for agronomic purposes (Moore et al. 1998). Several studies have shown that P concentration in a water extract resembles the P concentration in surface runoff water from agricultural land (Yli-Halla et al. 1995; Pote et al. 1996). Extraction with deionised water has been carried out at different water-to-soil ratios, a variable that substantially impacts the results. The higher the solution-to-soil ratio, the more P is desorbed from soil to the solution and P from the soil surfaces is increasingly depleted, resulting in higher results as expressed in mg:kg basis. Results obtained with high solution-to-soil ratios (e.g. 2000:1) are particularly applicable in the assessment of P loading potential of soil material (e.g. Yli-Halla et al. 1995), while low solution to soil-ratios, such as 60:1, have been used for the estimation of soil P status from the agricultural point of view (Sissingh 1971). The shortcomings of water as an extractant for P include very low concentrations in soils poor in P, and dispersion of soil, owing to low ionic strength, which makes it necessary to use expensive membranes for filtering the extract. The dilute $CaCl_2$ solution even more resembles the soil solution and usually yields even lower concentrations of P in the extract than deionised water does. However, the low amounts of P extracted by both methods, H_2O and $CaCl_2$, in soils with low P-concentrations cause detection problems and makes the results vulnerable to even small contaminations. Thus, both extractants can be conveniently performed only with soils of medium or high P status. In Switzerland, CO_2-satured water has long been used as a reference method for available soil P (Frossard et al. 2004). Saturating water with CO_2 causes a weakly acid soil pH, which is, however, quickly buffered by the soil, so that the actual extraction takes place at a pH value which is typical of the soil under investigation. ^{33}P-marked soils will show a similar specific activity as soils extracted with normal water do, showing that P-CO_2 and P-water solubilise similar forms of P (Demaria et al. 2005). Similarly as the $CaCl_2$- extract, this method gives information on the amount of very easily available soil P (Frossard et al. 2004; Demaria et al. 2005).

Another category of methods utilises anion exchange resins (AER), anion exchange membranes (AEM) or iron impregnated filter papers as a strong sink for P to simulate P uptake from the plant. The use of anion exchange resin (AER) beads was first introduced by Amer et al. (1955) who suggested that they remove P from the solution without releasing other P ions in exchange. The method was simplified by Sibbesen (1977) by putting the beads into nylon bags to facilitate the separation of the resin from the soil suspension. Originally, the soil had to be finely ground before separating it from the resin by sieving; this procedure was now unnecessary (Sibbesen 1977). Unfortunately, the nylon bag also caught fine root material and soil particles which had a negative impact on the analytical procedure (Fernandes and Warren 1996). Thus, the development of AEM further simplified the extraction procedure since it was no longer difficult to separate the soil from the resin, and no unwanted soil material was trapped by the membrane. Furthermore, it is also possible to apply the AEM under field conditions (Saunders 1964; Fernandes and Warren 1996; Bissani et al. 2002).

The Pi-method uses FeO-impregnated paper strips which function as a P sink in contact with wet soil samples. The amount of P which is sorbed on the oxide over an extended period of time is then considered as an index for P availability (Bissani et al. 2002). The Pi-method was originally developed as a simpler alternative for the resin methods by Sissingh in 1977. The water method, which is the common extraction method used for fertiliser recommendations in the Netherlands, turned out to be unsuitable for tropical soils because it often extracted only very low P-amounts which lead to analytical problems (Chardon et al. 1996). To overcome this problem, FeO-strips were used as a P sink during the extraction, and water was replaced by $CaCl_2$ since the former caused too much dispersion of the soil and contamination of the FeO-paper strip with soil particles (Chardon et al. 1996).

Those methods which apply P sinks use water or a dilute salt solution, such as 0.005 M $CaCl_2$ as the medium for desorption. As soon as P is desorbing from the soil, it is adsorbed by the sink. In this manner, phosphate concentration in the solution is kept at a very low level, which promotes effective desorption from soil. Eventually, P collected into the sink is determined. Without such a sink, P desorption from soil is stopped as soon as a (semi)equilibrium between soil P and solution P concentration is reached. Even though more laborious than the conventional soil test methods, these methods yield measurable results also from soils which are low in P. The process taking place resembles the depletion of solution P upon nutrient uptake by plant roots and therefore correlates closely with P uptake by plants in experiments (Aura 1978; Sibbesen 1983). These methods have also been used as surrogates for algal tests for turbid runoff water samples in estimating the potential P supply to algae from eroded soil material (Ekholm et al. 1991; Uusitalo and Ekholm 2003).

Similar to anion exchange membranes (AEM) or resins (AER), the DGT (= diffusive gradient in thin films)-method also uses a P-sink to mimic the plant root uptake of P from soils. A ferrihydrite gel-binding layer acts as an infinite sink for P after the solved particles passed a hydrogel, which works as a diffusion layer. The function of the diffusion layer is the limitation of the maximum P flux from the soil solution to the binding layer, comparable to the plant controlling the P influx (Zhang et al. 1998; Degryse et al. 2009; Six et al. 2013a, b). After a defined deployment time, the binding layer is eluted and the mass of solutes that accumulated on the binding layer is measured (Mason et al. 2005). The concentration in the solution is then calculated by using the measured mass of solutes and Fick's first law of diffusion (Zhang et al. 1998). Similar to the actual plant uptake, the transfer of P from the soil solution to the sink lowers the P concentration in the soil solution. This, again, induces the re-supply from the solid phase. The P-flux assessed by DGT resembles the flow of P from the soil solution to the plant and thus can be seen as a direct surrogate for plant-available P-fractions (Zhang et al. 1998; Six et al. 2013a, b). Compared to conventional extraction methods or the AEM-method, the DGT-method has its advantages because P is measured in saturated soil and not in soil slurry, thus resembling the actual field situation more accurately. Secondly, the standard extraction methods include vigorous shaking to mix the soil with the solution. This may induce the abrasion of soil particles exposing reaction surfaces, which would originally not be accessible by plant roots (Barrow 2008; Six et al. 2013a, b).

Another desorption technique by using water as a medium is the electro-ultrafiltration (EUF) technique. It has been successfully applied to predict the short- and long-term availability of nutrients in soils in Europe, Asia and Africa (Ziadi et al. 2001; Akinrinde et al. 2006; Teklić et al. 2009) and is currently practised as a routine method in Germany (mainly by advisory services for sugar beet producers such as Bodengesundheitsdienst Ochsenfurt, see www.bodengesundheitsdienst.de) as well as in Austria, Hungary, China and the countries of former Yugoslavia (Akinrinde et al. 2006). The extraction is based on an electric field induced by platinum electrodes, with nutrient ions in the soil suspension moving to either anode or cathode and passing an ultra-membrane filter in the process. Near the electrodes, the ions are collected in an aqueous eluate, in which they can be analysed later on by standard analytical methods. Thus, nutrients are continuously removed from the solution. By varying the duration of the extraction, as well as the voltage of the electric field and the water temperature during extraction, nutrient fractions of different solubility can be collected. For advisory purposes, commonly two fractions are separated: the first after 30 min at 200 V and 20 °C represents easily or directly available nutrients, while the second after 5 more minutes at 400 V and 80 °C is interpreted as nutrient reserve (Nemeth 1982; Horn 2006). Some authors differentiate additional fractions, which allows them to assess nutrient availability in soils more comprehensively, e.g. EUF (0–10 min) as a measure of the intensity factor, EUF (0–35 min) as quantity factor, EUF (30–35 min) as buffer capacity and the quotient EUF (5–10 min)/EUF (30–35 min) as fixation capacity of the soil (Akinrinde et al. 2006; with further references). The main advantage of the EUF method is that, in addition to P, all other macronutrients as well as a number of micronutrients can be determined by one extraction (two-phase) after only 35 min. If, in a third step, DTPA is added and the extraction is run for another 5 min at 80 °C, further micronutrients and heavy metals can be determined in a third fraction as well (Nemeth 1982; Horn 2006).

Another alternative way to gain information on the amount of available P in the soil solution, the amount of P that will exchange into the soil solution within 3 months and also on the capacity of the soil to buffer changes of the P content in the soil solution is the determination of isotopically exchangeable P (IEP) (Frossard et al. 2004). The method bases on the addition of the radioactive P-isotopes ^{32}P and ^{33}P to the soil. Once in the soil, the labelled P starts to mix with and will be diluted by the ^{31}P of the soil solution and equilibrates (Amer et al. 1955; Di et al. 1997; Frossard et al. 2011). This phenomenon was already noted by McAuliffe et al. (1947) (cited after Frossard et al. 2011) who observed a decreasing concentration of ^{32}P added to a soil-solution system with time as a result from the exchange and dilution between ^{32}P in soil solution and ^{31}P in the soil solid. According to the authors, this indicated the presence of two different P pools in the soil: one is assumed to contain the fraction of readily exchangeable P forms, while the other one contains the fraction of less-exchangeable P (Frossard et al. 2011).

The suitability of isotopic exchange was verified by Larsen (1952) who concluded by means of two complementary experiments that isotopically exchangeable P reflects the amount of available soil P. Amer et al. (1955) attributed the suitability of isotopic exchange to the conditions during the measurements, which the authors considered as "not unnatural". No acids, bases, or complexing or precipitating reagents, but water is used and thus, the behaviour of P in the soil is mirrored (Amer et al. 1955). As mentioned before, conventional acidic or alkaline extraction methods might also extract P-fractions, which are not available to plants and also only parts of those P-forms which are available (Fardeau 1996). However, it has been elaborated that isotopic measurements have shortcomings for those soils containing high amounts of P-fixing compounds, such as reactive iron oxides. In that case, an application of the IEP-method might lead to an overestimation of exchangeable P in the soil, especially when carrier-free radioisotopes [i.e. they are added without ^{31}P (Bünemann et al. 2004)] are used (Amer et al. 1969). Further constraints for the use of ^{32}P and ^{33}P are their short half-lives (14.3 and 24.4 days, respectively), which limits the information on the behaviour of P in the soil to a short time span, and the high energy β-radiation emitted by ^{32}P which is a considerable hazard and thus restricts studies to the greenhouse or climate chambers (Di et al. 1997; Frossard et al. 2004).

Different Methods—Different Results

Different methods can result in significantly different numerical values of STP in a given soil. One of the most popular topics in soil fertility research has been comparing different methods for P extraction and reporting the correlations or regression equations between the results of different methods within the soil material studied. This activity has been practiced in almost all countries, and often the value of methods has been assessed against results of P uptake by plants in pot experiments. Probably one of the most comprehensive comparisons was made within the COST 832 action "*Quantifying the Agricultural Contribution to Eutrophication*" where 135 soil samples from 12 countries were analysed for P with 13 different methods that are in common use in Europe (Neyroud and Lischer 2003). In the order of P concentrations extracted from soil, the different soil test methods gave the following median values (mg kg^{-1}):

$$\begin{aligned}&P_{total}(546;1) > P_{AL}(76{-}42;5) > P_{M3}(30;1) > P_{AAAc-EDTA}(25;2)\\&\quad > P_{DL}(35{-}20;2) > P_{CAL}(28{-}26;2) > P_{Olsen}(9{-}19;3) > \mathrm{Pi}\ (8;1)\\&\quad > P_{AAAc}(7;1) > P_{Morgan}(6;1) > P_{w}(4.0{-}1.7;3) > P_{CO_2}(0.5;2)\\&\quad > P_{CaCl_2}(0.3{-}0.0;2)\end{aligned}$$

The first numbers in the parentheses give the median or the range of medians obtained by the different laboratories carrying out the analysis, and the last number expresses the number of laboratories involved in the respective determination. The material of 135 soils had a median total P of 546 mg kg^{-1}, and the AL method was the most efficient and extracted 8–14 % of the total P, as the median of results, whereas the least efficient one, $CaCl_2$, extracted 0–0.05 % of the total.

Within the framework of the EU-project *Baltic Manure*, 9 national P-extraction methods were applied to 183 soil samples from 4 different countries to investigate the possibilities of comparison and interpretation of results obtained by different standard extraction methods (Shwiekh et al. 2014). As expected, significant differences in the extraction force of the different methods were found. The extraction force decreased in the order: $P_{AR} > P_{AL}$, $P_{M3} \geq P_{AAAc+EDTA}$, $P_{DL} \geq P_{CAL} \geq P_{Olsen}$, $P_{AAAc} \geq P_W$, which is in good agreement with the findings of Neyroud and Lischer (2003; see also Fig. 2).

As mentioned before, extractants with a higher acidity or alkalinity than the soil solution will also extract P forms with a low plant-availability (Self-Davis et al. 2000), which is reflected by the fact that all of them yield higher results than the water extraction (see previous chapter for further explanation). Furthermore, Shwiekh et al. (2014) observed in their study that the addition of EDTA to the AAAc-extract increases the amount of extracted P significantly in comparison to the AAAc-extract alone. This supports the findings of Lakanen and Erviö (1971) and Neyroud and Lischer (2003) who referred to a higher extraction force of AAAc + EDTA than AAAc alone.

As already discussed in the chapter "Determination of plant available P", the differences in extraction force between the various methods can be mainly attributed to different extraction mechanisms which base on individual active components in the extract. The three methods with the highest amounts of extracted P, AL, M3 and AAAc-EDTA, all contain one or more strong chelating agents such as lactate, EDTA, NH_3F and acetate (the latter, however, according to Eriksson (2009), displaying only mild chelating properties), which are able to release P from

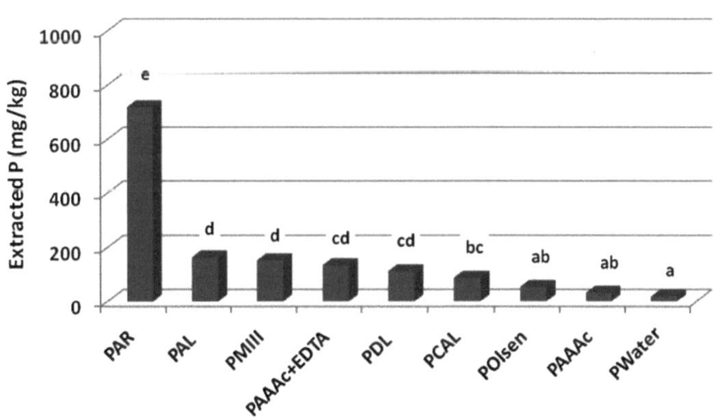

Fig. 2 Extractable P (mean; mg kg^{-1}) determined by 9 different methods. Significant differences between groups were determined by Tukey post hoc test ($p < 0.05$) and are denoted by different letters (Shwiekh et al. 2014)

Al- and Fe-P-compounds. In addition, AL and M3 have a low pH (<4), leading to the hydrolysis of P in insoluble Al-humic-P substances and the dissolution of sparingly soluble Ca-P compounds (Ottabong et al. 2009). In contrast, the two methods with medium strengths, DL and CAL, both contain only one chelating agent, lactate, combined with Ca as the active cation. Ca^{2+} may precipitate some of the extracted P in the extraction solution leading to lower P-values compared to those extractants containing NH_4^+ as the active cation, like AL (Eriksson 2009). The weaker extractants, Olsen and water, only hydrolyse those Fe-/Al-P compounds, which are easily soluble. Furthermore, the alkaline Olsen-extract releases P from Ca-oxides and some of the P from Ca-phosphates (Olsen et al. 1954). The extraction force of AAAc is mainly based on its content of acetic acid. Therefore it can also be included into the group of weak P extractants.

According to Neyroud and Lischer (2003), the differing extraction forces of the various methods are useful for different purposes and ask for different interpretation strategies for each method. The authors assume that weak extraction methods, which tend to reflect immediately available P, may be useful for immediate fertilisation recommendations, while the stronger extractants are rather related to those P species which are more strongly bound and will become available on a medium to long-term basis, and are thus essential for planning fertilisation on a mid to long-term basis. Regarding the assessment of the actual risk of P loss and environmental pollution (eutrophication of ground and surface waters), the weak extraction methods appear to be more suitable (Shwiekh et al. 2014).

Besides the active components in the extract, factors such as solution-to-soil ratio as well as duration and power of shaking the samples influence the extraction force of the different methods (Eriksson 2009). This was clearly reflected in the results published by Neyroud and Lischer (2003) who reported that the median of the results for P-AL analysed in 135 soils in five laboratories varied from 42 to 76 mg kg^{-1}, and those for Olsen-P, analysed in three laboratories, varied from 9 to 17 mg kg^{-1}. They assumed that the differences in results between the laboratories were attributable to local adaptations of the original method, such as differences in solution-to-soil ratio and extraction time in the first place. For example, according to Saarela et al. (2004) extending the extraction time from 30 to 60 min for Olsen-P in soils of Finland increased the results by 20 %. Extending the extraction time for the Bray 1 test from 1 to 5 min increased the amounts extracted in 30 acidic mineral soils by 14 % (Aura 1978). Results of water extraction are very sensitive to the solution-to-soil ratio due to the strong buffering against solution P concentration around the soil particles. This phenomenon was demonstrated by Hartikainen (1986) who found that the solution P concentration in a water extract from soils, which had received 60 kg P ha^{-1} a^{-1} for 7 years decreased only from about 0.090 to 0.054 mg l^{-1} when the water-to-soil ratio increased from 2:1 to 400:1. From these results it can be calculated that P desorption increased from 0.2 to 22 mg kg^{-1}, simply by adding more water around the soil particles. Similarly, Saarela (1992) measured increasing P_w-values (2.0, 10.1 and 27.1 mg dm^{-3} of soil) in 21 acidic soils with increasing water-to-soil ratios of 10:1, 60:1 and 300:1, respectively. Drying of the soil samples before the extraction can also lead to increased

P_{W-}-values. Koopmans et al. (2006), for instance, observed a 1.5–2.7-fold increase of P_W after drying of the soil.

Finally, the actual determination of P has an impact on the results. In 99 soils from the Baltic region, determination of solution P with ICP yielded 19 % higher results than the colorimetric method (Eriksson 2009). Even bigger differences between these methods of determination have been reported in soil material with a higher abundance or organic material (on average 36 %; Ziadi et al. 2009). While the colorimetric method determines orthophosphate only, the results obtained with ICP also comprise dissolved organic P as well as non-orthophosphate inorganic P compounds (Shwiekh et al. 2013). These examples indicate that whenever comparing soil test results obtained in different studies, all the details of the extraction should be checked and the real comparability of the results has to be evaluated on this basis.

Even though the extraction methods and their modifications yield widely different numerical values for plant available P, the results of many methods have proved to correlate, indicating that to a considerable extent, they extract P from the same pools. Neyroud and Lischer (2003) found that, based on the diverse material of 135 soil samples from 12 countries, the results of the strongly acidic extractants containing lactate (AL, DL, CAL) correlated closely among each other, and also with the results of the AAAc-EDTA method. Quite naturally, the two acetate methods, Morgan and AAAc, showed a high correlation. So did the results of water, CO_2 and $CaCl_2$. It is slightly surprising that also the results of the rather different methods Mehlich III and Bray, with very low extraction pH, correlated closely with the results of the Olsen test, carried out at pH 8.5, and with those of the Pi method. Within the results of these four groups of methods, the correlation coefficient was close to or higher than 0.90.

In a pot trial with maize and rice as test crop, 8 soil P tests (Olsen, Colwell, Bray 1, Mehlich III, $CaCl_2$, DGT, AEM and ammonium oxalate) were applied and their results compared. Correlation analysis revealed that the methods could be separated into two different groups. The first group comprised ammonium oxalate, Olsen, Colwell, Bray-1, Mehlich III and AEM-methods which base on chemical extractions and are considered to be P quantity measures, i.e. they indicate the number of sorption sites in the solid phase. However, the two methods of the second group, $CaCl_2$ extractions and the DGT-method, are intensity based measures which sample P from the soil solution plus a small fraction of labile P from the solid phase which buffers the P concentration in solution when the equilibrium is disturbed (Six et al. 2013a, b). It can be assumed that this resupply is more pronounced when a zero-sink is used which removes all P from the solution and thus disturbs the equilibrium between solution and solid phase more drastically. The authors consider AEM to be a measure of intensity and quantity, depending on the P buffering capacity of a soil. Prior experiments showed that for soils with a high phosphorus sorption index (PSI), a significant fraction of the non-labile, and thus unavailable, pool of P can be removed by the AEM method (Six et al. 2013a, b).

Studies on the comparability of results produced by either AEM or AER came to different conclusions. Sibbesen (1977) observed only slightly higher amounts of extracted P when resin beads were put into nylon bags. Similarly, loose resin beads

(AER) and resin membranes (AEM) behaved in a very similar manner when applied to a loam at low, medium and high soil P status, indicating that they would both lead to equally good assessments of available soil P (Saunders 1964). In contrast to this, Fernandes and Warren (1996) reported clear differences between AER (in bags) and AEM. In their study, AER generally extracted more P than AEM.

Applying different IEP-methods on the same soils often resulted in the determination of significantly different E-, A- and L-values, which even lead to contradictory information regarding the fertility of the soil (Fried 1964; cited after Fardeau et al. 1996). According to Fardeau et al. (1996) these contradictions might be explained by differences in the mixing procedure of the isotopes or the addition of a carrier, which might modify the system. Furthermore, experimental errors due to very low P-concentrations in the soil solution have to be considered.

Difficulties in the Comparison of Results Across Borders

As mentioned above, the choice of a standard soil P test usually depends on the suitability of the extractant for the majority of soils in that particular country. Since methods to extract STP differ from one country to another, it is often difficult to compare the soil P status across the borders. For this reason, it is equally difficult to compare the level of P fertilisation recommendations based on soil testing in different countries that use different methods to assess the amounts of available P in the soil. This is underlined by the findings of Shwiekh et al. (2014) who compared exemplarily the standard P extraction methods and the respective classification systems for Estonia, Germany, Poland and Sweden on the basis of the analyses of 183 soil samples (Table 2). Since such different reference methods and classification systems are applied in each country, it is clear that the assessment of the P supply of agricultural soils will vary significantly between countries (Fig. 3). While the German classification tends to group sites into lower (deficient) classes, they are assessed as being sufficiently or even excessively supplied with P if another national procedure is applied. Since international research cooperation is expanding, there is an urgent need to harmonise the interpretation of results obtained from

Table 2 Country-specific differences in the classification of the P-supply of soils

Country	National extraction method	Classification (mg P kg^{-1} soil)				
		Strongly deficient	Deficient	Sufficient	Excessive	Strongly excessive
Estonia[a]	M3	<1.5	1.5–4.0	4.1–9.5	9.6–20.5	>20.6
Germany	CAL	≤2.2	2.2–3.9	4.9–7.2	7.3–10.4	≥10.5
Poland	DL	<2.2	2.2–4.4	4.4–6.5	6.5–8.7	>8.7
Sweden	AL	<2	2–4	4–8	8–16	>16

[a]<2 % humus content

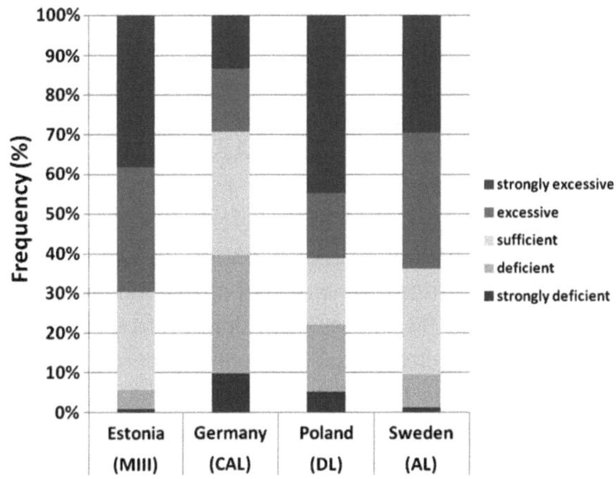

Fig. 3 Percentage distribution of P-classes obtained by applying different national standard extraction methods and the corresponding classification system (see also Table 10; n = 183; Shwiekh et al. 2014)

different soil P tests to formulate common fertiliser recommendations and to facilitate international research and environmental control. However, it is questionable whether soil P tests are comparable at all and can be transferred into one another (Ottabong et al. 2009).

In the past, several studies have investigated the comparability of different standard extraction methods and the possibility to "translate" one into another. For the results of Bray 1, Mehlich I, Mehlich III and Olsen methods, the most commonly used soil test methods in the USA, conversion equations were developed on the basis of non-calcareous soil samples from 19 south eastern, north central and north eastern states (Wolf and Baker 1985). The R^2 values ranged from 68 to 97 % and it is noteworthy that also in this material of non-calcareous soils the correlation between Olsen-P with the results of Mehlich and Bray 1 methods (involving acidic extractants) was as close as the correlations between the Mehlich and Bray methods. Conversion equations were also developed for the results of Mehlich III, Morgan and Bray 1 in Alaska (Michaelson et al. 1991). In a study involving 99 soil samples from Sweden, Estonia, Latvia and Lithuania, conversion equations for Olsen, AL, DL and Mehlich III methods were developed, including also soil properties such as carbon content, pH, iron and aluminium, and R^2 values ranging from 0.75 to 0.94 were obtained (Eriksson 2009).

Based on the results of 9 different extraction methods used as standard soil P tests in the Baltic countries, Schick et al. (2013) provided regression equations for the "translation" of these national soil P tests into one another. The aim was to allow the comparison and interpretation of the values obtained by the different methods to enable a reliable assessment of P supply and turnover in agricultural soils under different environmental conditions. The authors showed that it was generally possible to transform data from one soil P test into another. However, the quality of the resulting values depended on the pair of soil tests at question. All soil P tests applied in the study were significantly correlated with each other, however, only CAL, AL, M3, AAAc-EDTA and water extracts showed strong correlations and consequently allowed for the calculation of highly significant regression equations with a strong coefficient of determination. In some cases, simple regressions already yielded a coefficient of determination >80 %, in other cases additional soil parameters (such as soil-pH, Zn_{WH}, Fe_{WH}, Al_{WH}, Ca_{WH}, and C_{total}; WH stands for 0.43 M HNO_3 extraction) were included to achieve this level of accuracy. Extractions with sodium bicarbonate (Olsen), DL, and AAAc displayed somewhat weaker correlations. The weakest relations with other methods were found for the AAAc extract. Accordingly, no satisfactory regression equations (i.e. with R^2 > 80 %) could be produced for the latter set of methods. To evaluate and validate the models, the best regression equations for each pair of methods was applied to an independent data set produced from 93 soil samples derived from Germany (Shwiekh et al. 2014). The samples were selected to represent a wide range of available P concentrations and soils comparable to those used for the calculation of the regression equations. All regression equations displaying coefficients of determination R^2 > 80 % were used to predict available soil P in the set of independent samples. Then, the predicted values were plotted against the original measured values and regression equations were calculated to evaluate their quality. Out of the 26 equations tested, 8 produced predictions with a coefficient of determination R^2 < 80 %, i.e. a transformation of STP using these equations would not yield a statistically satisfactory result (data not shown here; see Shwiekh et al. 2014). Thus it can be concluded that even regression equations derived with a high level of determination (R^2 > 80 %) may not necessarily perform well when applied to another set of soil samples. As major obstacles for a reliable transformation of values obtained by different soil tests the authors identified differences in chemical composition, acidity and extraction force between methods as well as differences in the soil sample sets the equations were derived from and those they were applied to. The results of the studies indicate that usually there is a rough agreement between the results obtained with different extraction methods in non-calcareous soils, but in calcareous soils the results of acidic and basic extractants usually have a poor correlation. However, all studies involving comparisons of STP results obtained with different methods indicate that there are outliers, which calls upon caution in the attempt of converting results between methods and raises doubts about the accuracy of such conversions.

Establishing the Biological Significance of STP Values and Formulation of Fertiliser Recommendations

Only the plant itself is able to measure the available amount of P in the soil. The amount of P dissolved/desorbed from soil does not imply how suitable the method is to predict the P supply to plants. Thus, chemical soil P tests can only be evaluated by correlating the extracted P with plant parameters such as yield, P-uptake or P-content and not on the basis of absolute amounts extracted or taken up by the plant (Fitts and Nelson 1956; cited after Kamprath and Watson 1980; Olsen and Khasawneh 1980; Sibbesen 1983).

For example, in 12 phosphate-rich soils investigated by Delgado and Torrent (1997) the actual quantity of plant available P was estimated to be about three-fold compared to the result of the Olsen soil test. Thus, the results should be rather understood as relative values and it has to be ensured that the STP-values are closely related to the crop response (Beegle 2005). In the past, regression analyses for STP values and plant uptake or yield were carried out for most conventional extraction methods in most countries with a multitude of different soils and crops, producing so-called yield-response curves which allowed to predict the likelihood of plant response to fertilisation and develop soil and crop specific fertilising recommendations (see Fig. 1 and related text). This is done by identifying a so-called critical STP level above which little or no plant response to fertilizer P is observed. When STP levels range around this critical level, it is recommended to add just the amount of fertiliser P, which will maintain the adequate soil P or provide a starter effect (Kamprath and Watson 1980). In other words, the amount of P fertilisation should match plant uptake. Below the critical level, the recommendation will be to feed the plants and at the same time build up some P reserves with fertilisation, while above it, no additional P fertilisation is recommended. To evaluate the suitability of different standard P-tests with different extraction mechanisms for correctly estimating the amount of plant available P, Sibbesen (1983) statistically investigated the correlation of P-test values with the particular plant P uptake of various pot experiments. It turned out that the P-tests can be divided into three groups: (i) Best group: anion-exchange resin method; (ii) intermediate group: water and sodium bicarbonate methods, and (iii) worst group: all 'acid' methods.

Similarly, Fixen and Grove (1990) investigated the results of a multitude of studies dealing with the correlation of several STP-tests and the results of pot and field trials. They elaborated that the correlation coefficients varied widely among all the tests investigated (Bray 1 and 2, Olsen, AB-DTPA, Mehlich I + II, Water or dilute salts, resin) from no significant correlation with the availability to explaining more than 90 % of the variability of P-availability (r^2). However, since the lowest r-value for the resin methods was 0.69, it is indicated that these methods are suitable over a broader range of soil properties than the other methods. According to Fixen and Grove (1990) the results indicate that nearly any soil test will fail if it is applied on inappropriate soils.

As pointed out by Morel et al. (2000), major sources of P taken up or mobilised by plant roots are P ions in solution or P attached to soil constituents but ready to replenish and buffer the soil solution. Therefore, the authors emphasize that methods describing P ion transfer between soil constituents and solution, such as EUF or isotopic exchange studies, have a high potential to improve the accuracy of soil P testing. Akinrinde et al. (2006) and Akinrinde and Obigbesan (2005), comparing EUF to conventional STP methods (water, AAAc, 0.1 M HCl, $CaCl_2$ and Bray-1) using sorghum cultivated in some tropical soils as a test plant, found only weak or insignificant correlations between EUF and conventional tests. However, EUF showed better correlations with crop response than the conventional methods, which underlines the suitability of the method for studying nutrient availability. Interestingly, the authors report that in contrast to conventional STP methods, the largest proportion of nutrients desorbed by EUF came from the soil organic fraction.

The superiority of DGT and resin P in comparison to standard chemical extraction methods to assess the amount of available P in the soil has been reported by several authors. Most studies which investigated the suitability of DGT to assess the soil P status for different soils reported significant correlations between the amount of plant available P in the soil determined by DGT and plant response (e.g. dry matter yield or P uptake), while other standard extraction methods (Olsen, Colwell, Bray-1) predominantly displayed weaker or no correlations at all (Menzies et al. 2005; Mason et al. 2010; Tandy et al. 2011; Six et al. 2013a, b). In contrast to this, however, the yields of rice in a pot trial were best predicted by Mehlich III, Bray-1, Olsen and resin P in comparison to DGT and $CaCl_2$. The authors concluded that the main factor explaining the P uptake for rice is not the diffusion of P in the soil which is measured by DGT (Six et al. 2013a, b).

In the past, the suitability of AEM and AER to accurately predict the amount of plant available P in the soil, also in comparison to other STP-methods, was proven by various studies: Basing on the comparison of Resin-, Colwell- and Olsen-P with yield data from a field trial (Saggar et al. 1999) and a pot trial (Menzies et al. 2005), the suitability of resin-P to assess the amount of plant available P in soils was emphasised, while Bray-P and the Colwell-P did not provide accurate estimates. Similarly, Bissani et al. (2002) suggested that both resin methods lead to a better assessment of plant available P in the soil than Mehlich I and Pi. Besides the superiority of the resin methods to reflect the P uptake of ryegrass in a pot trial compared to conventional extraction methods (Olsen, Colwell, Bray I + II and Mehlich), Naidu et al. (1991) found that IEP showed no correlation with plant uptake. The authors attributed this to the difficulties in measuring P in the solution of soils with a very high P-sorption capacity.

The problem of the overestimation of labile P by isotopic measurements in soils with high amounts of P-fixing compounds (i.e. Fe-oxide) had been already brought up by Amer et al. (1969). Furthermore, Cabeza et al. (2013) assumed an overestimation of plant available P by IEP when the equilibrium concentration in the soil is either the same or below the P concentration at which no net P plant uptake occurs. In contrast to these findings, Fardeau et al. (1996) postulated, with reference to the results of different studies, that in all soil types the isotopically exchangeable P-ions in the soil solution are bioavailable and thus can be detected by the isotopic exchange method.

The Pi-method showed a good correlation with plant parameters such as dry matter yield or plant uptake in field studies and greenhouse experiments. It turned out that the Pi-test is independent of the soil type and can be used on alkaline, acidic and neutral soils. Furthermore, it also seems to be suitable for soils, which have been fertilised with water-soluble P-fertilisers or phosphate rock. In contrast, traditional acidic soil tests such as Bray, Truog or Mehlich are not suitable for these soils since they overestimate the P availability, whereas alkaline extractants such as Olsen tend to underestimate it (Menon et al. 1997; with further references).

Conclusions

A multitude of methods aiming at the determination of plant available P in the soil have been developed over the past 50–100 years. None of them can claim to be universally accepted all over the world. Rather, each country has selected those of them as their standard reference methods which fitted best to the majority of the particular countries' soils. While in the first place, these methods were developed for agricultural purposes, many of them are also useful in the field of environmental protection, since it has been demonstrated that high STP values contribute to a higher dissolved P concentration in runoff water, and to a higher P release during the erosion process and in the eventual water body. The variety of chemical extraction mechanisms as well as differences in analytical procedures offer both advantages and disadvantages. While the use of different standard methods and analytical procedures in different countries makes international comparisons rather difficult, turning the formulation of valid cross-border fertiliser recommendations into a very challenging, if not impossible task, differences in the extraction forces of the multiple methods can also be very useful to distinguish between various soil P fractions of different agricultural as well as environmental relevance. In the first place, P fractions, which are immediately available can be distinguished from those which may only become available in the medium or in the long term. This knowledge, in turn, can be used to formulate accurate fertiliser recommendations to ensure optimum yields and to prevent P-surpluses as well as an increased risk of P losses from the soil at the same time. Furthermore, modern methods basing on desorption techniques such as AEM/AER, DGT, IEP or EUF even allow to estimate additional parameters relevant for environmental monitoring and risk assessment, such as the amount of directly/easily soluble P, or the P buffer and fixation capacity of a soil. Thus, the selection of a suitable method depends not only on the soil under study, but also on the key issues addressed in the particular investigation.

Plenty of efforts have been made to provide transfer factors or equations allowing to "translate" the results of two or more methods into one another. However, the quality of the translation equations, the parameters needed to include, as well as the particular resulting coefficients depend very much on the characteristics of the soil samples used for their development, and transfer to other, non-related soil samples remains at least questionable. So far, no universally valid equations can be produced, and calculated values are only rough estimates at best. Future investigations should systematically include samples of different soil types and characteristics, resulting in individual translation equations valid only under clearly defined circumstances (i.e. a particular soil type, soil pH a.s.o.).

References

Akinrinde EA, Obigbesan GO (2005) Comparison of six extractants for assessing available phosphorus, zinc and manganese in selected Nigerian soils. Glob J Pure Appl Sci 11:179–184

Akinrinde EA, Obigbeson GO, Gaiser T (2006) Electro-ultrafiltration technique in relation to conventional methods of soil testing for the determination of available P, Ca, Mg and NO_3-N in some tropical soils. J Agron 5:375–381

Amer F, Bouldin DR, Black CA, Duke FR (1955) Characterization of soil phosphorus by anion exchange resin adsorption and P^{32} equilibration. Plant Soil 6:391–408

Amer F, Mahdi S, Alradi A (1969) Limitations in isotopic measurements of labile phosphate in soils. J Soil Sci 20:91–100

Aura E (1978) Determination of available soil phosphorus by chemical methods. J Sci Agric Soc Finland 50:305–316

Barrow NJ (2008) The description of sorption curves. Eur J Soil Sci 59:900–910

Beegle DB (2005) Assessing soil phosphorus for crop production by soil testing. In: Sims JT, Sharpley AN (eds) Phosphorus: agriculture and the environment. Agronomy Monograph 46. American Society of Agronomy, Inc., Madison, pp 123–143

Bissani CA, Tedesco MJ, de Camargo FAO, Miola GL, Gianello C (2002) Anion-exchange resins and iron oxide-impregnated filter paper as plant available phosphorus indicators in soils. Comm Soil Sci Plant Anal 33:1119–1129

Bray RM, Kurtz LT (1945) Determination of total, organic and available forms of phosphorus in soils. Soil Sci 59:39–45

Bünemann EK, Steinebrunner F, Smithson PF, Frossard E, Oberson A (2004) Phosphorus dynamics in a highly weathered soil as revealed by isotopic labelling techniques. Soil Sci Soc Am J 68:1645–1655

Cabeza R, Steingrobe B, Römer W, Claassen N (2011) Effectiveness of recycled P products as P fertilizers as evaluated in pot experiments. Nutr Cycl Agroecosys 91:173–184

Cabeza R, Steingrobe B, Römer W, Claassen N (2013) Plant availability of isotopically exchangeable and isotopically nonexchangeable phosphate in soils. J Plant Nutr Soil Sci 179:688–695

Carmo Horta M, Torrent J (2007) The Olsen P method as an agronomic and environmental test for predicting phosphate release from acid soils. Nutr Cycl Agroecosyst 77:283–292

Chardon WJ, Menon RG, Chien SH (1996) Iron oxide impregnated filter paper (Pi test): a review of its development and methodological research. Nutr Cycl Agroecosys 46:41–51

Colwell JD (1963) The estimation of phosphorus fertiliser requirements for wheat in southern New South Wales by soil analysis. Aust J Exp Agric Anim Husb 19:337–343

Cottenie A, Camerlynck R, Verloo M, Dhaese A (1979) Fractionation and determination of trace elements in plants, soils and sediments. Pure Appl Chem 52:45–53

Daly K, Casy A (2005) Environmental aspects of soil phosphorus testing. Irish J Agr Food Res 44:261–279

Degryse F, Smolders E, Zhang H, Davison W (2009) Predicting availability of mineral elements to plant with the DGT technique: a review of experimental data and interpretation by modelling. Environ Chem 6:198–218

Delgado A, Torrent J (1997) Phosphate-rich soils in the European Union: estimating total plant-available phosphorus. Eur J Agron 6:205–214

Demaria P, Flisch P, Frossard E, Sinaj S (2005) Exchangeability of phosphate extracted by four chemical methods. J Plant Nutr Soil Sci 168:89–93

Di HJ, Condron LM, Frossard E (1997) Isotope techniques to study phosphorus cycling in agricultural and forest soils: a review. Bio Fertil Soils 24:1–12

Dirks B, Scheffer H (1930) Der Kohlensäure-Bikarbonatauszug und der Wasserauszug als Grundlage zur Ermittlung der Phosphorsäurebedürftigkeit der Böden. Landwirtschaftl Jahrb 71:73–99

Egnér H (1954) Neue Beiträge zur chemischen Bodenuntersuchung unter besonderer Berücksichtigung der Laktatmethode. Landwirt Forsch 6 SH:28–32
Egnér H, Köhler G, Nydahl F (1938) Die Laktatmethode zur Bestimmung leichtlöslicher Phosphorsäure in Ackerböden. Annalen der landwirtschaftlichen Hochschule Schwedens 6:253–298
Egnér H, Riehm H, Domingo WR (1960) Untersuchungen über die chemische Bodenanalyse als Grundlage für die Beurteilung des Nährstoffzustandes der Böden. II. Chemische Extraktionsmethoden zur Phosphor – und Kaliumbestimmung. Kungl Lantbrukshögsk Ann 26:199–215
Ekholm P, Yli-Halla M, Kylmälä P (1991) Availability of phosphorus in suspended sediments estimated by chemical extraction and bioassay. Verhandlungen—Internat Verein Theor Angewe Limnol 24:2994–2998
Eriksson AK (2009) Phosphorus in agricultural soils around the Baltic Sea—comparisons of different laboratory methods as indices for phosphorus leaching to waters. Swedish University of Agricultural Sciences. Division of Water Quality Management. Seminarier och examensarbeten 64, 33 pp
Fardeau JC (1996) Dynamics of phosphate in soils. An isotopic outlook. Fert Res 45:97–100
Fardeau JC, Guiraud G, Marol C (1996) The role of isotopic techniques on the evaluation of the agronomic effectiveness of P fertilizers. Fert Res 45:101–109
Fernandes MLV, Warren GP (1996) Comparison of resin beads and resin membranes for extracting soil phosphate. Fert Res 44:1–8
Fitts JW, Nelson WL (1956) The determination of lime and fertilizer requirements of soils through chemical tests. Adv Agron 8:241–282
Fixen PE, Grove JH (1990) Testing soils for phosphorus. In: Westermann RL (ed) Soil testing and plant analysis. SSSA Book Ser. 3, Madison, pp 141–180
Fried M (1964) "E", "L" and "A" values. Trans 8th Int Congr Soil Sci IV:29–39
Frossard E, Julien P, Neyroud JA, Sinaj S (2004) Phosphor in Böden. Standortbestimmung Schweiz. Schriftenreihe Umwelt Nr. 368, BUWAL, Bern
Frossard E, Achat DL, Bernasconi SM, Bünemann EK, Fardeau J-C, Jansa J, Morel C, Rabeharisoa L, Lalajaona R, Sinaj S, Tamburini F, Oberson A (2011) The use of tracers to investigate phosphate cycling in soil-plant systems. In: Bünemann EK, Oberson A, Frossard E (eds) Phosphorus in action. Springer, Berlin, pp 59–91
Gallet A (2001) Phosphorus availability and crop production in seven Swiss field experiments. Dissertation, Zürich
Gassner A (2003) Factors controlling the spatial specification of phosphorus in agricultural soils. Landbauforsch Völk SH 244, Braunschweig
Hartikainen H (1986) Phosphorus concentration in soil-water extract as a function of extraction ratio. Trans 13th Congr Int Soc Soil Sci 2:325–326
Hesketh N, Brookes PC (2000) Development of an indicator for risk of phosphorus leaching. J Environ Qual 29:105–110
Horn D (2006) Bestimmung von Mikronährstoffen und Schwermetallen in Böden mit dem Verfahren der Elektro-Ultrafiltration (EUF) durch Zugabe von DTPA. Determination of micronutrients and heavy metals in soils using electro-ultrafiltration (EUF) technique by addition of DTPA. J Plant Nutr Soil Sci 169:83–86
Houba VJG, Novozamsky I, van Dijk D (1998) Certification of an air-dry soil for pH and extractable nutrients using one hundredth molar calcium chloride. Commun Soil Sci Plant Anal 29:1083–1090
Jaakkola A, Hartikainen H, Lemola R (1997) Effect of fertilization on soil phosphorus in a long-term field experiment in southern Finland. Agric Food Sci Finland 6:313–322
Jones CA, Cole CV, Sharpley AN (1984) A simplified soil and plant phosphorus model: I documentation. Soil Sci Soc Am J 48:800–805
Kamprath EJ, Watson ME (1980) Conventional soils and tissue tests for assessing the phosphorus status of soils. In Khasawneh FE (ed) The role of phosphorus in agriculture, American Society of Agronomy, Madison, pp 433–469

Knisel WJ (1980) CREAMS: a field scale model for chemicals, runoff, and erosion from agricultural management systems. USDA-ARS Conservation research report, no. 26. U.S. Government Printing Office, Washington, DC

Koopmans GF, Chardon WJ, Dekker PHM, Römkens PFAM, Schoumans OF (2006) Comparing different extraction methods for estimating phosphorus solubility in various soil types. Soil Sci 171:103–116

Lakanen E, Erviö R (1971) A comparison of eight extractants for the determination of plant available micronutrients in soils. Acta Agr Fenn 123:223–232

Larsen S (1952) The use of ^{32}P in studies on the uptake of phosphorus by plants. Plant Soil 4:1–10

Maguire RO, Chardon WJ, Simard RR (2005) Assessing potential environmental impacts of soil phosphorus by soil testing. In: Sharpley AN (ed) Sims JT. Phosphorus, Agriculture and the environment, pp 145–180

Mason S, Hamon R, Nolan A, Zhang H, Davison W (2005) Performance of a mixed binding layer for measuring anions and cations in a single assay using the diffusive gradients in thin film technique. Anal Chem 77:6339–6346

Mason S, McNeill A, McLaughlin MJ, Zhang H (2010) Prediction of wheat response to an application of phosphorus under field conditions using diffusive gradients in thin-films (DGT) and extraction methods. Plant Soil 337:243–258

McAuliffe CD, Hall NS, Dean LA, Hendricks SB (1947) Exchange reactions between phosphates and soils: hydroxylic surfaces of soil minerals. Soil Sci Soc Am Proc 12:119–123

McIntosh JL (1969) Bray and Morgan soil test extractants modified for testing acid soils from different parent materials. Agron J 61:259–265

Mehlich A (1953) Determination of P, Ca, Mg, K, Na, and NH4. North Carolina Soil Test Division (Mimeo). Raleigh, NC

Mehlich A (1984) Mehlich 3 soil test extractant: a modification of the Mehlich 2 extractant. Commun Soil Sci Plant Anal 15:1409–1416

Menon RG, Chien SH, Chardon WJ (1997) Iron oxide-impregnated filter paper (P_i test): II. A review of its application. Nutr Cycl Agroecosys 47:7–18

Menzies NW, Kusumo B, Moody PW (2005) Assessment of P availability in heavily fertilized soils using the diffusive gradient in thin films (DGT) technique. Plant Soil 269:1–9

Michaelson GJ, Gavlak RG, Ping CL (1991) Relating Mehlich 3 extractable P to Morgan and Bray 1 extractable P for Alaska soils. University of Alaska Fairbanks. Agricultural and Forestry Experiment Station. School of Agriculture and Land Resources Management. Circular 81. 6 pp

Moore PA Jr, Joern BC, Provin TL (1998) Improvements needed in environmental soil testing for phosphorus. In: Sims JT (ed) Soil testing for phosphorus. Environmental uses and implications (Chap. 3). Southern Cooperative Series Bulletin No. 389, pp 21–30

Morel C, Tunney P, Plenet D, Pellerin S (2000) Transfer of phosphate ions between soil and solution: perspectives in soil testing. J Environ Qual 29:50–59

Morgan MF (1941) Chemical soil diagnosis by the universal soil testing system. Conn Agric Exp Stn Bull 450

Naidu R, Syers JK, Tillman RW, Kirkman JH (1991) Assessment of plant-available phosphate in limed, acid soils using several soil-testing procedures. Fert Res 30:47–53

Nemeth K (1980) The availability of nutrients in soils as determined by electro-ultrafiltration (EUF). Adv Agron 31:155–187

Nemeth K (1982) Electro-ultrafiltration of aqueous soil suspension with simultaneously varying temperature and voltage. Plant Soil 64:7–23

Neyroud J-A, Lischer P (2003) Do different methods used to estimate soil phosphorus availability across Europe give comparable results? J Plant Nutr Soil Sci 166:422–431

Olsen SR, Khasawneh FE (1980) Use and limitations of physical-chemical criteria for assessing the status of phosphorus in soils. In: Khasawneh FE, Sample EC, Kamprath EJ (eds) The role of phosphorus in agriculture. American Society of Agronomy, Madison, pp 361–410

Olsen SR, Sommers LE (1982) Phosphorus. In: Page AL, Miller RH, Keeney DR (eds) Methods of soil analysis, part 2. Agronomy Monographs No. 9. American Society of Agronomy, Soil Science Society of America, Madison, pp 403–430

Olsen SR, Cole CV, Watanabe FS, Dean LA (1954) Estimation of available phosphorus in soils by extraction with sodium bicarbonate. USDA Circular 939. U.S. Government Printing Office, Washington DC

Ottabong E, Börling K, Kätterer T, Mattsson L (2009) Comparability of the ammonium lactate (AL) and sodium carbonate (Olsen) methods for determining available phosphorus in Swedish soils. Acta Agr Scand, Section B Soil Plant Sci 59:373–378

Paasikallio A, Häkkinen U (1977) Acid ammonium acetate and acid ammonium acetate/EDTA as extractants for phosphorus-32, aluminium and iron in soils. Ann Agric Fenn 16:227–237

Pote DH, Daniel TC, Sharpley AN, Moore PA Jr, Edwards DR, Nichols DJ (1996) Relating extractable soil phosphorus to phosphorus losses in runoff. Soil Sci Soc Am J 60:855–859

Riehm H (1942) Bestimmung der laktatlösischen Phosphorsäure in karbonathaltigen Böden. Phosphorsäure 1:167–178

Riehm H (1958) Die Ammoniumlaktatessigsäure-Methode zur Bestimmung der Leichtlöslichen Phosphorsäure in Karbonathaltigen Böden. Agrochimica 3:49–65

Saarela I (1992) A simple diffusion test for soil phosphorus availability. Plant Soil 147:115–126

Saarela I (2002) Phosphorus in Finnish soils in the 1900s with particular reference to the acid ammonium acetate soils test. Agr Food Sci Finland 11:257–271

Saarela I, Sippola J (1990) Inorganic leaf phosphorus and soil tests as indicators of phosphorus nutrition in cereals. Commun Soil Sci Plant Anal 21:1927–1943

Saarela I, Järvi A, Hakkola H, Rinne K (2004) Phosphorus status of diverse soils in Finland as influenced by long-term P fertilisation 2. Changes of soil test values in relation to P balance with references to incorporation depth of residual and freshly applied P. Agric Food Sci 13:276–294

Saggar S, Hedley MG, White RE, Perrott KW, Gregg PEH, Cornforth IS, Sinclair AG (1999) Development and evaluation of an improved soil test for phosphorus, 3: field comparison of Olsen, Colwell and Resin soil P tests for New Zealand pasture soils. Nutr Cycl Agroecosys 55:35–50

Saunders WMH (1964) Extraction of soil phosphate by anion-exchange membrane. New Zeal J Agr Res 7:424–431

Schick J, Kratz S, Shwiekh R, Rückamp D, Haneklaus S, Schnug E (2013) Comparison and inter-calibration of different soil P tests used in the Baltic Sea Countries. Knowledge Report

Schnug E, Hanotiaux G, Long M, Vanoverstraeten M (1999) Handbook of analytical methods for soils used by the IMPHOS Western European Research Network on Phosphorus

Schüller H (1969) Die CAL-Methode, eine neue Methode zur Bestimmung der pflanzenverfügbaren Phosphates im Boden. Z Pflanz Bodenk 123:48–63

Self-Davis ML, Moore PA Jr, Joern BS (2000) Determination of water- and/or dilute salt-extractable phosphorus. In: Pierzynski GM (ed) Methods of phosphorus analysis for soils, sediments, residuals, and waters. Kansas State University, Manhattan, pp 24–26

Sharpley AN (1995) Dependence of runoff phosphorus on extractable soil phosphorus. J Environ Qual 24:920–926

Sharpley AN, Smith SJ, Menzel RG (1982) Prediction of phosphorus losses in runoff from southern plains watersheds. J Environ Qual 11:247–251

Shwiekh R, Kratz S, Schick J, Kammerer H, Ahmed SS, Schnug E (2013) Determination of inorganic and organic P dissolved in water and Olsen extracts by inductively coupled plasma optical emission spectroscopy (ICP-OES) and colorimetry. Landbauforsch 64:303–306

Shwiekh R, Schick J, Kratz J, Rückamp D, Schnug E (2014) Comparative investigation and inter-calibration of different soil P tests (Unpublished)

Sibbesen E (1977) A simple ion-exchange resin procedure for extracting plant-available elements from soil. Plant Soil 46:665–669

Sibbesen E (1983) Phosphate soil tests and their suitability to assess the phosphate status of soil. J Sci Food Agric 34:1368–1374

Simard RR, Tran TS (1993) Evaluating plant-available phosphorus with the electro-ultrafiltration technique. Soil Sci Soc Am J 57:404–409

Sims JT (1993) Environmental soil testing for phosphorus. J Prod Agric 6:501–507

Sims JT (2000) Soil test phosphorus: Olsen P. In: Pierzynski, GM (ed) Methods of phosphorus analysis for soils, sediments, residuals, and waters. Kansas State University, Manhattan, pp 20–21
Sims JT, Hodges S, Davis J (1998) Soil testing for phosphorus: current status and uses in nutrient management programs. In: Sims JT (ed) Soil testing for phosphorus—environmental uses and implications. Southern Cooperative Series Bulletin No. 389, pp 13–20
Sissingh HA (1971) Analytical technique of the Pw method used for the assessment of the phosphate status of arable soils in the Netherlands. Plant Soil 34:438–446
Six L, Smolders E, Merckx R (2013a) The performance of DGT versus conventional soil phosphorus tests in tropical soils—maize and rice responses to P application. Plant Soil 366:49–66
Six L, Smolders E, Merckx R (2013b) Testing phosphorus availability for maize with DGT in weathered soils amended with organic materials. Plant Soil 376:177–192
Soltanpour PN, Schwab AP (1977) A new soil test for simultaneous determination of macro and micro-nutrients in alkaline soils. Commun Soil Sci Plant Anal 8:195–207
Tandy S, Mundus S, Yngvesson J, de Bank TC, Lombi E, Schjoerring JK, Husted S (2011) The use of DGT for prediction of plant available copper, zinc and phosphorus in agricultural soils. Plant Soil 346:167–180
Tattari S, Bärlund I, Rekolainen S, Posch M, Siimes K, Tuhkanen H-R, Yli-Halla M (2001) Modelling sediment yield and phosphorus transport in Finnish clayey soils. Trans Am Soc Agric Eng (ASAE) 44:297–307
Teklić T, Vukadinović V, Bertić B, Lončarić Z (2009) Comparability of electroultrafiltration and ammonium acetate-lactate extraction data for phosphorus and potassium in pseudogley soils. Comm Soil Sci Plant Anal 40:1–6
Thomas GW, Peaslee DE (1973) Testing soils for phosphorus. In: Walsh LM, Beaton JD (eds) Soil testing and plant analysis (revised edition). Soil Science Society of America Inc., Madison, pp 115–132
Torrent J, Delgado A (2001) Using phosphorus concentration in the soil solution to predict phosphorus desorption to water. J Environ Qual 30:1829–1835
Truog E (1930) The determination of the readily available phosphorus in soils. J Am Soc Agron 22:874–882
Uusitalo R, Ekholm P (2003) Phosphorus in runoff assessed by anion exchange resin extraction and an algal assay. J Environ Qual 32:633–641
van der Paauw F, Sissingh HA, Ris J (1971) Een verbeterede methode van fosfaatextractie van grond met water: het Pw-getal; with a summary: an improved method of water extraction for the assessment of availability of soil phosphate: PW value. Verslagen van landbouwkundige onderzoekingen, vol 749
van der Zee SEATM, Fokkink LGJ, van der Riemsdijk WH (1987) A new technique for assessment of reversibly adsorbed phosphate. Soil Sci Soc Am J 51:599–604
Vanderdeelen J (1994) Phosphate immobilization in an uncropped field experiment on a calcareous soil. Plant Soil 171:209–215
VDLUFA (2012) VDLUFA-Methodenbuch Band I: Die Untersuchung von Böden. 1.-6. Teillieferung. Darmstadt
Vuorinen J, Mäkitie O (1955) The method of soil testing in use in Finlad. Agrogeol Publ 63:1–44
Watson M, Mullen R (2007) Understanding soil tests for plant-available phosphorus. Online: http://ohioline.osu.edu/agf-fact/pdf/Soil_Tests.pdf. Last access: 22 July 2013
Werner W (1974) Der Einfluß des Kalkgehaltes auf das Lösungsvermögen der konventionellen Lactatlösungen für Düngerphosphate und deren Umwandlungsprodukte. Die Phosphorsäure 30:189–198
Williams JR, Sharpley AN (1989) EPIC—Erosion/productivity impact calculator: 1. Model documentation. Technical Bulletin No. 1768. Washington, DC: USDA Agricultural Research Service
Wolf AM, Baker DE (1985) Comparisons of soil test phosphorus by Olsen, Beay P1, Mehlich I and Mehlich III methods. Commun Soil Sci Plant Anal 16:467–484

Yli-Halla M, Hartikainen H, Ekholm P, Turtola E, Puustinen M, Kallio K (1995) Assessment of soluble phosphorus load in surface runoff by soil analysis. Agric Ecosys Environ 56:53–62
Yli-Halla M, Tattari S, Bärlund I, Tuhkanen H-R, Posch M, Siimes K, Rekolainen S (2005) Simulating processes of soil phosphorus in geologically young acidic soils of Finland. Trans Am Soc Agric Eng (ASAE) 48:101–108
Zhang H, Davison W, Gadi R, Kobayashi T (1998) In situ measurement of dissolved P in natural waters using DGT. Anal Chim Acta 370:29–38
Ziadi N, Simard RR, Tran TS, Allard G (2001) Soil available phosphorus as evaluated by desorption techniques and chemical extractions. Can J Soil Sci 81:167–174
Ziadi N, Bélanger G, Gagnon B, Mongrain D (2009) Mehlich 3 soil phosphorus as determined by colorimetry and inductively coupled plasma. Comm Soil Sci Plant Anal 40:132–140

Assessing the Plant Phosphorus Status

Silvia H. Haneklaus and Ewald Schnug

Abstract Plant analysis is an approved method to assess the nutrient status of crops. Diverging values for critical nutrient values for one crop species are related among others to growth stage, sampled plant part and interpretation method. This makes it difficult for scientists and farmers alike to decide on correct values. A metadata analysis of existing symptomatological, critical nutrient, no effect and toxicological values for phosphorus with respect to different crop plants has been carried out. Values have been clustered in different groups of crops based on morphogenetic and physiological characteristics and phosphorus concentrations have been agglomerated in three major categories: severely deficient, sufficient and toxic. A total P concentration of 0.175–0.25 % proved to denote severe P deficiency in all crops. Phosphorus contents between 0.30 and 0.34 % P indicate a sufficient P supply of graminoids, legumes and oil crops. Toxic effects may occur if the phosphorus content exceeds 1 % P. Factors which affect the phosphorus uptake of crop plants such as root architecture and features which are closely related to the phosphorus status, for example resistance against diseases are outlined.

Keywords BOLIDES (BOundary Line Development System) · Critical nutrient value · Manure · No effect value · Plant analysis · Remote sensing · Root anatomy · Symptomatological value · Toxicological value · Variable rate fertilization

Introduction

Phosphorus (P) is an essential nutrient for all living organisms. A sufficient P supply is important to warrant crop productivity. Non-renewable P reserves will be depleted statistically in about 100 years (Perspectives Agriculture 2011). Though

S.H. Haneklaus (✉) · E. Schnug
Federal Research Centre for Cultivated Plants, Institute for Crop and Soil Science, Julius Kühn-Institute (JKI), Bundesallee 50, 38116 Brunswick, Germany
e-mail: silvia.haneklaus@julius-kuehn.de

E. Schnug and L.J. De Kok (eds.), *Phosphorus in Agriculture: 100 % Zero*,
DOI 10.1007/978-94-017-7612-7_6

the P resources are manifold higher than the actual P reserves, it can be assumed that access to mineral P fertilizers will be restricted increasingly by economic constraints. Yet another problem is the export of P in cheap feedstuff and energy crops from developing countries as increasing stock market prices favor P mining in these areas of the world. In contrast, the excessive use of farmyard manure on big livestock enterprises is the main reason for disproportionately high P levels in soils of industrialized countries (Haneklaus et al. 2016). A global concept is urgently required on how to deal with limited P resources in a socially just way as are strategies for balanced P fertilization that are purely driven by the P demand of the crop (Ulrich 2014; Schnug and Haneklaus 2016).

A truly sustainable P use avoids equally P mining and P surplus thus transcribing the concept of 'P: 100 % Zero' to production fields (Schnug and Haneklaus2016). Indispensable prerequisites for the successful implementation of the concept are the access to reliable critical nutrient values for different crop species and the determination of optimum nutrient ranges in soils, which are required for accomplishing the site-specific maximum yield. These are the basis for evaluating the plant P status and developing algorithms for P fertilization that match the small-scale spatial variation of plant available P in soils with variable rates of organic and mineral fertilizers. It can be expected that P recycling and thus the use of recycled P fertilizer products will play a key role in a sustainable P management. An important criterion for using such products is the speciation of P in the recycled product as it is imperative that the materials contain P in a plant available form.

This section evaluates methods for assessing the plant P status, discusses strategies for a balanced P fertilization, and addresses almost unnoticed aspects of P nutrition in agricultural practice. Hereby, focus was put on presenting and interpreting scientific data in such way that they can be transcribed into management practices on farms.

Soil Phosphorus Reserves

About 30–70 % of the total P content in agricultural soils is organically bound (Li et al. 2014). On low-fertility grassland, peat and forest soils this percentage can be as high as 80–95 % (Li et al. 2014). The median value for the total P content in northern European soils is 690 mg kg^{-1} P in top soils and 370 mg kg^{-1} P in bottom soils (Reimann et al. 2003). The minimum value was 50 mg kg^{-1} P and the maximum value 4080 mg kg^{-1} in top soils (Reimann et al. 2003).

It is possible that absolute P deficiency occurs on soils, but addressing transient P deficiency on agricultural soils will be the real challenge under conditions of a limited P input. Vitousek et al. (2010) summarized the bio-geo-chemical factors causing absolute and transient P deficiency and forecasted the horizons for their actual occurrence. P will be mineralized completely irrespective of the parent material if infinite periods of time are supposed. Absolute P deficiency is expected on extremely old and leached soils, which will by then reach a so-called terminal steady state (Walker and Syers 1997). However, the time frame for such a development is estimated to take several million years (Vitousek et al. 2010). It is important to localize such endangered sites, as otherwise changes in structure and functionality of the ecosystem will occur (Vitousek et al. 2010). Only anthropogenic activity and geological interferences may cause changes in soils, which have achieved their terminal steady state.

In contrast to absolute P deficiency induced by a pedogenetically complete P loss, transient P deficiency results from processes such as barriers in soil, which prevent access of plant roots to P, parent material with extremely low P content, accumulation of P in depressions, and induced P deficiency by extremely high nitrogen (N) application rates. The estimated period of time varies between decades and tens of thousands of years (Vitousek et al. 2010). The authors emphasize that also in low-grade ecosystems P losses will occur after steady state has been achieved, which are balanced by supply through erosion on a minimum level. It is important to follow up changes in the P status of soils and plants over time in order to evaluate the progress of the processes mentioned previously. Besides this it is necessary to determine biomass production, N/P ratio in long-term field experiments in order to capture changes in the site-specific productivity whilst taking geographic and climatic constraints into account. Anthropogenic activities cause continuously increasing N loads, which change entire areas and increase the risk of P deficiency (Vitousek et al. 2010). From agricultural point of view it is relevant for balancing P fertilization that occluded P is dynamic on a time scale of decades (Richter et al. 2004) and not inaccessible for organisms infinitely as suggested by Crews et al. (1995).

Factors Affecting Phosphorus Acquisition of Crop Plants

Genetic and morphological characteristics of the plant, parent material and pedogenesis of soils influence besides climatic factors and fertilizer management P acquisition of crop plants. There is an urgent need to explore agro-technical measures in order to improve a resource-efficient use of P in agriculture. Chances and challenges are highlighted next.

Root Anatomy

When agriculture began about 13,000 years ago plant species with a low resource requirement for growth prevailed (White et al. 2013). Root vigor was crucial to acquire water and minerals on soils with low fertility (White et al. 2013). Today, intensive agricultural production systems prevail so that modern cultivars likely lost their beneficial root traits for low input systems (Wissuwa et al. 2009). There are indications that breeding has altered cereal growth patterns towards shallower or lighter root systems in modern cultivars (Mac Key 1988; Siddique et al. 1990). Root size of modern cultivars proved to be small when compared to landraces (Waines and Ehdaie 2007). In general, resource efficient approaches to breed high-yielding crops with high P use efficiency on soils with restricted P availability will have to focus on improving P acquisition and root architectural traits rather than a higher utilization efficiency (White et al. 2012). This approach has been successfully accomplished in crops cultivated on soils with low P status in developing countries by proliferation of lateral roots, the production of long root hairs, colonization with mycorrhizal fungi, development of cortical aerenchyma, increasing P uptake capacity of root cells, and exudation of organic acids and phosphatases into the rhizosphere (White et al. 2013). In particular, P acquisition will increase with length and longevity of root hairs rather than a higher density (Brown et al. 2013). White et al. (2013) summarized the most important characteristics or root specifications for efficient P acquisition: early root vigor, large root biomass or root/shoot ratio, more cortical aerenchyma, large root surface in top soil (lateral rooting, root hairs), high root length density, proliferation in patches of high P phytoavailability, mycorrhizal associations, greater exudation of protons, organic compounds and phosphatases, and last but not least higher phosphate uptake capacity of root cells.

Under conditions of a limited P supply plants may improve P acquisition by changes in root growth and anatomy, increased production of phosphatases and RNases and altered activity of enzymes of the glycolytic pathway (Liu et al. 1998; George et al. 2011). Changes in root architecture in relation to the soil P supply is controlled by crosstalk between exogenous and endogenous hormones and signaling molecules such as reactive oxygen species (ROS), sugars and microRNAs (Niu et al. 2013; Jung and McCouch 2013). Root architecture is regulated by hormone-controlled pathways (Jung and McCouch 2013). Under conditions of high phosphate availability in soil interaction with auxin promotes primary root growth while that of lateral roots is reduced (Jung and McCouch 2013). In comparison, a low soil phosphate pool interacts with auxin, cytokinins, ethylene, gibberellins and endogenous strigolactones expressed by a shallow and highly branched root system (Jung and McCouch 2013; Niu et al. 2013).

Mycorrhizal symbioses contribute significantly to the P supply of angiosperms on soils with low P status (Vance 2011). An overview of the significance of mycorrhiza for the P nutrition of crop plants is provided for instance by Shah (2014), Naher et al. (2013), Bünemann et al. (2011), Garg and Chandel (2010) and Koltei and Kapulnik (2010).

Farm Management Practices

The farm management system, here crop rotation, polycultures and intercropping have been shown to improve the nutrient supply distinctly (Fang et al. 2011; White et al. 2013). In case of P only minimum beneficial effects were observed in polycultures as only few roots were sufficiently close to benefit from exudates of a neighboring plant when simulation modeling was applied (Postma and Lynch 2012). In contrast, Li et al. (2007) reviewed existing intercropping experiments with the result that overyielding was 46 % for maize and 26 % for faba bean in a four years field intercropping experiment. In comparison to monocultures, intercropping enhanced P acquisition by faba bean and maize. An association with P mobilizing plant species facilitates mobilization of organic P by microbial or root-released phosphatases and inorganic P by releasing carboxylates or protons on alkaline soils (Li et al. 2014).

Eichler-Löbermann et al. (2008) tested various catch crops (oil radish, buckwheat, serradella, ryegrass and phacelia) with view to their ability to solubilize P in soils and contribute to the P nutrition of the main crop. The results of a three-year field experiment revealed that phacelia contributed strongest to the P supply of the main crop, since it increased the P uptake as well as the P content in soil significantly. In contrast, ryegrass reduced the plant available P pool in the soil.

Farm management practices that may interfere with the mineral acquisition and utilization of plants are the use of pesticides. Glyphosate (N-phosphonomethylglycine) is an herbicide that reduced nutrient uptake and shoot biomass of glyphosate-resistant soybean under field conditions in relation to rate and growth stage (Zobiole et al. 2012). Glyphosate uptake by plants increased with the orthophosphate concentration in the soil (Denis and Delrot 1993). Denis and Delrot (1993) concluded from their experiments that glyphosate is absorbed partly by a phosphate transporter of plants. These data comply with those of Cakmak et al. (2009) who showed that simulated drift of glyphosate increased the P content in young leaves of non-resistant soybean varieties while that of Ca, Mg and Mn was reduced. At the same time the risk of glyphosate being leached from agricultural soils is higher if the plant available P pool is high (Simonsen et al. 2008).

Another example for the impact of farm management practices on nutrient acquisition is the use of the fungicide phosphite. For farmers it is important to distinguish strictly between phosphite and phosphate as the first will not contribute to the P nutrition of the crop (Araujo et al. 2013; Schroetter et al. 2006). In these and other studies phosphite unfolded negative growth effects when the P nutritional status was below optimum. Visible macroscopic effects under greenhouse conditions were a stunted growth of the plants up to a complete die off if P was applied exclusively as phosphite (Schroetter et al. 2006). The physiological reason is simply that plants and microorganisms cannot metabolize phosphite (Lopez-Arredondo and Herrera-Estrella (2012). Consequently, phosphite is no suitable P source for agricultural crops (Thao and Yamakawa 2009).

Regulatory Aspects of Phosphorus Uptake by Crop Plants

In a soil pH dependent manner, P is present in the soil solution as $H_2PO_4^-$, HPO_4^{2-} and H_3PO_4. P uptake as $H_2PO_4^-$ is highest at soil pH 5–6 (Schachtman et al. 1998; George et al. 2011). Plant P requirements are high while the P concentration in the soil is low with at maximum 10 μM (Bieleski 1978). This means that crop plants require efficient transporters at the soil/root interface in order to have access to sufficient amounts of P.

P acquisition is characterized by epidermal uptake and apoplastic transport within the plant (Liu et al. 1998). Kinetic and molecular data showed that crop plants have multiple transporters across membranes (Schachtman et al. 1998). There exist two P uptake systems, one with high affinity and activity, which is regulated by the external P supply while the other has a lower activity and affinity (Schachtman et al. 1998).

After transport of P into cells it is used basically in three pathways: in the cytoplasm where it plays a major role for P homeostasis, in mitochondria and plastids for the synthesis of organic P compounds and in the vacuole for storage of P (Sano and Nagata 2008).

LePT1 transporter was expressed mainly in roots but also in leaves, stems and petioles while LePT2 was exclusively expressed in roots und conditions of severe P deficiency. Transcript levels of both genes increased after 24 h and peaked after 5 days if the P concentration in the nutrient solution was <100 μm P (Liu et al. 1998). Gene expression of the transporters proved to be reversible when P was resupplied to the plants (Liu et al. 1998). The authors suggested that signals for increased P uptake by roots under conditions of severe P deficiency are triggered by internal changes in the P concentration. Only if the plant P status falls below a certain threshold P transporters in the roots will be expressed (Liu et al. 1998).

Zhou et al. (2008) showed that overexpression of OsPHR2 genes in rice mutants mimics P starvation inducing changes of root architecture such as root elongation and proliferation of root hairs which finally result in an accumulation of P in shoots though P was sufficiently supplied. The identification of regulatory mechanisms behind OsPHR2 is a promising approach to improve P efficiency of plants on marginal soils.

Genotypic differences between the highest and lowest P content of wheat cultivars varied between 10 and 31 % and reflect differences in uptake efficiency (Schnug and Strampe 1988). Westermann (2007) reports such genotypic differences for alfalfa, cotton, corn, tomato, rice, wheat and common bean. Leiser et al. (2014) found distinct differences in P uptake of sorghum cultivars whereby landraces showed a higher uptake efficiency on P-deficient soils than bred lines, which had a higher utilization efficiency. Waines and Ehdaie (2007) came to similar results and explained this observation by differences in root morphology (see above). Improving the P uptake efficiency is usually the prime target of breeding programs while the P utilization efficiency in terms of translocation and remobilization deserves intensified research in order to improve the overall P efficiency (Wang et al. 2010).

Synergisms and antagonisms in nutrient uptake influence the nutritional status of crops and unbalanced fertilization will favor nutritional imbalances. Three forms of synergistic and antagonistic relationships can be distinguished which comprise nutrient uptake by roots, transport and translocation within the plant and elemental functions (Finck 1976). An increasing P supply enhanced iron (Fe) and manganese (Mn) uptake while it decreased that of N, potassium (K), calcium (Ca) and magnesium (Mg) in tomato (Gunes et al. 1998). Antagonistic effects between excessive P and micronutrients, here Fe, Mn, Zn and copper (Cu) are a common phenomenon (Ranade-Malvi 2011). For instance Fe is precipitated as Fe-phosphate in lateral roots, which hampers translocation into shoots (Wanasuria and Kühn 1978). Obviously P fertilization interferes with translocation of Zn into shoots, which was most pronounced on alkaline soils (Fageria 2001). Further recommended readings, which deal with interactions of phosphorus with other plant nutrients are provided by Adams (1980), Sumner and Farina (1986) and Fageria (2001).

Functions of Phosphorus in Crop Plants

An element needs to comply with the criteria defined by Arnon (1950) in order to be declared as being essential for plant growth. Firstly, the element must be essential to complete the life cycle of the plant. Secondly, the element must have specific functions. Thirdly, the element must have a direct nutritional impact on plant nutrition. In comparison, minerals are rated as beneficial for plants if they promote growth and yield, replace the function of essential nutrients in plant metabolism, or strengthen the natural resistance of plants against abiotic and biotic stress. *George Ville of France* showed in 1861 that phosphorus (P) is an essential macronutrient (Epstein 2000).

After uptake, P remains in its oxidized form. Inorganic P can be found as orthophosphate and pyrophosphate in the plant tissue (Maathuis 2009). In comparison, organic P is bound to hydroxyl groups of sugars and alcohols by esterification (Bünemann et al. 2011). Formation and disruption of pyrophosphate bonds is a key mechanism for energy homeostasis in plants and signal transduction pathways (Raghothama 2007). P activates organic compounds by phosphorylation (Finck 1976). Phosphorylation of proteins is a signal for protein/protein interactions (Raghothama 2007). P is an integral component of nucleic acids, phospholipids and ATP (Schachtman et al. 1998; White and Hammond 2008). Phytin is the most important reserve fraction in cereal grains with about 80 % of P bound as phytin (Finck 1976). P is involved in the regulation of enzyme reactions and metabolic pathways (Theodorou and Plaxton 1993). As a constituent of phospholipids P influences cell membrane characteristics (Raghothama 2007; White and Hammond 2008). It is the concentration of P in chloroplasts that regulates the transport of phosphorylated sugars and starch synthesis (Raghothama 2007).

According to Vance (2011) and George et al. (2011) plants may exert two basic approaches to improve P acquisition and utilization if the soil P supply is limited. The first aims at a sparing internal P use: the growth rate is decreased, whilst the growth per unit of P uptake is increased; carbon metabolism bypasses steps that require P, changes in membrane biosynthesis and alternative respiratory pathways occur. The second focuses on an increased P mobilization of soil reserves by inducing changes in root architecture and release of mobilizing exudates (see above).

Phosphorus Supply and Plant Health

Plant diseases are a major limitation to improved production efficiency and crop quality as they reduce nutrient availability, uptake, distribution, or utilization by the plant (Huber and Haneklaus 2007).

The mineral nutrient supply is supposedly the primary and pivotal barrier against infection, which influences the course of pathogenesis. In general, the greatest benefit to the plant in terms of health can be expected when full nutrient sufficiency is provided; however, the response to a particular nutrient may be different when going from deficiency to sufficiency than from sufficiency to excess (Huber and Haneklaus 2007).

The availability of mineral fertilizers has brought about the demise of many diseases through improved plant resistance, disease escape, or altered pathogenicity. Fertilization aims among others at preventing biotic plant stress thus strengthening the natural resistance against pathogen attack, and at promoting the synthesis of compounds involved in resistance against fungal diseases (Haneklaus et al. 2007, 2009). Essential plant nutrients influence plant, pathogen, and microbial growth so that they are an important factor in disease control; the interaction with nutrition is dynamic and all essential nutrients are reported to influence the incidence or severity of some diseases (Huber and Haneklaus 2007). An overview of current knowledge on nutrient-induced resistances against plant disease is provided by Datnoff et al. (2007).

Reuveni et al. (1992, 1995, 1998, 2000) reported about induced systemic resistance against fungal diseases by P, which has been applied to the soil or leaf. Compared to other nutrients, studies on the impact of the plant P status on plant disease are limited and often conflicting (Prabhu et al. 2007). Under conditions of an insufficient P supply P fertilization may increase resistance against diseases in younger and older plants. In the first case P improves plant vigor, among others by development of new roots, which promote disease escape (Huber 1991). In the latter case it enhances the maturity process (Prabhu et al. 2007). Reuveni et al. (2000) observed that P fertilization significantly increased the Ca content of the plant. In this context Gottstein and Kuc (1989) suggested that phosphates induce sequestering of apoplastic Ca, which increases for instance the activity of polygalacturonases. During this process oligogalacturonides are released, which trigger signaling pathways that induce resistance mechanisms (Vallarino and Osorio 2012). The same authors showed that the mechanisms of P-induced systemic resistance were different when P was soil-applied before establishment of powdery mildew (*Sphaerotheca fuliginea*) in cucumbers, or after infection by foliar P sprays.

A higher P supply has been linked to increased levels of fungistatic components such as phenolic compounds and flavonoids in different plant parts. The size of the plant available P pool in the soils depends among others on microbial activity, root exudates and soil pH (see above). Amino acid exudation by roots has been identified as one factor controlling infection of wheat with take-all (Graham and Menge 1982; Duffy et al. 1997). Apparently disease severity was reduced by a higher plant available P pool and subsequently higher P uptake of the plant, which reduced the amount of root exudates. Root exudates influence pathogen activity (Prabhu et al. 2007). The indirect impact of the P nutritional status in the form of an increased growth rate after P application seems to outreach the significance of a direct fungistatic effect by increased synthesis of phenolic compounds for resistance against fungal pathogens (Prabhu et al. 2007).

Symptomatology of Phosphorus Deficiency and Toxicity

The cytoplasmic P concentration ranges from 5 to 10 mM, is fairly constant and constitutes 1–5 % of the total inorganic P in plants. In contrast, vacuolar P concentrations may vary widely in relation to the external P supply, from below the detection limit at P starvation to up to 25 mM (Schachtman et al. 1998). P deficiency affects photosynthesis, glycolysis and respiration in such way that organic acids, starch and sucrose are accumulated (George et al. 2011). Under conditions of P starvation intracellular phosphatases and nucleases are synthesized in order to remobilize P from cellular compounds and nucleic acids (George et al. 2011; Raghothama 2007). Phospholipids are replaced by sulfolipids and galactolipids, and the RNA level in cells is reduced (George et al. 2011). The reddish purple color due to enrichment of anthocyanins is generally promoted under conditions of high photosynthetic activity. The phenomenon is initialized by the enrichment of carbohydrates in the cells following the inhibition of the protein metabolism (Raghothama 2007). To avoid physiological disorders caused by too high carbohydrate concentrations, plants detoxify these in anthocyanins derived from the reaction with cell-borne flavonoids (Schnug and Haneklaus 1998). Morphological changes comprise an increased root/shoot biomass and changes in root morphology. Cluster root formation is a highly specialized adaptation of generally non-mycorrhizal plants to low P in soils (Shane and Lambers 2005). Some species of Proteaceae show symptoms of P toxicity after the storage capacity for P is exceeded though the external P supply is low (Shane and Lambers 2005).

Macroscopic P deficiency symptoms are not very specific and can be easily mismatched with an excessive N supply under field conditions. Deficiency of both nutrients causes stunted growth of plants. Retarded growth and a dark green, later

Fig. 1 Macroscopic symptoms of P deficiency expressed by chlorosis and stunted growth (*left*) together with anthocyanin enrichments (*right*) in maize. *Photos* E. Schnug and J. Schick, JKI Braunschweig

bronze violet color of leaves and stalks is a common description of severe P deficiency. Visual symptoms of macroscopic P deficiency in shoots of crop plants comprise a retarded leaf development, which affects the main and axillary shoot so that the habitus appears stunted (Vance 2011). Besides that flowering and ripening are delayed (Vance 2011; Finck 1976). Dwarfish growth accompanied by chlorosis may occur, followed by necrosis under conditions of continuous P starvation. Brown and dark brown necrotic spots are visible on older leaves as are dark brown leaf edges before leaves die off (Bergmann 1983). In faba beans, potatoes and sugar beets these spots can be brown-black or black because of melanin synthesis (Bergmann 1983). Symptoms can be observed first on older leaves as P is mobile within the plant. As a consequence leaf senescence is accelerated. Root system development and tillering both are humble (Finck 1976; Larcher 1995). Typical symptoms of macroscopic P deficiency in maize at growth stage BBCH 39 (Stauss et al. 1994) are shown in Fig. 1. Interestingly, P deficiency in conifers yields automatically necrosis without previous chlorosis (Larcher 1995).

Not only P deficiency, but also P toxicity induces characteristic symptoms in agricultural crops. An excessive P supply accelerated ripening and decreased the lignin content of wheat. In addition, it favored the induction Ca, Zn, B, Cu and Mn deficiency (Bergmann 1983). Bergmann (1983) defined an excess of P when plants show five to 10-fold higher P concentrations than the average, which equals P concentrations of >3 %. These data are in agreement with those found by Silber et al. (2002), however, significantly higher than those found in various other studies (see Table 2). P toxicity results in growth inhibition, causes interveinal chlorosis and necrosis of leaves, and accelerates senescence (Silber et al. 2002). Although the mechanism of P toxicity in plants is poorly understood, it is accepted that P toxicity is associated with P-Zn interactions, either in the soil or in the plant (Loneragan and Webb 1993). As mentioned previously, P toxicity symptoms may emerge after the storage capacity for P is exceeded (Shane and Lambers 2005).

An overabundant P supply induced dark brown zones parallel to the leaf veins of poplar (Schönnamsgruber 1955). Clover and soybean reacted with white to pale grey necroses along the leaf edges, which proceeded basipetally (Rossiter 1952; Bergmann 1983). In oats and barley necrosis started at the tips of older leaves without prior chlorosis (Bergmann 1983).

Evaluation of Diagnostic Phosphorus Values in Crop Plants

The ideal plant is characterized by high nutrient and water use efficiency, resistance against abiotic and biotic stress, high productivity and quality of the harvest products. Particularly, higher nutrient utilization efficiency will contribute to higher crop yield under limiting conditions such as in organic farming and developing countries.

The major advantages of plant analysis when compared to soil testing are that it is a direct measure for the nutritional status, a real parameter for the relationship to crop yield, a method for verifying severe and latent nutrient deficiency and assessing reliably the nutrient demand, a direct proof of fertilizer efficiency, and last but not least the method of choice for determining nutrient ratios in crop plants. A clear disadvantage is that plant analysis is usually a *post mortem* (Aldrich 1973).

Scientists and farmers need equally reliable crop-specific P concentrations indicating symptomatological value, critical nutrient value, no effect value and toxicological value. The former in order to adjust and control the P supply in the growth medium or field accordingly, the latter to identify a P disorder.

During plant growth morphological changes occur, which imply the translocation of nutrients within the plant. Thus changes in the nutrient concentration are not only related to fluctuations in their supply, but also to the plant part and plant age. These factors need to be taken into account when interpreting and comparing results of plant analysis (Haneklaus et al. 2006). Basically, time of sampling and analyzed plant part are simply a convention, but there are some practical reasons, which deserve consideration: (a) photosynthetically active leaves show the highest P concentrations of all plant organs, and P is mobile in plants so that macroscopic deficiency symptoms appear first on older leaves; (b) sampling earlier in the vegetative state during main growth of a crop allows more time to correct P deficiency by fertilization. Relevant in this context is that plant analysis is a reliable tool to evaluate the P nutritional status, but usually is not applicable as a diagnostic tool on production fields because of a generally insufficient practicability.

In dicotyledonous crops younger, fully expanded leaves are the strongest sinks for P and available during vegetative growth so that they are preferably suited for tissue analysis (Haneklaus et al. 2006). Oilseed rape for instance delivers suitable leaves for tissue analysis until one week after flowering and sugar beet until the canopy covers the ground and the storage roots start to extend. For the analysis of gramineous crops, either whole plants (1 cm above ground) after the appearance of the first and before the appearance of the second node, or flag leaves are suited best (Haneklaus et al. 2006). In all cases care has to be taken to avoid contamination of tissue samples with P for instance from foliar fertilizers or P containing pesticides.

Analytical Methods

P occurs in plants in different chemical forms and besides the total P concentration, P extracted by water, weak acids and salt solution have been employed to determine the nutritional status (Westermann 2007). In addition, enzymatic methods have been tested to assess the P supply of crop plants (Westermann 2007). Most promising seemed the plant acid phosphatase activity, which increases under conditions of an insufficient P supply and reflected visual symptoms of P deficiency (Westermann 2007). However, a relationship between acid phosphatase activity and P content/crop yield has not been shown so far. In this context the studies of Yan et al. (2001) question a link between acid phosphatase activity and P acquisition efficiency.

Yet another approach to determine the small-scale spatial variation of the plant nutrient status within fields is by employing remote sensing techniques. Next, the major obstacles for using spectral analysis on-the-go for nutritional analysis are summarized. Changes in the supply with essential plant nutrients will result in changes of the chlorophyll content and spectral signature under conditions of severe deficiency and toxicity (Haneklaus and Schnug 2006). According to present knowledge this applies to the visible region of the spectrum, the UV and IR range. Here, negative impacts on plant turgor and morphological leaf structures, which are characteristic of extreme S, K, Ca and B deficiency result in changes in the spectral signature apart from changes in the chlorophyll content. Individual nutrients have no specific effect on the chlorophyll content or the spectral signature so that it can be concluded that their specific impact cannot be inferred from the chlorophyll content. Other symptoms of severe P deficiency such as the stunted growth and enrichment of anthocyanins are not element specific and can for instance be observed under conditions of severe S supply, too (Haneklaus et al. 2006).

As N imposes the strongest impact on the chlorophyll content, N sensors advanced to the major field of application. However, the causal chain spectral signature—green color—chlorophyll content—protein content—total N content—N status has been extended irrespective of scientific basic knowledge by the element N supply of soil N pool (Haneklaus and Schnug 2006). There are many different nutritional disorders caused by S, K, Mg, Mn, Zn and Cu deficiency, which influence the chlorophyll content and spectral signature. Under the constraints of nutrient mining on arable farms it can be expected that P will add to the list in due time. Abiotic factors are extreme temperatures and stagnant soil water.

The total P content is most frequently used for the evaluation of the P nutritional status. Yet, Bergmann (1983) estimates the nutrient ratios P/Fe, P/Zn, P/Al to be of prime interest for a balanced nutrition due to their antagonism. Particularly, high P fertilizer rates may influence the uptake unfavorably. Bergmann (1983) provided some key data for P/Zn ratios in crop plants whilst stressing that besides an optimum nutrient ratio both elements need to be sufficiently supplied (Table 1).

Higher N and P concentrations in seeds than shoots together with a low N/P ratio in seeds suggests that P deficiency affects reproductive output more than vegetative

Table 1 Threshold values for P/Zn ratios based on mass ratios (mg kg^{-1}) in shoots of crop plants at main growth

Crop plant	Severe Zn deficiency	Hidden Zn deficiency	Sufficient Zn supply	Excessive Zn supply
Maize	>300	>200–300	50–200	<25
Beans	>300	>200–300	80–200	<40
Sunflower	>400	>300–400	80–200	<50
Potatoes	>300	>200–300	100–200	<60

Bergmann (1983)

growth (Güsewell 2004). However, the P supply affects biomass and grain yield equally so that there is obviously no specific role for N and P (Allison 2002).

Güsewell (2004) reviewed factors affecting variations of N/P ratios in different plant species of natural and semi-natural vegetation. She states that according to Liebig's Law of the Minimum there is a minimum value for the N/P ratio where only N is limiting and a maximum value where only P is limiting. P deficiency will occur if plant P concentrations are <0.10 % P (Güsewell 2004). The average N/P ratio varies between 12 and 13 with a range from 1 to 100. N is supposed to be limited at a N/P ratio <13.5 and P if the ratio is >16 (White and Hammond 2008). The data are in agreement with those found in agricultural experimentation where the optimum N/P ratio in wheat plants was 12:1 at flowering (Boldyrev 1972). When the N/P supply ratio is varied 10-fold, this will result in N/P plant ratios, which vary only two to threefold (Güsewell 2004). In relation to P fertilization the N/P ratio may vary up to 50-fold.

There is no doubt that balanced nutrient ratios in plant tissues are essential for crop productivity, quality and plant health, but the strongest argument against using nutrient ratios to assess the nutritional status is that it can result from totally different N, P and Zn concentrations in the plant tissue. Surplus with one element may therefore falsely be interpreted as a deficiency with the other (Finck 1970). The suitability of nutrient ratios as a diagnostic criterion also implies a constancy. Additionally, it requires the determination of two elements and thus is more laborious and costly.

Assessment of Critical Nutrient Values and Ranges

Critical values are indispensable for evaluating the nutritional status of a crop. Important threshold markers are: the symptomatological value, which reflects the P concentration below which macroscopic deficiency symptoms become visible; the no effect value (NEV), which stands for the P concentration above which the plant is sufficiently supplied with P for achieving the maximum potential yield, the critical nutrient value which corresponds with P concentrations that reduce yield by 5, 10 and 20 % (Reuter and Robinson 1997), and the toxicological value, which indicates the P concentration above which toxicity symptoms can be observed. But there is no one exclusive critical nutrient value for each crop as it depends on biotic and abiotic factors, experimental growth conditions, the developmental stage of the plant at sampling, the collected plant part, the targeted yield and mathematical approach for calculating it (Smith and Loneragan 1997). As the mathematical procedure itself to derive a critical value has a significant impact it is only possible to define ranges, which characterize different nutritional levels (Smith and Loneragan 1997). This also implies that it is more or less impossible to compare results from different experiments.

Numerous, widely differing critical P values and sufficiency ranges exist for a lot of crops, which have been compiled for instance by Reuter and Robinson (1997) for cultivated plants including forest plantations.

In this section an attempt was made to compile and categorize available, individual data based on studies with varying experimental conditions from literature. Critical values for P concentrations were clustered in different groups of crops in order to facilitate an easy and appropriate evaluation of the P supply. Plant groups were assembled by morphogenetic and physiological features. Because of the wide heterogeneity of results for similar classes of P supply and for a better comparability of results, concentrations were agglomerated in three major categories: severely deficient, sufficient and toxic irrespective of the sampled plant part during vegetative growth (Table 2). A sufficient supply comprises P concentrations, which are equal or higher than the critical nutrient value. Finck (1976) defines all nutrient

Table 2 Mean critical values and ranges of P nutrition for different groups of agricultural crops

P nutritional status			Total P content (%)
Severely deficient	Sufficient	Noxious	
Poaceae: barley (*Hordeum vulgare*), corn (*Zea mays*), oats (*Avena sativa*), rice (*Oryza sativa*), rye (*Secale cereale*), sorghum (*Sorghum vulgare*), wheat (*Triticum aestivum*; *Triticum durum*)			
0.20	0.34	0.92	Median
0.15	0.25	0.70	25 % quartile
0.28	0.50	1.34	75 % quartile
0.068–0.700	0.07–1.60	0.2–3.00	Range
173	368	20	(n)
Legumes: chickpea (*Cicer arietinum*), cowpea (*Vigna unguiculata*), Faba bean (*Vicia faba*), (field) pea (*Pisum sativum*), lupin (*Lupinus angustifolius, Lupinus albus, Lupinus cosentinii*), navy, bush, snap, green, dwarf, French beans (*Phaseolus vulgaris*), soybean (*Glycine max*)			
0.20	0.34	1.135	Median
0.145	0.26	0.80	25 % quartile
0.29	0.43	2.10	75 % quartile
0.04–0.67	0.063–1.50	0.56–3.20	Range
184	346	22	(n)

(continued)

Table 2 (continued)

P nutritional status			Total P content (%)
Severely deficient	Sufficient	Noxious	
Oil crops I: mustard (*Brassica juncea*) , oilseed rape, spring and winter varieties (*Brassica napus; Brassica campestris*)			
0.17	0.31	0.70	Median
0.108	0.26	–	25 % quartile
0.243	0.548	–	75 % quartile
0.06–0.50	0.15–1.00	0.65–0.70	Range
25	50	3	(n)
Oil crops II: cotton (*Gossypium hirsutum*), linseed (*Linum usitatissimum*), peanut (*Arachis hypogaea*), sunflower (*Helianthus annuus*)			
0.175	0.30	2.08	Median
0.13	0.23	–	25 % quartile
0.23	0.40	–	75 % quartile
0.05–0.39	0.03–1.28	0.41–3.24	Range
58	159	3	(n)
Root crops: potato (*Solanum tuberosum*), sugar beet (*Beta vulgaris*)			
0.17	0.40	–	Median
0.11	0.25	–	25 % quartile
0.23	0.545	–	75 % quartile
0.02–0.50	0.075–1.40	0.45–1.00	Range
39	87	2	(n)
Fodder crops/Pastures: alfalfa (*Medicago sativa*), bermudagrass (*Cynodon dactylon*), buffel gras (*Cenchrus ciliaris*), centro (*Centrosema pubescens*), Kentucky bluegrass (*Poa pratensis*), orchard grass (*Dactylis glomerata*), perennial ryegrass (*Lolium perenne*), red clover (*Trifolium pratense*), subterranean clover (*Trifolium subterraneum*), tall fescue (*Festuca arundinacea*), white clover (*Trifolium repens*)			
0.21	0.32	0.70	Median
0.195	0.25	0.70	25 % quartile
0.25	0.40	1.00	75 % quartile
0.07–0.50	0.065–1.00	0.45–1.00	Range
57	178	7	(n)
Brassica vegetables: broccoli (*B. oleracea* var. *italica*), brussels sprouts (*Brassica oleracea* var. *gemmifera*), cabbage (*Brassica oleracea*), cauliflower (*Brassica oleracea*)			
0.205	0.50	–	Median
0.11	0.35	–	25 % quartile
0.40	0.70	–	75 % quartile
0.07–0.69	0.20–1.07	–	Range
8	94	–	(n)
Non-brassica vegetables: asparagus (*Asparagus officinalis*), bush bean (*Phaseolus vulgaris*), carrot (*Daucus carota*), celery (*Apium graveolens*), cucumber (*Cucumis sativus*), garlic (*Allium sativum*), lettuce (*Lactuca sativa spp.*), onion (*Allium cepa*), spinach (*Spinacea oleracea*), tomato (*Lycopersicon esculentum*)			

(continued)

Table 2 (continued)

P nutritional status			Total P content (%)
Severely deficient	Sufficient	Noxious	
0.25	0.45	1.17	Median
0.20	0.30		25 % quartile
0.33	0.603		75 % quartile
0.07–0.65	0.014–1.50		Range
50	241	1	(n)

Compiled from data and references cited in Bergmann (1983), Finck (1976), Haneklaus et al. (1998), Klages (2011), Mills and Jones (1997), Reuter and Robinson (1997), Sanchez (2007), Schnug and Haneklaus (1992, 2008)

concentrations that exceed the NEV as luxury consumption. A subdivision made prior, which took additionally criteria such as crop species, sampled plant part and growth stage at sampling into consideration proved to be not feasible as the variation of results was so high that no clear ranges, let alone threshold values could be assigned for individual classes of P supply. The compilation of the data made in Table 2 suggests, however, that the sampled plant part during the main vegetative development (for example from stem extension to start of flowering in case of cereals and oilseed rape) seems to be of minor relevance for generally addressing the P nutritional status. Similar results were determined for the major plant nutrient sulfur (Haneklaus et al. 2006).

The data presented in Table 2 reveal that the three classes of P supply are overlapping for all plant groups. A total P concentration of 0.175–0.25 % proved to denote severe P deficiency in all crops. Referring to the median values, P contents between 0.30 and 0.34 % P indicate a sufficient P supply of graminoids, legumes and oil crops. In comparison, root crops and vegetables have a somewhat higher P demand as the optimum content varied between 0.40 and 0.50 % P (Table 2). With respect to the 75 % quartiles, severe P deficiency has been attributed if the P content was lower than ~0.25 % P for all crops except vegetables, which are in the severe deficiency range at concentrations of 0.33–0.40 % P. Consequently, a sufficient P supply is provided if the 75 % quantil is about 0.46 % for poaceae, legumes, oil crops, root crops, fodder crops and 0.65 % P for vegetables (Table 2). The database for excessively high P concentrations, which unfold negative impacts on crop yield is strictly limited so that a value of >1 % P may be rated as indicative for exceeding a toxicological threshold. In Poaceae, legumes and oil crops the corresponding value may be distinctly higher with about 3 % (Table 2).

The major criticism of critical values for the interpretation of tissue analysis is the small experimental basis, which often consists of not more than a single experiment (Vielemeyer et al. 1983). Besides the lack of data, the method of interpretation may also yield erroneous results. Methods based on regression analysis, like the 'broken stick method' by Hudson (1966) and Spencer and Freney (1980), or the 'vector analysis' by Timmer and Armstrong (1987) investigate mathematical, but not necessarily causal interactions between the nutrient content

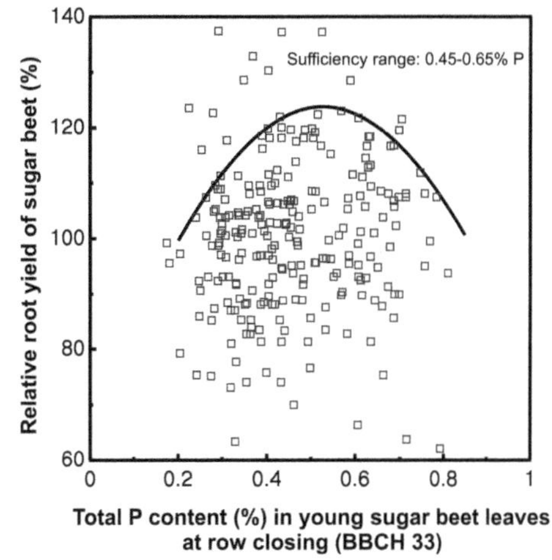

Fig. 2 Scattergram of total P content (%) in younger leaves of sugar beet at row closing (BBCH 33) and yield data (n = 278), and upper boundary line calculated by *BOLIDES* (data extracted from Haneklaus and Schnug 1998)

and yield, because the dictate of minimizing the sum of squared distances aims only to find a function which fits best across the data set. Like the method of Cate and Nelson (1965, 1971), these methods have been designed primarily for the investigation of small data sets and plants grown under *ceteris paribus* conditions, where only the response to variations in the nutrient supply varied. Another quite important disadvantage of critical values and critical ranges (Bergmann 1983; Mills and Jones 1997; Baier 1995), or NEVs is that they ignore the non-linearity of the *Mitscherlich* function describing the relationship between growth factors and yield (Wallace 1990). The ideal basis for critical values for the interpretation of tissue analysis are large sets of yield data and nutrient concentrations in defined plant organs which cover a wide range of growth factor combinations. The data may include samples from field surveys, field and pot experiments if the reference yield of 100 % was obtained in all cases under optimum growth conditions. In Fig. 2 a corresponding example is given for the total P concentration in younger leaves of sugar beet that have been collected on production fields in northern Germany and eastern Juetland, Denmark.

Comprehensive data sets like that presented in Fig. 2 allow the accurate calculation of so-called upper boundary line functions, which describe the highest yields observed over the range of nutrient concentrations measured. Boundary lines describe the 'pure effect of a nutrient' on crop yield under *ceteris paribus* conditions (Moeller-Nielsen 1973; Moeller-Nielsen and Frijs-Nielsen 1976; Evanylo and Sumner 1987; Walworth et al. 1986). Data points below this line relate to samples where some other factor limited the crop's response to the nutrient. Commonly, the boundary lines show for all essential plant nutrients a steep increase at the beginning, which corresponds with the range of severe deficiency. In case of micronutrients the boundary line continues in a steep slope, the range of sufficient supply being usually narrow. In comparison, the yield response to increasing macronutrient concentrations processes asymptotically over a wide range when leaving the severe deficiency zone towards the value above which no further yield increase (NEV) is to be expected from increasing P concentrations (Fig. 2). Above total P concentrations of 0.65 % P in foliar tissue of sugar beet no further yield increases by increasing tissue P concentrations are to be expected (NEVs; Fig. 2). But still in this range of the response curve there is no linearity between tissue P levels and yield. The non-linearity of the boundary lines reveals once more the limited value of critical values.

An overview of the scientific background and development of upper boundary lines is given by Schnug et al. (1996). The BOundary LIne DEvelopment System (BOLIDES) is a statistical software, which fits upper boundary line functions and evaluates optimum nutrient values and ranges. A similar approach has been chosen by Banneheka et al. (2013). BOLIDES is based on a 5-step algorithm (Fig. 3; Schnug et al. 1996). For the identification of outliers, cell sizes are defined for nutrient and yield values together with an optional number of data points per cell (Fig. 3a). The cell size can be chosen variably with proposed values for X (nutrient content) corresponding to the standard deviations and for Y (yield) with the coefficient of variation. If another variable, often a stable soil feature such as organic matter or clay content, is having a significant effect on the response to the nutrient, its presence is indicated by two or more distinct concentrations of points, each with its own boundary line response to the nutrient (Fig. 3b). The data can be classified on the basis of this third variable and the boundary line can be determined separately for each class. Next, a boundary step function is calculated for each class starting from the minimum nutrient content up to the point of maximum yield as well as from the maximum nutrient content up to the maximum yield (Fig. 3c). Then the boundary line, usually a 4th order polynomial function, is fitted according to the least square method (Fig. 3d). The first derivative of the fitted polynomial gives predicted yield response to fertilization in relation to the nutrient content (Fig. 3). The last step is the classification of the nutrient supply to determine optimum nutrient levels and/or optimum nutrient ranges. The optimum nutrient value corresponds with the zero of the first derivative of the upper boundary line and the sign of the second derivative at this point. For the determination of the optimum ranges i.e. the range of nutrient concentration which gives 95 % of the maximum yield, standard, numerical root-finding procedures are used for real polynomials of degree four with constant coefficients (Fig. 3).

The relationship between P concentration in plant tissue and yield reflects the physiological patterns in the internal nutrient utilization, which is specific for each plant species and can be best established by boundary lines. In comparison, the relationship between fertilizer dose and P concentration in plant tissue is much less dependent on physiological factors, but strongly influenced by factors affecting the physical mobility of P in soils. This transfer-function bears the largest part of insecurity for the effectiveness of P fertilization. Thus, for the derivation of fertilizer recommendations the common relationship between fertilizer dose and yield is best split into two partial relationships: fertilizer dose versus nutrient uptake and nutrient uptake versus yield (Janssen et al. 1990). If tissue analysis shall be used for fertilizer recommendations, concentrations need to be calibrated against P doses.

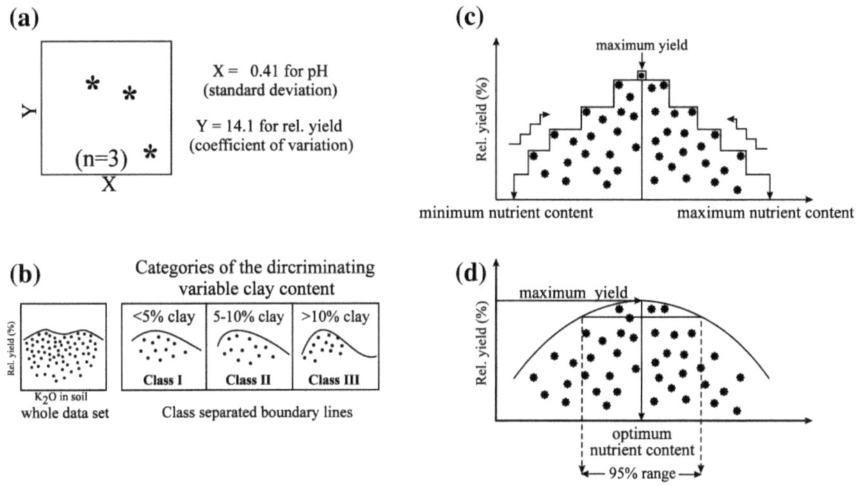

Fig. 3 Structure of BOLIDES for the determination of upper boundary line functions and optimum nutrient values and ranges in plants and soils (extracted from Schnug and Haneklaus 2008). **a** Identification of outliers. **b** Discrimination against a third variable. **c** Calculation of step functions. **d** Determination of the upper boundary line and calculation of optimum nutrient values and ranges

With PIPPA (Professional Interpretation Program for Plant Analysis), another software has been written, which not only evaluates the status of individual, essential plant nutrients, but which also appraises results from multiple elemental analyses (Schnug and Haneklaus 2008). In PIPPA, both boundary line and transfer functions are integrated for each element so that the yield limiting effect is calculated for each specified nutrient and finally fertilizer recommendations are given.

Phosphorus Fertilization

The key objectives of agricultural production are to warrant food security and food quality, to acknowledge the requirements of soil protection and to achieve a harmonization with ecological demands. So far a severe discrepancy exists regularly between agronomic and ecological requirements. Agriculture is the largest user of P and also the largest source for losses of P by environmental dispersion (surplus enrichment in soils, erosion, discharge into water bodies) and irreversible fixation (meat bone meals and ashes). Sustainability in P fertilization can only be achieved if point and non-point P losses are reduced to an unavoidable minimum and if the access to this finite resource and essential plant nutrient is warranted on a global scale. The Perspectives *Sustainable agriculture in the Baltic Sea Region in times of peak phosphorus and global change* are dedicated to the problem of how to deal with limited P resources in agricultural production (Perspectives Agriculture 2011).

Spatial Variation of Soil and Crop Parameters

While it is feasible to control point losses of P, it is a major challenge to efficiently reduce the discharge of diffuse losses from agriculture. A simple reduction of P fertilizer rates will not reduce ecological problems as soils are neither static, nor homogenous in space and time so that the standard approach of a reduced uniform application rate will still result in a side by side of over and under supply. It is the small-scale variability of soil and crop features that must be addressed and consequently transformed into variable rate applications in order to efficiently reduce environmental impacts of agricultural production.

The variability of soil, plant or yield is neither completely random, nor completely deterministic. Therefore, in generating a continuous model of a field related structure, this means that the value (e.g. the plant available P content in the soil) for an unknown point in the field can be estimated from neighboring samples whereby the unknown point is not completely dependent upon the values of the known points. The semi-variogram analysis is a method to determine the variability between data points in relation to their geographic distance from each other. The maximum distance up to which data points are related is called the range (a[h]) and the variance (ϒ[h]) at this point is called the sill. Relevant to note is that the determination of spatial correlations within a single field will not give a representative estimation of the scale of the spatial correlation for different fields (Haneklaus and Schnug 2006). Ranges for the major plant nutrients Ca, K, Mg and S were in different landscapes in the same range of about 115 m, while the ranges for P were at least 20–75 m longer (Haneklaus et al. 1997). The generally larger ranges for plant concentrations can probably be explained by compensatory regulation mechanisms during P uptake and translocation (see above). Mean ranges in different landscapes showed a smaller variation than those observed on different fields of one farm. In the case of plant available P in the soil, ranges were as large as 364 m but were on average, around 150 m (Haneklaus et al. 1997). Panten et al. (2016) provide a comprehensive overview of the spatial variability of P in soils and plants.

Relevant with view to soil P is the spatial speciation of nutrients. It was shown that different environmental factors resulted in the spatial speciation of P (Gassner 2003). Gassner (2003) was able not only to separate total, soluble and labile P pools, but also extracted different environmental factors that resulted in the formation of these pools. In the studies of Gassner (2003) the soluble P fraction was related to the adsorption capacity of the soil and in particular by the variation in soil texture. Topography and soil erosion processes were reflected in the distribution of total P, whereas the labile P fraction was influenced by the recent fertilizer history.

Variable Rate Fertilization

Field experimentation is considered as one of the pillars for developing fertilization recommendation schemes, but in the past it consistently aimed at averaging results by arranging replicates across the variability within a field (Haneklaus and Schnug 2006). Related to this problem is the calibration of soil and plant data. Consequently the authors strictly recommend on-farm experimentation employing Precision Agriculture (PA) technologies in order to set up site-specific critical nutrient ranges for the supply with essential plant nutrients and to assess response curves.

Variable rate application of fertilizers matches the small-scale variation of the nutrient demand of crops in the field. It is intrinsic that a site-specific nutrient management complies with a sustainable input of resources that favors crop growth and minimizes negative impacts on the environment. Algorithms for the continuous, variable rate application of straight and compound mineral fertilizers, and manure have been elaborated and tested on production fields (Haneklaus and Schnug 2000, 2006; Haneklaus et al. 2016). The algorithms for mineral compound fertilizers with flexible nutrient ratios target at a balanced nutrient input of all major nutrients within one crop rotation. A mere linear optimization of the N/P/K ratio will not lead to a balanced nutrient input as the discrepancy between actual P and K fertilizer demand and applied P and K rates favors an imbalanced input and thus increases surplus or deficit in the field. The example given in Fig. 4 illustrates the problem: the optimum N rates for each grid cell of the field were sorted in ascending order and the lines (in fact small dots) indicate the P rates applied with varying N/P ratios of 0.7–1. The mean P/K ratio in the field was 0.83. In comparison the markers reflect the optimum P rates in the individual grid cells. It is obvious that the mean N/P ratio is associated with a more or less equal deficit and surplus of P while with a N/P ratio of 1 surpluses are avoided but deficit zones increase. In comparison a N/P ratio of 0.71 avoids deficits but strongly promotes surpluses. This means that whatever N/P ratio is chosen, it will be associated with an inevitable surplus and/or deficit of P (see Fig. 4). This means that a linear shifting of the N/P ratio will not meet the spatial variability of the actual P requirement. And fertilizer strategies for compound fertilizers cannot focus on “the” optimum nutrient ratio but rather have to use nutrient ratios, which avoid surpluses in combination with the use of single nutrient products. A balanced input of N, P and K using compound fertilizers can for instance be realized with an alternative use of compound and single nutrient fertilizers (Haneklaus and Schnug 2000).

Besides addressing the spatial variability of soil and crop features, it is the monitoring of temporal changes in the soil and plant nutrient supply that can be realized easily in PA. Remote sensed images and yield maps can be used for

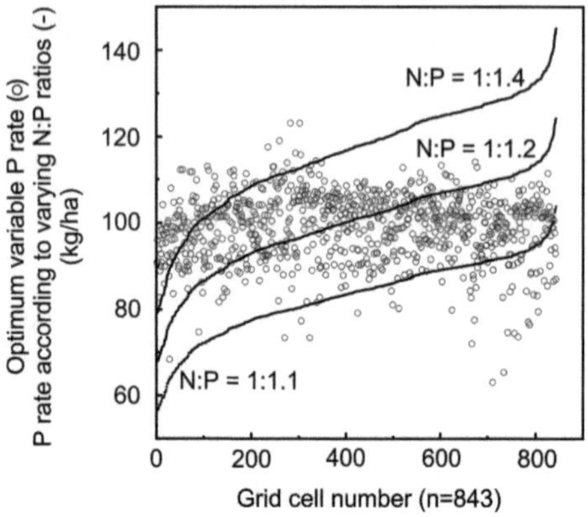

Fig. 4 Discrepancy between optimum variable P rates for the crop rotation and P rates applied with varying N:P ratios which follow the variation of the N demand of the crop on a field (9°28.944′ E; 52°34.040′ N) in lower Saxony (adapted from Haneklaus and Schnug 2000)

developing directed sampling strategies, and hereafter for fixing a minimum number of control plots, so-called monitor pedo cells (MPCs) in the field, which comprise the whole range of variation of soil properties and their temporal fluctuations (Haneklaus et al. 2000; Haneklaus and Schnug 2006). MPCs are selected for long-term investigations, which enable a fast adaptation of fertilizer algorithms to actual needs. Thus enrichment and depletion of soils with P can be followed up timely.

If soils are sufficiently supplied with P, the P rates may equal P off-take. Important to note is that it must be obligatory that P is fully plant available, meaning that the entire P in a fertilizer must be completely water soluble or at least soluble in neutral ammonium citrate solution (Schnug and Haneklaus 2016). Only then P utilization will be 100 % on a long-term basis (Schnug and Haneklaus 2016). Djodijic et al. (2005) came to the same conclusion as a result of a long-tem field experiment in Sweden. Recycled P fertilizer products and rock phosphates have to be evaluated critically as a significant amount of P will not take place in the soil P cycle and thus will not contribute to the P nutrition of agricultural crops (Schick 2010). In case of manure P is fully plant available (Hansen 2006).

Manure—Resource or Removal?

Application rates of manure are commonly not based on a documented demand of nutrients, but fully exhaust the maximum legal input. Studies of Eghball et al. (2005) showed that already in the first year of application P accessibility can be as high as 100 %. Thus it is fair to assume that the entire P in manure is plant available on a long-term basis, which makes it an ideal P source. Advanced PA technologies are likely to be objected by farmers as they demand changes in the production and recycling chain of liquid and solid manure. The variable rate application of manure is an important step towards a targeted input of nutrients. Modern spreading

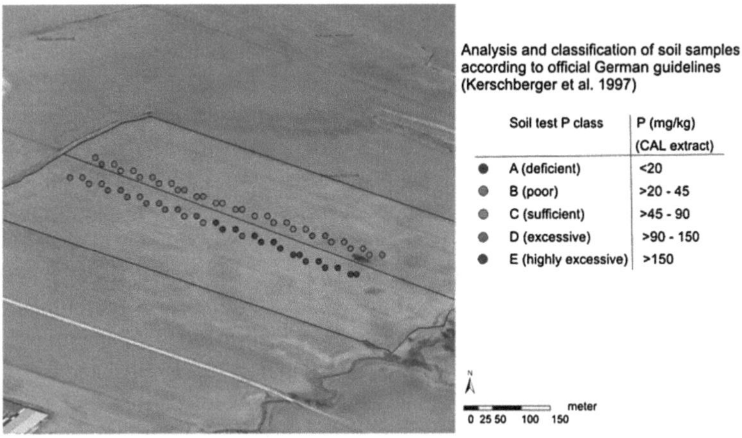

Fig. 5 Categorized P content along transects of neighboring fields (the northerly field received exclusively mineral fertilizers and the southerly field exclusively chicken manure as P supply in the past; adapted from Haneklaus et al. 2016)

machines automatically control manure application on a volumetric basis so that the nutrient output can be adjusted (Saeys et al. 2008). A basic rule of balanced fertilization is that the nutrient input matches the demand in order to avoid equally surpluses and undersupply. Here, the problem arises that many soils of livestock farms are already overloaded with P (see Fig. 5). The comparison of the soil P status along transects of neighboring fields that received historically either mineral fertilizers or manure showed that the P supply was regularly higher on livestock farms (Haneklaus et al. 2016; Fig. 5).

With a view to the N/P ratio in manure of about 6.4 in cattle, 4 in pig and 3.5 in poultry manure this implies that maximum loads must be based on the P rather than the N content of manure. Then with 22 kg ha^{-1} P (mean P off-take), on an average 141, 88 and 77 kg ha^{-1} N would be applied with cattle, pig and poultry manure, respectively. In contrast, maximum application rates of manure equaling 170 kg ha^{-1} N (as commonly practiced) exceed the mean P off-take of agricultural crops by 95 to >120 % if pig and poultry manure is used. If cattle manure is applied the surplus is minor with approximately 2 %.

A P-based regulation of manure rates is the only option for a sustainable use of the resource and reduction of nutrient losses to water bodies. A mandatory limitation of the P input to 22 kg ha^{-1} P as practiced in Sweden implies that alternative ways for the utilization of manure are required or livestock densities are reduced. The first involves co-fermentation in biogas plants, technological separation and processing of liquid and solid phases in order to preserve the fertilizer value of the product (Popovic 2012). The extension of the acreage where manure is applied is another option (Powers and van Horn 2001).

For maximum utilization of manure variable rates should follow the spatial variability of the plant available P content in soils as an indicator for the P demand of crops. The problem is that not only the ranges, which characterize different levels of the soil P status proved to be not congruent in different countries of the Baltic Sea region, but also recommended fertilizer rates based on soil analytical data deviated highly (Haneklaus et al. 2016). In other words, a farmer sending his soil samples to laboratories in Germany and Sweden for instance will receive results deviating by up to two classes of P supply. The authors analyzed all samples with one official national extraction method in one laboratory. In addition to the diverging evaluation of the soil P status the recommended fertilizer rates may deviate by up to 28 % for grain crops and 37 % for sugar beet if the P supply is rated optimum in both countries. In the worst case, three and four times higher application rates to sugar beet and cereals are recommended to German farmers (Haneklaus et al. 2016). Tóth et al. (2014) compared the P fertilizer recommendation systems in the UK and Hungary with distinct disparities in the advised fertilizer doses. These findings stress the urgent need for a harmonization of analytical methods, interpretation and recommendation procedures.

At this point a brief retrospective of the adoption of soil analysis in agriculture as control quantity for fertilization seems appropriate. Soil analysis, which is meanwhile the standard method in agriculture, faced unanimous skepticism at the start of 20th century (Uekötter 2006). This early mistrust gradually faded away though the uncertainties of the method did not disappear (Uekötter 2006). The causes of error and imprecision are numerous and the spatial variability of soil features is definitively one major factor while many other components contribute to the inaccuracy (Uekötter 2006). A reliable and tailor-made approach is that of on-farm experimentation if precision agriculture technologies are available as truly site-specific threshold values and response curves to nutrient input and nutrient mining can be determined (Haneklaus and Schnug 2006). Upper boundary lines are calculated using geocoded soil and/or plant data, and yield information from online yield monitoring systems. Accordingly, these statistics are transcribed into algorithms for variable rate applications of fertilizers (Haneklaus and Schnug 2006).

Nutrient Mining and Phosphorus Phytomining—Dealing with Extremes

Nutrient mining is a synonym for exhausting the mineral pool of agricultural soils by relinquishing to replenish nutrients by fertilization as a measure to reduce variable costs, or because of inaccessibility of fertilizers. It is fair to assume that the acreage that is affected by P mining will increase together with increasing commodity prices after peak P. Such management practice will become critical when crop productivity is affected. Here, timely countermeasures need to be taken such as the cultivation of landraces with a higher P uptake efficiency and introduction of intercropping systems where no access to mineral P fertilizers is provided (see above). It is extremely important to follow up changes in the plant available P supply of soils and the plant P status of crops in order to maitain crop yields.

Yet, another problem is soils, which are overloaded with P. Such sites can be found in the vicinity of big livestock enterprises and intensive agriculture of cash crops. In Fig. 5 the soil P status of neighboring fields from arable and livestock farm in Finland is shown. Excessive and strongly excessive P concentrations (CAL extract) corresponded with P concentrations of >90–150 and >150 mg kg^{-1} P (Haneklaus et al. 2016). Such extraordinarily high plant available P concentrations (Olsen extract) of 100 mg kg^{-1} were found also on soils where vegetables were grown in China (Yan et al. 2013).

Phytomining is a procedure to extract valuable metals, for instance gold and cobalt from plant material (Brooks and Robinson 1998), and Sharma and Sahi (2012) proposed to apply it in order to recycle P for fertilizer purposes whilst cleaning up charged soils. The authors rated P concentrations of >1 % to be sufficiently high to implement the technology. With reference to the data presented in Table 2 such P concentration is on the verge of toxicity so that negative impacts on yield biomass may occur. An enforcement of renewable energy in agriculture sounds like a promising concept to cope with global change, scarcity of energy and resources. And phytomining of P would add to the overall theory. It is intrinsic that the competition between food production, fuel and resources such as rock phosphates based on industrial crops or P-rich crops will be enforced with the expansion of the acreage of cultivated, arable land. Rather soils, which are not suitable for the production of food products, seem to be predestined for the cultivation of crop plants for industrial purposes. Such areas include for instance reclaimed mining sites, roadsides and agricultural soil overcharged with P, for example as a result of continuous long-term application of excessive loads of manure on big livestock enterprises.

References

Adams F (1980) Interactions of phosphorus with other elements in soils and in plants. In: Khasawneh FE, Sample EC, Kamprath EJ (eds) The role of phosphorus in agriculture. American Society of Agronomy, Crop Science Society of America, Soil Science Society of America, Madison, pp 655–680

Aldrich SR (1973) Plant analysis: problems and opportunities. In: Walsh LM, Beaton JD (eds) Soil testing and plant analysis. SSSA, Madison, pp 213–221

Allison VJ (2002) Nutrients, arbuscular mycorrhizas and competition interact to influence seed production and germination success in *Achillea millefolium*. Funct Ecol 16:742–749

Araujo JL, Faquin V, de Avila FW, Pedroso TQ (2013) Effect of phosphate-phosphate interaction on growth and on phosphorus nutrition of common bean in nutrient solution. Revista Brasileirs de Ciencia do Solo 37:482–490

Arnon DL (1950) Criteria for essentiality of inorganic micronutrients for plants. Loytsia 3:31–38

Baier J (1995) Computer program for foliar fertilization. In: Proceedings of IAOPN symposium, Cairo, pp 23–28

Banneheka BMSG, Dhanushika MP, Wijesuriya W, Herath K (2013) A linear programming approach to fitting an upper quadratic boundary line to natural rubber data. J Nat Sci Found Sri Lanka 41:13–20
Bergmann W (1983) Ernährungsstörungen bei Kulturpflanzen. Gustav Fischer Verlag, Stuttgart, 614 p
Bieleski RL (1978) Phosphate pools, phosphate transport, and phosphate availability. Annu Rev Plant Physiol 24:225–252
Boldyrev NK (1972) Komplexmethode der Blattdiagnostik für Ernährung, Höhe und Qualität des Ertrages landwirtschaftlicher Kulturen. 3. Europ. Kolloquium Ernährung der Kulturpflanzen, Budapest
Brooks RR, Robinson BH (1998) The potential use of hyperaccumulators and other plants for phytomining. In: Brooks RR (ed) Plants that hyperaccumulate heavy metals. CAB International, New York, pp 327–356
Brown LK, George TS, Dupuy L, White PJ (2013) A conceptual model of root hair ideotypes for future agricultural environments: what combination of traits should be targeted to cope with limited P availability? Ann Bot 112:317–330
Bünemann EK, Oberson A, Frossard E (2011) Phosphorus in action. Springer, Heidelberg, 483 p
Cakmak I, Yazici A, Tutus Y, Ozturk L (2009) Glyphosate reduced seed and leaf concentrations of calcium, manganese, magnesium, and iron in non-glyphosate resistant soybean. Eur J Agron 31:114–119
Cate BR Jr, Nelson LA (1965) A rapid method for correlation of soil test analyses with plant response data. North Carolina State University Agricultural Experiment Station Bull, vol 1. Int. Soil Testing Series
Cate RB, Nelson LA (1971) A simple statistical procedure for partitioning soil test correlation data into two classes. Soil Sci Soc Am Proc 35:658–660
Crews TE, Kitayama K, Fownes J, Herbert D, Mueller-Dombois D, Iley RH, Vitousek PM (1995) Changes in soil phosphorus and ecosystem dynamics across a long soil chronosequence in Hawaii. Ecology 76:1407–1424
Datnoff LE, Elmer W, Huber DM (2007) Mineral nutrition and plant disease. APS Press, St. Paul, MN, p 278
Denis MH, Delrot S (1993) Carrier-mediated uptake of glyphosate in broad bean (Vicia faba) via a phosphate transporter. Physiol Plant 87:569–575
Djodijic F, Bergstrom L, Grant C (2005) Phosphorus management in balanced agricultural systems. Soil Use Manag 21:94–101
Duffy BK, Ownley BH, Weller DM (1997) Soil chemical and physical properties associates with suppression of take-all of wheat by *Trichoderma koningii*. Phytopathology 87:1118–1124
Eghball B, Wienhold BJ, Woodbury BL, Eigenberg RA (2005) Plant availability of phosphorus in swine slurry and cattle feedlot manure. USDA-ARS/UNL Faculty, Paper 1186
Eichler-Löbermann B, Kohne S, Kowalski B, Schnug E (2008) Effect of catch cropping on phosphorus bioavailability in comparison to organic and inorganic fertilization. J Plant Nutr 31:659–676
Epstein E (2000) The discovery of the essential elements. In: Kung S-D, Yang S-F (eds) Discoveries in plant biology, vol 3. World Scientific, Singapore, pp 1–16
Evanylo GK, Sumner ME (1987) Utilization of the boundary line approach in the development of soil nutrient norms for soybean production. Commun Soil Sci Plant Anal 18:1355–1377
Fageria VD (2001) Nutrient interactions in crop plants. J Plant Nutr 24:1269–1290
Fang S, Deng XG, Chen X, Liao H (2011) Crop root behavior coordinates phosphorus status and neighbors: from field studies to three-dimensional in situ reconstruction of root system architecture. Plant Physiol 155:1277–1285
Finck A (1970) Die Pflanzenanalyse als Hilfsmittel zur Ermittlung des Düngerbedarfes. Sonderdruck aus Chemie und Landwirtschaftliche Produktion, pp 183–188
Finck A (1976) Pflanzenernährung in Stichworten. Hirt Verlag, Kiel, p 200
Garg N, Chandel S (2010) Arbuscular mycorrhizal networks: process and functions. A review. Agron Sustain Dev 30:581–599

Gassner A (2003) Factors controlling the spatial specification of phosphorus in agricultural soils. Ph.D. thesis, TU Braunschweig, Germany, p 126
George TS, Fransson A-M, Hammond JP, White PJ (2011) Phosphorus nutrition: rhizosphere processes, plant response and adaptations. In: Bünemann EK, Oberson A, Frossard E (eds) Phosphorus in action. Springer, Heidelberg, pp 245–271
Gottstein HD, Kuc J (1989) Induction of systemic resistance to anthracnose in cucumber by phosphates. Phytopathology 79:176–179
Graham JH, Menge JA (1982) Influence of vesicular-arbuscular mycorrhizae and soil phosphorus on take-all disease of wheat. Phytopathology 72:95–98
Gunes A, Aplaslan M, Inal A (1998) Critical nutrient concentrations and antagonistic and synergistic relationships among the nutrients of NFT-grown young tomato plants. J Plant Nutr 21:2035–2047
Güsewell S (2004) N:P ratios in terrestrial plants: variation and functional significance. New Phytol 164:243–266
Hammond JP, White PJ (2008) Diagnosing phosphorus deficiency in crop plants. In: White PJ, Hammond JP (eds) The ecophysiology of plant-phosphorus interactions. Springer, Heidelberg, pp 225–246
Haneklaus S, Schnug E (1998) Evaluation of critical values for soil and plant analysis of sugar beets by means of boundary lines applied to field survey data. Aspects Appl Biol 52:87–94
Haneklaus S, Schnug E (2000) Decision-making strategies for the variable-rate application of compound fertilizers. Commun Soil Sci Plant Anal 31:1863–1873
Haneklaus S, Schnug E (2006) Site specific nutrient management—objectives, current status and future research needs. In: Srinivasan A (ed) Precision farming—a global perspective. Marcel Dekker, New York, pp 91–151
Haneklaus S, Ruehling I, Schroeder D, Schnug E (1997) Studies on the variability of soil and crop fertility parameters and yields in different landscapes of Northern Germany. In: Stafford JV (ed) Precision agriculture, vol II. IOS Scientific Publishers Ltd., UK, pp 785–792
Haneklaus S, Knudsen L, Schnug E (1998) Minimum factors in the mineral nutrition of field grown sugar beets in northern Germany and eastern Denmark. Aspects Appl Biol 52:57–64
Haneklaus S, Schnug E, Panten K (2000) Evaluation of structural coincidences of patterns in remote sensing images and yield maps for the identification of sampling locations. Aspects Appl Biol 60:37–44
Haneklaus S, Bloem E, Schnug E, De Kok L, Stulen I (2006) Sulphur. In: Barker AV, Pilbeam DJ (eds) Handbook of plant nutrition. CRC Press, Boca Raton, pp 183–238
Haneklaus S, Bloem E, Schnug E (2007) Sulfur and plant disease. In: Datnoff L, Elmer W, Huber D (eds) Mineral nutrition and plant diseases. APS Press, St. Paul, MN, pp 101–118
Haneklaus S, Bloem E, Schnug E (2009) Plant disease control by nutrient management: sulphur. In: Walters D (ed) Disease control in crops. Biological and environmentally-friendly approaches. Wiley, New York, p 280
Haneklaus S, Schick J, Kratz S, Rückamp D, Schnug E (2016) Variable rate application of manure —gain or pain? Landbauforschung—Appl Agri Forstry Res 66(1):1–9
Hansen DJ (2006) Manure as a nutrient source. In: Haering KC, Evanylo GK (eds) Mid-atlantic nutrient management handbook. MAWP 06-02, pp 207–225
Huber DM (1991) The use of fertilizers and organic amendments in the control of plant disease. In: Pimentel D (ed) CRC handbook of pest management in agriculture. CRC Press, Boca Raton, Florida, pp 405–495
Huber DM, Haneklaus S (2007) Managing nutrition to control plant disease. Landbauforschung Völkenrode 57:313–322
Hudson DJ (1966) Fitting segmented curves whose join points have to be estimated. J Am Stat Assoc 61:1097–1129
Janssen BH, Guiking FCT, Van der Eijk D, Smaling EMA, Wolf J, Van Reuler H (1990) A system for quantitative evaluation of the fertility of tropical soils (QUEFTS). Geoderma 46:299–318
Jung JKH, McCouch S (2013) Getting to the roots of it: genetic and hormonal control of root architecture. Front Plant Sci 4:1–32

Kerschberger M, Hege U, Jungk A (1997) Phosphordüngung nach Bodenuntersuchung und Pflanzenbedarf. VDLUFA Standpunkt. (Online) http://www.vdlufa.de/joomla/Dokumente/Standpunkte/0-4-phosphor.pdf, last access 21 Nov 2012
Klages S (2011) Evaluation of the mineral status of organically grown cotton in Egypt. Ph.D. thesis, TU Braunschweig, Germany, p 259
Koltei H, Kapulnik Y (2010) Arbuscular mycorrhizas: physiology and function. Springer, Heidelberg, p 400
Larcher W (1995) Physiological plant ecology. Springer, Heidelberg, p 506
Leiser WL, Rattunde HFW, Weltzien E, Haussmann BIG (2014) Phosphorus uptake and use efficiency of diverse West and Central African sorghum genotypes under field conditions in Mali. Plant Soil 377:383–394
Li L, Li SM, Sun JH, Zhou LL, Bao XG, Zhang HG, Zhang FS (2007) Diversity enhances agricultural productivity via rhizosphere phosphorus facilitation on phosphorus-deficient soils. Proc Nat Acad Sci, USA 104:11192–11196
Li L, Tilman D, Lambers H, Zhang F-S (2014) Plant diversity and overyielding: insights from belowground facilitation of intercropping in agriculture. New Phytol 203:63–69
Liu C, Muchhal US, Uthappa M, Kononowicz AK, Raghotama KG (1998) Tomato phosphate transporter genes are differentially regulated in plant tissues by phosphorus. Plant Physiol 116:91–99
Loneragan JF, Webb MJ (1993) Interactions between zinc and other nutrients affecting the growth of plants. In: Robson AD (ed) Zinc in soils and plants. Kluwer Academic Publishers, Dordrecht, pp 119–134
Lopez-Arredondo DL, Herrera-Estrella L (2012) Engineering phosphorus metabolism in plants to produce a dual fertilization and weed control system. Nat Biotechnol 30:889–893
Maathuis FJM (2009) Physiological functions of mineral macronutrients. Curr Opin Plant Biol 12:250–258
Mac Key J (1988) Shoot: root interrelations in oats. In: Mattsson B, Lyhagen R (eds) 3rd international Oat conference. Svalöv, Sweden, pp 340–344
Mills HA, Jones JB Jr (1997) Plant analysis handbook II. Micro Macro Publications, Athens, Georgia, p 422
Moeller-Nielsen J (1973) Kornplanters erbaerinstilstand vurderet og reguleret udfra planternes kemiske sammensaetning. Diss. Kgl. Veterinaerog Landbohoiskole, Kopenhagen
Moeller-Nielsen J, Frijs-Nielsen B (1976) Evaluation and control of the nutritional status of cereals. II. Pure effect of a nutrient. Plant Soil 45:339–351
Naher UA, Othman R, Panhwar QA (2013) Beneficial effects of mycorrhizal association for crop production in the tropics—a review. Int J Agric Biol 15:1021–1028
Niu YF, Chai RS, Jin GL, Wang H, Tang CX, Zhang YS (2013) Responses of root architecture development to low phosphorus availability: a review. Ann Bot 112:391–408
Panten K, Godlinski F, Schroetter S, Hofmeier M (2016) Variability of P off-take by plants. This volume
Perspectives Agriculture (2011) Sustainable agriculture in the Baltic Sea region in times of peak phosphorus and global change. Baltic 21 Series no. 4/2011
Popovic O (2012) Pig slurry separation and storage. Effects on P, Cu and Zn distribution between separation fractions and particle size classes. Ph.D. thesis, Copenhagen, Denmark
Postma J, Lynch JP (2012) Complementarity in root architecture for nutrient uptake in ancient maize/bean/squash polycultures. Ann Bot 110:521–534
Powers WJ, van Horn HH (2001) Nutritional implications for manure management planning. Appl Eng Agric 17:27–39
Prabhu AS, Fageria NK, Berni RF, Rodrigues FA (2007) Phosphorus and plant disease. In: Datnoff LE, Elmer W, Guber DM (eds) Mineral nutrition and plant disease. APS Press, St. Paul, MN, pp 45–55
Raghothama KG (2007) Phosphorus and plant nutrition: an overview. In: Sims JT, Sharpley AN (eds) Phosphorus: agriculture and the environment. Agronomy monograph No. 46. American Society of Agronomy, Wisconsin, pp 355–378

Rajala A (2003) Plant growth regulators to manipulate cereal growth in northern growing conditions. Ph.D. thesis, University of Helsinki, Finland, p 47

Ranade-Malvi U (2011) Interaction of micronutrients with major nutrients with special reference to potassium. Karnataka J Agric Sci 24:106–109

Reimann C, Siewers U, Tarvainen T, Bityukova L, Eriksson J, Gilucis A, Gregorauskiene V, Lukashev VK, Matinian NN, Pasieczna A (2003) Agricultural soils in northern Europe: a geochemical atlas. Geologisches Jahrbuch, Sonderhefte, Reihe D, Mineralogie, Petrographie, Geochemie, Lagerstättenkunde, H. SD 5. Schweizbart'sche Verlagsbuchhandlung, Stuttgart, p 279

Reuter DJ, Robinson JB (1997) Plant analysis—an interpretation manual. CSIRO Publishing, Collingwood, Australia, p 572

Reuveni R, Reuveni M, Agapov V (1992) Systemic resistance against norther leaf blight and common rust in maize by foliar spray of phosphates. Phytopathology 82:1179

Reuveni M, Agapov V, Reuveni R (1995) Induced systemic protection to powdery mildew in cucumber by phosphate and potassium fertilizers: effects of inoculums and post-inoculation treatment. Can J Plant Pathol 17:247–251

Reuveni R, Dor G, Reuveni M, Tuzun S (1998) Local and systemic control of powdery mildew (*Leveillula taurica*) on pepper plants by foliar spray of mono-potassium phosphate. Crop Prot 17:703–709

Reuveni R, Dor G, Reuveni M, Tuzun S (2000) Systemic resistance against *Sphaerotheca fuliginea* in cucumber plants exposed to phosphate in hydroponics system, and its control by foliar spray of mono-potassium phosphate. Crop Prot 19:355–361

Richter DD, Allen HL, Li J, Markewitz D, Markewitz D, Raikes J (2004) Bioavailability of slowly cycling soil phosphorus: major reconstructuing of soil phosphorus fractions over four decades in an aggrading forest. Oecologia 150:259–271

Rossiter RC (1952) Phosphorus toxicity in subterranean clover and oats grown on Muchea sand, and the modifying effects of lime and nitrate-nitrogen. Austr J Agric Res 3:227–243

Saeys W, Deblander J, Ramon H, Anthonis J (2008) High-performance flow control for site-specific application of liquid manure. Biosyst Eng 99:22–34

Sanchez CA (2007) Phosphorus. In: Barker AV, Pilbeam DJ (eds) Handbook of plant nutrition. CRC Press, Boca Raton, pp 51–90

Sano T, Nagata T (2008) Phosphate uptake and transport to plant cells. In: Jaiwal PK, Singh RP, Dhankher OP (eds) Plant membrane and vacuolar transporters. CAB International, Wallingford, pp 131–147

Schachtman DP, Reid RJ, Ayling SM (1998) Phosphorus uptake by plants: from soil to cell. Plant Physiol 116:447–453

Schick J (2010) Untersuchungen zu P-Düngewirkung und Schwermetallgehalten thermochemisch behandelter Klärschlammaschen. Ph.D. thesis, Braunschweig, Germany, p 179

Schnug E (1988) Quantitative und qualitative Aspekte der Diagnose und Therapie der Schwefelversorgung von Raps (*Brassica napus* L.) unter besonderer Berücksichtigung glucosinolatarmer Sorten. D.sc. thesis, Kiel University, p 256

Schnug E, Haneklaus S (1992) PIPPA: un programme d'interprétation des analyses de plantes pour le colza et les céréales. Supplément de Perspectives Agricoles 171:30–33

Schnug E, Haneklaus S (1998) Diagnosis of sulphur nutrition. In: Schnug E, Beringer H (eds) Sulphur in agro-ecosystems, vol 2., Mineral Nutrition in EcosystemsKluwer Academic Publication, Dordrecht, pp 1–38

Schnug E, Haneklaus S (2008) Evaluation of the relative significance of sulfur and other essential mineral elements in oilseed rape, cereals, and sugar beet production. In: Jez J (ed) Sulfur: a missing link between soils, crops, and nutrition. B. Crops and sulfur. CSSA-ASA-SSSA, pp 219–233

Schnug E, Haneklaus S (2016) The enigma of fertilizer phosphorus utilization. This volume

Schnug E, Strampe U (1988) Sortentypische Unterschiede der Nährelementkonzentrationen bei Winterweizen. J Agron Crop Sci 160:163–172

Schnug E, Heym J, Achwan F (1996) Establishing critical values for soil and plant analysis by means of the boundary line development system (BOLIDES). Commun Soil Sci Plant Anal 27:2739–2748
Schönnamsgruber G (1955) Studien über den Phosphathaushalt von jungen Holzpflanzen, insbesondere von Pappeln. Mitt Württemb Forstl Versuchsanstalt 12:1–68
Schroetter S, Angeles-Wedlel D, Kreuzig R, Schnug E (2006) Effects of phosphite on phosphorus supply and growth of corn (*Zea mays*). Landbauforschung Völknerode 56:87–99
Shah MA (2014) Mycorrhizas: novel dimensions in the changing world. Springer, Heidelberg, p 87
Shane MW, Lambers H (2005) Cluster roots: a curiosity in context. Plant Soil 274:101–125
Sharma NC, Sahi SV (2012) Excess soil phosphorus: accelerated P transfer, water quality deterioration, and sustainable remediation strategies. Improving Crop Resist Abiotic Stress 1 (2):165–191
Siddique KHM, Belford RK, Tennant D (1990) Root: shoot ratios of old and modern semi-dwarf wheats in a Mediterranean environment. Plant Soil 121:89–98
Silber A, Ackerman A, Bar-Tal A, Levkovitch I, Matsevitz-Yosef T, Swartzberg D, Granot D (2002) Interrelationship between phosphorus toxicity and sugar metabolism in *Verticordia plumose* L. Plant Soil 245:249–260
Simonsen L, Fomsgaard IS, Svensmark B, Spliid NH (2008) Fate and availability of glyphosate and AMPA in agricultural soil. J Environ Sci Health—Part B: Pesticides, Food Contam Agric Wastes 43:365–375
Smith FW, Loneragan JF (1997) Interpretation of plant analysis: concepts and principles. In: Reuter DJ, Robinson JB (eds) Plant analysis—an interpretation manual. CSIRO Publishing, Collingwood, Australia, pp 3–26
Spencer K, Freney JR (1980) Assessing the sulfur status of field-grown wheat by plant analysis. Agron J 72:469–472
Stauss R, Bleiholder H, van der Boom T, Buhr L, Hack H, Hess M, Klose R, Meier U, Weber E (1994) Einheitliche Codierung der phänologischen Entwicklungsstadien mono- und dikotyler Pflanzen. Ciba, Geigy AG, Basel
Sumner ME, Farina MPW (1986) Phosphorus interactions with other nutrients and lime in field cropping systems. Adv Soil Sci 5:201–236
Thao HTB, Yamakawa T (2009) Phosphite (phosphorous acid): fungicide, fertilizer or bio-stimulator? Soil Sci Plant Nutr 55:228–234
Theodorou ME, Plaxton WC (1993) Metabolic adaptations of plant respiration to nutritional phosphate deprivation. Plant Physiol 101:339–344
Timmer VR, Armstrong G (1987) Diagnosing nutritional status of containerized tree seedlings: comparative plant analyses. Soil Sci Soc Am J 51:1082–1086
Tóth G, Guicharnaud R-A, Tóth B, Hermann T (2014) Phosphorus levels in croplands of the European Union with implications for P fertilizer use. Eur J Agron 55:42–52
Uekötter F (2006) Did they know what they were doing? An argument for a knowledge-based approach to the environmental history of the twentieth-century agriculture. GHI Bull Suppl 3:145–166
Ulrich A (2014) Peak phosphorus: opportunity in the making exploring global phosphorus management and stewardship for a sustainable future. Ph.D. thesis, ETH Zurich, Switzerland, p 170
Vallarino JG, Osorio S (2012) Signaling role of oligogalacturonides derived during cell wall degradation. Plant Signal Behav 11:1447–1449
Vance CP (2011) Phosphorus as a critical macronutrient. In: Hawkesford MJ, Barraclough P (eds) The molecular and physiological basis of nutrient use efficiency in crops. Wiley, Chichester, pp 229–264
Vielemeyer H-P, Neubert P, Hundt I, Vanselow G, Weissert P (1983) Ein neues Verfahren zur Ableitung von Pflanzenanalyse-Grenzwerten für die Einschätzung des Ernährungszustandes landwirtschaftlicher Kulturpflanzen. Archiv Acker- und Pflanzenbau und Bodenkde 27:445–453

Vitousek PM, Porder S, Houlton BZ, Chadwick OA (2010) Terrestrial phosphorus limitation: mechanisms, implications, and nitrogen-phosphorus interactions. Ecol Appl 20:5–15
Waines JG, Ehdaie B (2007) Domestication and crop physiology: roots of green-revolution wheat. Ann Bot 100:991–998
Walker TW, Syers JK (1997) The fate of phosphorus during pedogenesis. Geoderma 15:1–19
Wallace A (1990) Crop improvement through multidisciplinary approaches to different types of stresses-law of the maximum. J Plant Nutr 13:313–325
Walworth JL, Letzsch WS, Sumner ME (1986) Use of boundary lines in establishing diagnostic norms. Soil Sci Soc Am J 50:123–128
Wanasuria S, Kühn H (1978) Der Einfluß der Phosphaternährung auf die Aufnahme von Eisen, Mangan und Zink in Wasserkultur bei Sonnenblumen, Erbsen und Gurken. Die Bodenkultur 28:232–244
Wang XR, Shen JB, Laio H (2010) Acquisition or utilization, which is more critical for enhancing phosphorus efficiency in modern crops? Plant Sci 179:302–306
Westermann DT (2007) Plant analyses and interpretation. In: Sims JT, Sharpley AN (eds) Phosphorus: agriculture and the environment. Agronomy monograph no 46. American Society of Agronomy, Wisconsin, pp 415–436
White PJ, Hammond JP (2008) Phosphorus nutrition of terrestrial plants. In: White PJ, Hammond JP (eds) The ecophysiology of plant-phosphorus interactions. Springer, Heidelberg, pp 51–81
White PJ, Broadley MR, Gregory PJ (2012) Managing the nutrition of plants and people. Appl Environ Soil Sci. Article 104826
White PJ, George TS, Gregory PJ, Bengough AG, Hallett PD, McKenzie BM (2013) Matching roots to their environment. Ann Bot 112:207–222
Wissuwa M, Mazzola M, Picard C (2009) Novel approaches in plant breeding for rhizosphere-related traits. Plant Soil 321:409–430
Woodward EJ, Marshall C (1987) Effects of seed treatment with plant growth regulator on growth and tillering in spring barley (*Hordeum distichum*) cv. Triumph. Ann Appl Biol 110:629–638
Yan XL, Liao H, Trull MC, Beebe SE, Lynch JP (2001) Induction of a major leaf acid phosphatase does not confer adaptation to low phosphorus availability in common bean. Plant Physiol 125:1901–1911
Yan ZJ, Liu PP, Li YH, Ma L, Alva A, Dou ZX, Chen Q, Zhang FS (2013) Phosphorus in China's intensive vegetable production systems: overfertilization, soil enrichment, and environmental implications. J Environ Qual 42:982–989
Zhou J, Jiao FC, Wu ZC, Li YY, Wang XM, He XW, Zhong WQ, Wu P (2008) OsPHR2 is involved in phosphate-starvation signaling excessive phosphate accumulation in shoots of plants. Plant Physiol 146:1673–1686
Zobiole LHS, Kremer RJ, de Oliveira RS, Constantin J (2012) Glyphosate effects on photosynthesis, nutrient accumulation, and nodulation in glyphosate-resistant soybean. J Plant Nutr Soil Sci 175:319–330

P Solubility of Inorganic and Organic P Sources

Sylvia Kratz, Judith Schick and Anne Falk Øgaard

Abstract P solubility of mineral and organic fertilizers can be estimated by a variety of different chemical extraction methods. In Europe, the characterization of P solubility in fertilizers is regulated in the European fertilizer regulation 2003/2003 for commercial fertilizers, assigning different methods to the various fertilizer types. Relationships between chemical solubility and agricultural performance/P availability for plants have been documented in numerous pot and field trials. Non-commercial fertilizers like farmyard manures and slurries, on the other hand, as well as "new products" based on recycling materials such as sewage sludge ash, are not included in this regulation yet, and a number of methods, designed for mineral fertilizers as well as for P solubility in soils, have been tested over the last couple of years to adequately characterize these products. This review gives a critical overview of the chemical extraction methods currently practiced and/or tested in the European countries, and their performance as estimates for plant availability of fertilizer P from inorganic and organic P sources.

Keywords Agronomic efficiency · Biowaste compost · Chemical extraction methods · Commercial phosphorus fertilizers · Farmyard manure · Farmyard slurry · Organic phosphorus fertilizers · Phosphorus solubility · Sewage sludge

---

S. Kratz (✉) · J. Schick
Federal Research Centre for Cultivated Plants,
Institute for Crop and Soil Science, Julius Kühn-Institut (JKI),
Bundesallee 50, 38116 Brunswick, Germany
e-mail: sylvia.kratz@jki.bund.de

A.F. Øgaard
Bioforsk Soil and Environment, Frederik A. Dahls vei 20, 1430 Ås, Norway

E. Schnug and L.J. De Kok (eds.), *Phosphorus in Agriculture: 100 % Zero*,
DOI 10.1007/978-94-017-7612-7_7

Introduction

P solubility is one of the most important parameters to characterize the quality of P containing fertilizers, as it is a prerequisite for estimating plant availability and nutrient uptake efficiency. The European fertilizer regulation defines various chemical extraction procedures for the characterization of P contents and solubility of commercial (mineral and organo-mineral) fertilizers. Some countries, such as Germany, also have their own national fertilizer ordinance with some additional procedures, which have been kept till today for historical reasons. The applicability of the various extraction procedures depends on the origin and quality of the P fertilizer in question. However, there are challenges when using chemical extactions for estimating plant availability of P. The availability in soil is dependent on transport processes of P towards roots (diffusion and mass flow) and the capacity of soils and fertilizers to replenish the P concentration in the soil solution after plant P uptake. The higher the concentration in the soil solution, the more P is transported towards the roots through mass flow and diffusion. The P concentration in the soil solution is probably most important in the first period after seeding, and before the roots have been properly developed. Close to the roots the P availability is influenced by the plants' ability to extract P from soil and fertilizers by acidification of the rhizosphere and by exudation of low molecular organic acids, which desorb iron (Fe)- and aluminum (Al)-bound P by anion exchange, and by chelation of Fe and Al (Mengel and Kirkby 1987). Therefore, ideally, chemical extractions of P fertilizers should both reflect the P fertilizer's influence on soil solution P concentration and on the P fraction extractable by the root processes. Furthermore, soil characteristics (especially pH) influence the availability of P in the fertilizer product, and the fertilizer product may influence availability of P already in soil through influence on soil pH (e.g. limed fertilizer products) or through influence on soil's adsorption capacity for P (e.g. sewage sludge precipitated with Fe and/or Al). To sum it up, influences on the solubility of fertilizers can be classified into three categories (according to Hedley and McLaughlin 2005) viz. (i) Fertilizer characteristics: particle size and aggregate strength, P content and speciation, pH, electric conductivity (EC), carbon (C)/nitrogen (N) content and chemistry, Fe/Al content, amount of free carbonate (e.g. calcite, dolomite) and calcium phosphate, silica content (Thomas slag, fused phosphate), for apatite containing fertilizers (e.g. rock phosphates): isomorphic substitution of PO_4 by CO_3 in the apatite crystal, cementation of apatite minerals by silica mineral, degree of crystallization of the apatite (difference between sedimentary and magmatic rock phosphates); (ii) Soil parameters: sorption characteristics (pH, mineralogy, EC), bioturbation, porosity, bulk density, P status, content and quality of organic substance (activity of nitrifiers and C mineralizers), (iii) Management and environmental factors (placement and mixing of fertilizer, climate, crop type etc.). Therefore, it is clear that any result from chemical extraction may only provide a first estimate of plant availability of the investigated fertilizer and must be interpreted in the light of these parameters. Nevertheless, they should allow for a comparative evaluation of the potential solubility of different fertilizers which enables farmers to better adjust their fertilizer dressings and minimize any surplus in their individual P balance.

This chapter presents an overview of the extraction procedures currently used for commercial fertilizers in Europe and discusses (a) how reliable estimates of plant availability and nutrient uptake efficiency based on these procedures are (b) to which extent those methods can be applied to different kinds of recycled P-fertilizers and (c) if they are also suitable for organic fertilizers containing P as a nutrient, or, which other methods may be applied for organic fertilizers, respectively.

Criteria for the Evaluation of Agronomic Efficiency of P Fertilizers

Agronomic efficiency of fertilizers is usually evaluated by one or more of the following parameters: (i) Nutrient uptake/degree of utilization; (ii) Yield, e.g. calculation of relative yields or yield increase compared to a standard fertilizer: RAE (relative agronomic efficiency) (%) = 100 × yield (increase) by test fertilization/ yield (increase) by standard fertilization; (iii) For organic fertilizers: mineral fertilizer equivalent (MFE) i.e. efficiency of the organic fertilizer compared to that of an optimally applied mineral fertilizer, expressed as amount of P in kg ha^{-1} as mineral fertilizer, which—applied at the optimal time, results in the same fertilization effect as 100 kg P applied as organic fertilizer: MFE = P (kg ha^{-1}) in mineral fertilizer × 100/P (kg ha^{-1}) in organic fertilizer.

The utilisation rate of a nutrient is defined as the amount, which is taken up by the plant and is expressed in % of the total amount of nutrients applied (Finck 1992). The utilization rate can be referred to different time periods: The growing period of the fertilised plant, the after-effect for several years as long as it is detectable, and a time period of several decades (important for long-term considerations). Usually the utilization rate is referred to the growing period of the fertilized plant. However, it is essential to distinguish between the actual (or real) and the apparent (or effective) utilization rate (UR; Schnug et al. 2003a). The actual utilization rate considers only the utilisation rate in the year of application (%) and is calculated by the so-called "difference method": Actual UR = (total uptake − uptake from soil reserve) × 100/amount of nutrients supplied by fertilization. The actual utilization rate of P usually amounts to around 10–25 % of fertilizer P in the year of application. The after effects add up to 1–2 % per year. The apparent utilization rate is the total effect of the fertilizer in the year of application as well as in following years (Finck 1992; Schnug et al. 2003b): Apparent (effective) utilization (%) = utilization in the year of application + utilization of nutrients stored in the soil from preceding fertilization. In the long term, the apparent utilization can be up to 100 % of the applied nutrient. A major prerequisite of a high effective utilization is the ability of the fertilizer P to go into solution within the first vegetation period in order to be involved in the local P dynamics. However, it should always be considered that the "difference method" assumes that plant uptake of soil P will not be influenced by fertilization, which is not fully correct. In fact, this ignores the so-called “priming effect”, which means that application of fertilizer P may stimulate root growth and/or accelerate the turnover of organic soil substance, allowing for an increased uptake of soil P (Bingeman et al. 1953).

One alternative to assess the amount of P which is taken up by the plant is the application of the isotopic exchange method, which means that either the fertilizer (direct labelling) or the soil (indirect labelling) is labelled with the radioisotopes ^{33}P or ^{32}P, and the plant uptake of the fertilizer versus soil P is calculated from the specific activity of $^{33}P/^{31}P$ or $^{32}P/^{31}P$ in the plant and in the fertilizer or in the soil. However, this method is also limited since both isotopes have relatively short half lifes (^{32}P:14d, ^{33}P:25d), allowing only for short time studies, and the exchange of labelled isotopes in the soil by unlabelled ones cannot be controlled. Nevertheless, isotopically exchangeable P (IEP) may be a viable measure to estimate the amount of fertilizer that dissolves (in the soil water) and reacts with the soil (see below, Cabeza et al. 2011).

Chemical Extraction Procedures for Estimating P Solubility of Mineral Fertilizers

European and German Fertilizer Legislation

The EU regulation on fertilizers (VO (EG) 2003/2003 from 13.10.2003, Official Journal of the EC L 304, with modifications by VO (EG) No. 2076/2004, Official Journal of the EC L 359, pp 25ff. and VO (EG) No. 162/2007, EU-Amtsblatt L 51, pp 7ff.) defines the following chemical procedures for estimating P solubility in commercial fertilizers [the German fertilizer ordinance (DüMV from 26.11.2003, BGBl. I, p. 2373, modified on 03.11.2004, BGBl. I, S. 2767) lists some additional methods for products only marketed in Germany]: (1) Water soluble P, (2) P soluble in alkaline ammonium citrate according to Petermann (65 °C), (3) P soluble in alkaline ammonium citrate according to Petermann (20 °C), (4) P soluble in alkaline ammonium citrate according to Joulie (20 °C), (5) P soluble in water and neutral ammonium citrate according to Fresenius-Neubauer (only DüMV), (6) P soluble in neutral ammonium citrate (NAC), (7) P soluble in 2 % citric acid, (8) P soluble in 2 % citric acid and alkaline ammonium citrate (Petermann; only DüMV), (9) P soluble in 2 % formic acid, and (10) P soluble in mineral acid (HNO_3 + H_2SO_4 mix (with Cu sulfate), or, according to DüMV/VDLUFA-Methodenbuch II, 3 additional procedures: (a) with H_2SO_4, (b) with H_2SO_4 and Cu sulfate, (c) dry ashing with Ca carbonate, dissolution with HNO_3.

The inclusion of so many methods into the EU regulation is the result of a compromise by all EU countries, which was achieved in order to allow each country to stick with its nationally established procedures (Russell 1982).

Scope of the Procedures

The scope of each procedure is defined in the EU regulation on a European basis, and in the VDLUFA-Methodenbuch II (1995) on a national (German) basis (see Table 1). This illustrates that, except for water and mineral acid, none of these procedures was designed to be universally applied for all different types of P-containing fertilizers. Rather, each procedure extracts a different P fraction or compound. In the literature, the following associations between procedures and P compounds are found (Table 2).

Table 1 Scope of extraction procedures for P solubility of commercial fertilizers

Procedure	Scope according to VDLUFA and/or VO (EG) 2003/2003
Water soluble P	All types of P fertilizers
P soluble in alkaline ammonium citrate according to *Petermann* (65 °C)	Dicalcium phosphate dihydrate ($CaHPO_4 \cdot 2H_2O$)
P soluble in alkaline ammonium citrate according to *Petermann* (20 °C)	Fused phosphate[a] (VDLUFA: including mineral multinutrient fertilizers, Al-Ca-phosphate, special multinutrient suspensions)
P soluble in alkaline ammonium citrate according to *Joulie* (20 °C)	Al-Ca-phosphate
P soluble in neutral ammonium citrate	Superphosphate (SSP), Triple-Superphosphate (TSP)
P soluble in water and neutral ammonium citrate according *Fresenius-Neubauer*	Superphosphate, mineral multinutrient fertilizers
P soluble in 2 % citric acid	Thomas phosphate (basic slag) and multinutrient fertilizers made of basic slag
P soluble in 2 % citric acid and alkaline ammonium citrate (*Petermann*)	Mixtures of P containing slags and DCP and/or MCP
P soluble in 2 % formic acid	Partly digested and soft rock phosphates (RP) and their mixtures
P soluble in mineral acid:	
H_2SO_4	Fertilizers with no or little organic material content
Mix of HNO_3 and H_2SO_4 (VDLUFA: with Cu sulfate)	Standard method according to EU-VO, VDLUFA: for fertilizers with large share of organic material
H_2SO_4 and Cu sulfate	Fertilizers with large share of organic material, but without nitrates
Dry ashing with calcium carbonate, dissolution with HNO_3	Fertilizers with large share of organic material

[a]Mixed crystallization from calcium sodium phosphate $CaNaPO_4$ and calcium orthosilicate Ca_2SiO_4 (Scheel 1968a) or Ca-Na-silicophosphate (Finck 1992)

Table 2 P forms in inorganic P fertilizers extracted by various procedures

Extractant	Extracted P fraction/P compound	Reference
Water (W)	Mono calcium phosphate (MCP): $Ca(H_2PO_4)_2{\cdot}H_2O$, Ammonium phosphate	Scheel (1968b), Braithwaite (1987), Hignett and Brabson (1961)
Alkaline ammonium citrate (AAC)	Di calcium phosphate (DCP): $CaHPO_4$; or, "non water soluble, but plant available P fraction", i.e. including plant available Fe-Al-phosphates	Hignett and Brabson (1961), Werner (1967), Junge and Werner (1989)
Neutral ammonium citrate (NAC) or sequential extraction (W + NAC)	MCP, DCP, about 80 % of Al-/Fe-phosphates, basic Ca-phosphate (hydroxy apatite / tri calcium phosphate) or, rock phosphate residue which was not dissolved or recombined during ammonisation apatite, depending on carbonate substitution and share of free carbonate in the rock phosphate	Braithwaite (1987), Hammond et al. (1989), Hignett and Brabson (1961), Schmitt (1969a, b), Chien and Hammond (1978), Léon et al. (1986), (debated, contrasting opinion: Braithwaite 1987)
Citric acid (CA)	MCP, DCP, about 20 % of Al-/Fe-phosphates, about 20 % of undissolved RP residue in partly digested RP, P compounds containing silica, e.g. Ca-silico-phosphate (Silicocarnotite: $5CaO.P_2O_5.SiO_2$ mixed crystal)	Braithwaite (1987), Maercker (1895), Martens (1943), Gericke (1952, 1968)
Formic acid (FA)	Apatite, depending on carbonate substitution and share of free carbonate in the rock phosphate	Chien and Hammond (1978), Léon et al. (1986)
Mineral acid	Total phosphate	Finck (1992)

Plant Availability of Different P Compounds in Vegetation Trials

As shown above, some of the extraction procedures extract more than one P fraction. However, the plant availability of the different fractions is often different, as well. In the literature, the following classifications can be found:

(1) Mono calcium phosphate (MCP): completely and instantly plant available (Finck 1992; Schilling 2000),

(2) Di calcium phosphate (DCP), well available for plants (Quin 1982; for proof by vegetation trials see references cited by Hignett and Brabson 1961), but takes more time to become effective than MCP (Finck 1992),
(3) Ammonium phosphate ($NH_4H_2PO_4$ [MAP], $(NH_4)_2HPO_4$ [DAP]); completely and immediately plant available (Sikora and Mullins 1995; Schilling 2000),
(4) Silico carnotite/silica-containing phosphates (including Ca–Na-silico phosphate); well plant available, proven by numerous pot and field trials with Thomas slag (e.g. Maercker, 1895; Schleede et al. 1939; Scheffer 1956 (with further references), Gericke 1968), and Rhenania phosphates (comprehensive summary of pot and field trials see Werner 1967).
(5) Fe-/Al-Phosphates (e.g. $AlNH_4HPO_4F_2$, $FeNH_4(HPO_4)_2$ and $MgAl(NH_4)2H(PO_4)_2F_2$ in MAP (Sikora et al. 1989); $Fe_3(K, Na, H) H_8(PO_4)_6.6H_2O$ and $CaAlH(HPO_4)_2F_2.2H_2O$ in superphosphate (Prochnow et al. 2003): plant availability depends on type of compound and degree of crystallinity [amorphous Fe- and Al-compounds are often more soluble than crystalline ones (Sikora et al. 1989)]. MAP-impurities appear to be less available than MCP, but more soluble than DCP, variscite ($AlPO_4.2H_2O$) and strengite ($FePO_4.2H_2O$) (which are P compounds that typically precipitate in soils) (Mullins et al. 1990; Sikora et al. 1992). Some researchers assume that the above-mentioned Fe- and Al-phosphates may have a residual (long-term) availability (Sikora and Mullins 1995). However, a short term pot trial (6 weeks) with partly digested rock phosphates proved an adverse effect of Al_2O_3 and Fe_2O_3 on yield and P uptake, indicating a poor P availability. As Hammond et al. (1989) stated, a large share of these oxides will result in the transformation of water soluble P into Al- and Fe-P, which may or may not be soluble in NAC, over time. The solubility of Fe-Al-P-minerals generally depends on the pH of the soil (or extraction solution): since it increases with rising pH, the solubility of these minerals is for example better in NAC than in formic acid (Chien 1993). However, as Deeley et al. (1987) observed on acid sandy soils in SW-Australia, agronomic efficiency may—under special circumstances—be higher at low NAC solubility. In this case, superphosphates which were enriched with Fe and Al showed a better agronomic efficiency compared to conventional superphosphates, because Fe and Al protected P from rapid leaching which would otherwise have taken place in the acid sandy soils.
(6) Basic calcium phosphates/apatite: Low plant availability (see vegetation trials reported by Hignett and Brabson 1961); depending on the amount of free carbonate (e.g. calcite, dolomite) in the rock phosphate, which impairs solubility of the apatite due to the common ion effect (calcium). On the other hand, isomorphic substitution of PO_4 by CO_3 in the apatite crystal results in a structural instability, followed by an increasing reactivity/solubility of the rock phosphate (Chien 1993).

Correlations Between Chemical Extractions and Plant Uptake

Generally, the chemical extraction procedures were developed on empirical basis. Accordingly, vegetation trials for testing correlations between chemical solubility and plant uptake of one or more P fractions were usually designed specifically for a selected fertilizer or a particular extraction procedure. The following paragraphs report results from vegetation trials for the various EU-methods.

Water

Water extracts mainly primary and secondary alkali phosphates such as KH_2PO_4 and K_2HPO_4, primary earth alkali phosphates like $Ca(H_2PO_4)_2$, and their respective ammonia salts (Abrahamczik and Niedermaier 1968). There is wide agreement about the correlation between water solubility of superphosphate and its agronomic efficiency (e.g. Scheffer 1956, with further references). Good correlations were also established between water solubility and P uptake/yield for partly digested rock phosphates with different degrees of digestion (Hammond et al. 1980; Stephen 1985; further references see Stephen and Condron 1986). However, it was shown already in the beginning of superphosphate production in the middle of the 19th century that water-soluble P compounds will be transformed into non water-soluble forms in the soil within a short period of time. Other than undigested rock phosphate, these compounds still displayed a positive influence on plant growth (Terman et al. 1964). In order to differentiate between this plant available but not water-soluble P form and P from undigested rock phosphate, various extraction procedures with ammonium citrate were developed (Fresenius et al. 1871; Joulie 1873; Petermann 1880).

Neutral Ammonium Citrate (NAC)/Alkaline Ammonium Citrate (AAC)

The evaluation of commercial mineral P fertilizers (straight P and compound types) is often based on NAC extraction after Fresenius-Neubauer, not only in Europe, but worldwide. Different procedures (with or without preceding water extraction) are practiced depending on the country, however, they are all based upon the procedure suggested by Fresenius et al. (1871) (Schmitt 1969a). In the 1960s, some researchers started to question the viability of the NAC extraction, as it seemed to overestimate the amount of plant available P in fertilizers (Hignett and Brabson 1961; Gilliam and Wright 1964; Schmitt 1969a, b). Pot and field trials with Rhenania phosphate (fused phosphate; Niklas/Strobel/Scharrer/Schropp 1926, 1931, cited after Schmitt 1969a), ammoniated superphosphate (Gilliam and Wright 1964), and ammoniated NPK fertilizers (Brabson and Burch 1964) demonstrated good agreement between P contents of fertilizers extracted by alkaline ammonium citrate and their P effects. In vegetation trials with 15 soils supplied with superphosphate, Thomas slag, Rhenania phosphate, Kola apatite, Hyperphos (partly digested) and lab grade hydroxyapatite and tricalcium phosphate, Schmitt (1969a, b) showed that AAC was superior to NAC for differentiating easily soluble from sparingly soluble phosphates on slightly acid to slightly alkaline soils. This was explained by the fact that NAC did not only extract plant available P forms but also compounds with a poor effect on plant uptake, such as calcium phosphates (hydroxyapatite) which were undigested or formed during ammoniation of superphosphate (Hignett and Brabson 1961; Schmitt 1969a, b). In 1966, all German Agricultural Research Institutes (LUFAs) carried out a comparison between NAC after Fresenius-Neubauer and AAC after Petermann, analyzing a total of more than 1100 fertilizer samples. On average NAC extracted around 1.5 % P more than AAC, however, there were large variations depending on product and manufacturer (NN 1966). In a pot trial with 5 triple superphosphates, Mullins et al. (1990) confirmed again, that on a short-term basis (first vegetation period, less than 120 days), NAC overestimated the amount of plant available P content in a fertilizer. They assumed the reason being the formation of Fe- and Al-phosphates during fertilizer production, which are insoluble in water, but extractable by NAC. Similar results (i.e. a short term plant availability below that of lab grade MCP) were achieved by Bartos et al. (1991), who performed pot experiments with residues from mono-ammonium phosphates (MAP) which were insoluble in water, but at least partly soluble in NAC and were identified as $AlNH_4HPO_4F_2$, $FeNH_4(HPO_4)_2$ and $MgAl(NH_4)2H(PO_4)_2F_2$ by Sikora et al. (1989). A good correlation between NAC-extractable P and yield/P uptake in pot and field trials was found, however, for directly applied rock phosphates (Chien and Hammond 1978; Léon et al. 1986). For partly digested rock phosphates, contrasting results were achieved: Hammond et al. (1989) could not establish any correlation between P uptake and NAC-extractable P, while the sum of water- and NAC-extractable P showed a significant correlation with P uptake. Stephen (1985), on the other hand, observed a good correlation between P uptake and the NAC-extractable amount of P in finely granulated, partly digested rock phosphate from North Carolina. Probably, origin and nature of the raw material may influence the behavior of partly digested products.

Citric Acid

Dating back to the end oft he 19th century, a huge number of vegetation trials was carried out by VDLUFA (Association of German Agricultural Research Institutes) which established solid correlations between the solubility of thomas slag in citric acid and its agronomic efficiency (Maercker 1895). Accordingly, the extraction with 2 % citric acid after *P. Wagner* became the official laboratory procedure of the association in 1899 for the evaluation of Thomas slags (Gericke 1968). In the following decades, it was shown by a couple of hundred pot and field trials carried out by the LUFAs and other researchers that the citric acid soluble fraction of Thomas phosphate was equivalent to the water soluble fraction of superphosphate, i.e. the same yields were achieved at a given P level based on the respective soluble fraction (for a summary of trials see Scheffer 1956, or Gericke 1968). In contrast to

Thomas slag, rock phosphates did show no or only weak correlations between citric acid soluble P and P uptake/yield in vegetation trials (Schleede et al. 1939; Hofmann and Mager 1951). Similarly, only weak correlations between solubility in 2 % citric acid and agronomic efficiency were found for superphosphate (Stephen 1982). However, an adjustment of the extractant to pH = 5 brought some improvement. Evaluation of data from a couple of field trials in New Zealand indicated that the content of Al/Fe-oxides in superphosphate played an important role for the correlation between solubility in citric acid and agronomic efficiency (Quin 1982).

Formic Acid

Extraction with formic acid allows for a satisfactory discrimination between soft rock phosphates of different quality (Hofmann and Mager 1951; Müller 1968). A number of field trials demonstrated correlations betwen solubility of rock phosphates in formic acid and their plant availability (Gisiger and Pulver 1959; Chien and Hammond 1978). See Table 3 for a summary of correlations proven in vegetation trials between chemical extractability and agronomic efficiency for different types of fertilizers:

Table 3 Standard chemical exractants shown to correlate with agronomic efficiency of commercial P containing fertilizers in vegetation trials—literature survey (w.f.r. = with further reference)

Extractant	Fertilizer type	Reference
Water (W)	Superphosphate	Scheffer (1956) w.f.r.
	Partly digested rock phosphate	Hammond et al. (1980), Stephen (1985), Stephen and Condron (1986) w.f.r.
Alkaline ammonium citrate (AAC)	Rhenania phosphate	Schmitt (1969a, b)
	Ammoniated superphosphate	Gilliam and Wright (1964)
	Ammoniated NPK-fertilizer	Brabson and Burch (1964)
Neutral ammonium citrate (NAC)	Rock phosphate	Chien and Hammond (1978), Léon et al. (1986)
	Partly digested rock phosphate	Stephen (1985)
W + NAC	Partly digested rock phosphate	Hammond et al. (1989)
Citric acid (CA)	Thomas slag	Maercker (1895), Scheffer (1956) w.f.r., Gericke (1968)
Formic acid (FA)	Rock phosphate	Gisiger and Pulver (1959), Chien and Hammond (1978)

Requirements for a Chemical Extraction Procedure to Characterize Plant Availability of Fertilizer P

As this review shows, a large number of different extraction procedures have been developed until today in order to describe the plant availability of P from fertilizers. Often, the driving force for the creation of a new method was the demand of fertilizer producers for a method allowing for the best differentiation of their own product from other fertilizers on the market. With regard to the validity and comparability of the various procedures, many vegetation trials were published, some of them—as shown before—coming to contradictory results and/or conclusions.

All chemical extraction procedures described here share a common problem: They are static methods for determining a momentary solubility, which may depict the continuous process of P dissolution taking place under field conditions only as a kind of "snap shot" at best. None of these lab procedures will result in a real chemical equilibrium, therefore in a strict sense they do not allow physical or chemical statements about the "true" solubility of the fertilizer under study (Higgins 1982). As Terman et al. (1964) concluded, none of these chemical extractions will yield satisfactory results for a multitude of different fertilizers and growing conditions regarding the correlations between chemical solubility and behavior of test plants.

The question which extractant might be suited best to estimate the plant availability of a fertilizer should, however, also be dealt with on the background of the degree of potential utilization of the fertilizer. As mentioned earlier, crops usually do not take up more than 15–20 % of the added P in the year of application. The unused residue, minus losses by erosion and leaching, remains in the soil and can be used by the crops planted in the following years (so-called after-effect of fertilization). The low degree of utilization in the year of fertilizer application already indicates that crops satisfy a large share of their total demand by taking up transformed P compounds accumulated in the soil from earlier fertilizer applications. These transformation products mainly consist of organic P compounds, of phosphate anions sorbed to anorganic soil colloids (clay, Fe/Al-(hydr)oxides) and of precipitated Fe/Al/Ca-phosphates (Schachtschabel 1960). During the process of transformation, the phosphates, which are easily mobilized at first, turn into low-energy forms over a long period of time until they are finally thermodynamically stable under the given circumstances. This is associated with a decrease in solubility and thus with fixation of phosphates in the soil (Schnug et al. 2003a). However, the decrease in solubility does not mean an irrevocable transformation of the phosphate into a non-plant available form: as long as crops are still supplied with easily available phosphates, they still remain able to also take up soil phosphates which are more difficult to mobilize (Schachtschabel 1976, cited after Schnug et al. 2003a). A main prerequisite of a high effective utilization, i.e. a high total effect of fertilization is therefore that the P added by fertilization goes into solution within the first vegetation period and can thus be involved in the site specific P dynamics. What does this mean with regard to the requirements for a suitable extraction procedure to estimate plant availability of a P fertilizer? The crucial point is, as indicated by the above explanations, that the fertilizer-P is able to be transformed in the soil within an "adequate" period of time. The majority of agriculturally used soils in temperate European climates have a pH between 5.5 and 7.5. These chemical conditions are only mirrored by the extractants water and neutral ammonium citrate, while 2 % formic or citric acid have a pH around 2, and alkaline ammonium citrate has a pH around 10. The "effective" P content of a fertilizer, i.e. the content going into solution within the first vegetation period, is therefore, according to Schnug et al. (2003a), best determined by water extract, or, in (slightly) acid soils, per extraction with neutral ammonium citrate. A drawback of these neutral extracts, however, is that they cannot mirror the local impact of acids exsudated by plant roots. Thus, probably none of the known extracts will be able to make an unambiguous statement for every situation occurring at the soil-plant-fertilizer interface.

Applicability of Standard Procedures for Recycling Fertilizers

Since global P-reserves are finite and world market prices for P-fertilizers are increasing, P-recycling from secondary waste materials is urgently required to ensure food safety in the future as well as an independence from those few countries in which the global rock phosphate mines are located. At present, a multitude of methods to recycle P from secondary waste materials such as sewage sludge, manure or meat and bone meal have already been developed. However, one crucial point besides the concentration of inorganic and organic pollutants in those recycling products is the assessment of their fertilizing effect, especially with view to P. Thus, different standard extraction methods have to be applied to make statements about the P availability of these "new" products such as fertilizers based on processed sewage sludge ash.

A study about the quality of thermochemically treated sewage sludge ashes was conducted to learn more about their agronomic efficiency. To meet this aim, the P solubility of the ashes in 5 different extracts was assessed and compared with the results of growth experiments (dry matter yield, P uptake and P_{CAL}-concentration in the soil after harvest) with maize as test plant (Schick 2009). It became evident that neither the raw sewage sluge ash (SSA), nor the thermochemically (1050 °C) treated products contained an appreciable amount of water-soluble P forms (Table 4). Furthermore, the raw ash showed a higher P solubility in ammonium

Table 4 P_{AR}-concentration[a] (%) and P-solubilities (% of P_{AR}) of thermochemically treated sewage sludge ashes (SSA) and their original untreated ash (Schick 2009)

Product	P_{AR} (%)	P_{Water} (% of P_{AR})	P_{AAC} (% of P_{AR})	P_{W+NAC} (% of P_{AR})	P_{NAC} (% of P_{AR})	P_{CA} (% of P_{AR})	P_{FA} (% of P_{AR})
SSA (untreated)	10	0.03	26	14	40	42	28
SSA + $MgCl_2$	7.6	0.07	7.1	29	28	92	84
SSA + $CaCl_2$	7.2	0.04	0.39	16	13	90	82
SSA + $CaCl_2$ + H_2SO_4	6.5	34	40	45	37	100	78

[a]*AR* aqua regia

citrate (neutral and alkaline) than the processed ashes, i.e. the thermochemical treatment had a negative effect on this parameter, while the solubility in CA and FA was enhanced by the process. Among the treated ashes, the $MgCl_2$-variant contained a higher amount of soluble P forms than the $CaCl_2$-variant. A significant increase of the P solubility was obtained by a partly digestion of the ashes with H_2SO_4, the solubility in water was even increased up to 34 % of $P_{aqua\ regia}$.

The varying P-solubilities before and after processing, also with regard to the chlorine donator chosen, can be attributed to the re-formation of the P-phases during the thermochemical treatment (Fig. 1). Apparently, the formation of the

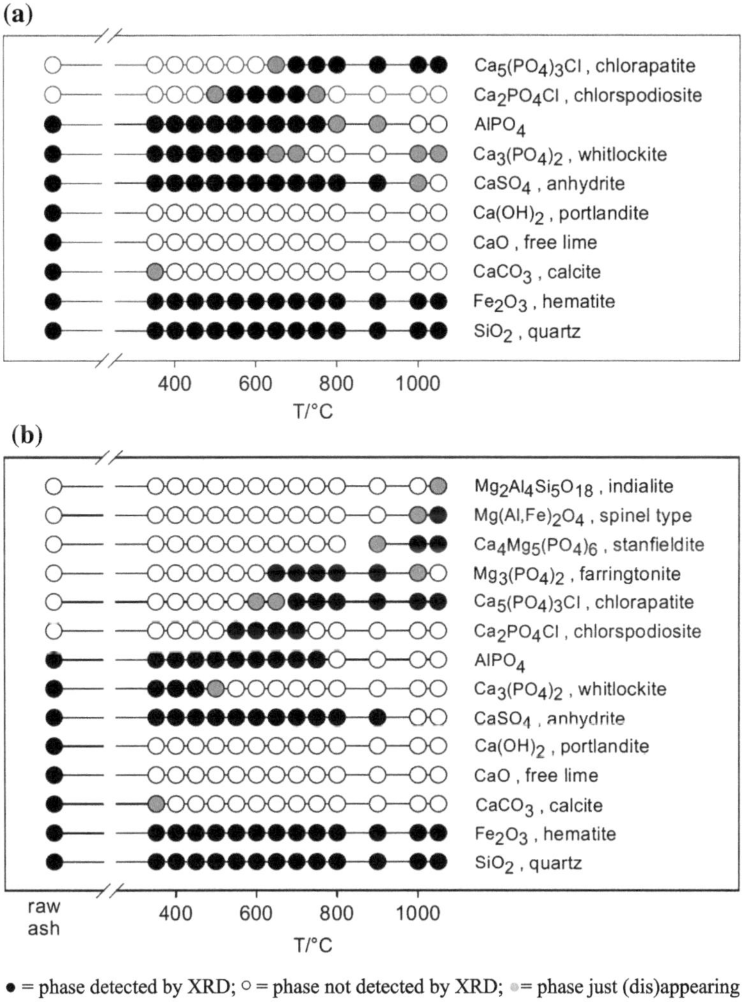

Fig. 1 Some of the mineral phases detected by XRD in a raw ash and in the same ash after being thermochemically treated with the chlorine-donor $CaCl_2$ (**a**) and $MgCl_2$ (**b**) Peplinski et al. (2009)

different phases is significantly influenced by temperature and chlorine donator used ($CaCl_2$ or $MgCl_2$). While the dominating P phases in the raw ash were $AlPO_4$ and $Ca_3(PO_4)_2$ (whitlockite) the thermochemical treatment using $CaCl_2$ at 1050 °C lead to the formation of $Ca_5(PO_4)_3Cl$ (chlorapatite). The use of $MgCl_2$ as a chlorine donator resulted in the formation of two P- phases, chlorapatite and $Ca_4Mg_5(PO_4)_6$ (stanfieldite; Peplinski et al. 2009). As the P solubility of the $CaCl_2$-variant was exceeded by the $MgCl_2$-variant, it can be assumed that stanfieldite is more easily dissolved by the extractants investigated and thus shows also a better plant availability than chlorapatite. The results of the pot trial (not shown here) confirm this assumption (Schick 2009).

Since thermochemically treated sewage sludge ashes did not contain water souble P forms, no significant correlation between this extraction method and the parameters dry matter yield, P uptake and the P_{CAL}-concentration was observed (Table 5). However, results of the growth tests showed strong and statistically significant correlations with P solubility in neutral ammonium citrate (with and without prior water extraction). Hence, the extraction with neutral ammonium citrate seems to be most suitable to characterize the P solubility of fertilizer products containing high amounts of apatite (Table 5). This is also confirmed by the results of Chien and Hammond (1978) as well as Léon et al. (1986) who found good correlations between NAC-extractable P and yield/P uptake in pot and field trials for rock phosphates.

Correlations between standard soil P extraction methods (water, alkaline ammonium citrate, water and neutral ammonium citrate according to Fresenius-Neubauer, neutral ammonium citrate without preceeding water extraction according to the EU regulation 2003/2003, citric acid, formic acid and mineral acid according to EU regulation 2003/2003) and plant P uptake were also investigated applying a number of different sewage sludge based and other recycling fertilizers (see Table 6, solubilities in Table 7) in a modified Neubauer trial (Kratz et al. 2010; Neubauer and Schneider 1923, Neubauer 1931). It should be noted that the Neubauer trial must be interpreted with care, since with its high root density and small amount of soil, it does create

Table 5 Correlation (Pearson) between P-solubilities (% of P_{AR}) and fertilizing effect (dry matter yield, P-utake, P_{CAL}-content in the test soil after harvest) of chemically treated sewage sludge ashes

Extract	Correlation coefficient (r)		
	Dry matter yield	P uptake	P_{CAL}
Water	0.645	0.443	0.469
Alkaline ammoncitrate	0.794*	0.631	0.662
Water + neutral ammoncitrate	0.885**	0.817*	0.930**
Neutral ammoncitrate	0.801*	0.869**	0.944**
Citric acid, 2 % (w/v)	0.952**	0.585	0.696
Formic acid, 2 % (w/v)	0.791*	0.409	0.545

*$p < 0.05$, **$p < 0.005$

Table 6 Recycling fertilizers tested by standard extraction procedures in the Neubauer trial by Kratz et al. (2010)

Organic NP from bone meal ($OrgNP_{BM}$)
Organic NP from meat and bone meal cat. 3 (Org NP_{MBM})
Ash from meat and bone meal cat. 1 (MBMA)
Organic NP from sewage sludge precipitation ($OrgNP_{SS}$)
MAP from waste water precipitation (Struvite)
Straight P from thermochemically treated sewage sludge ash ($P\text{-}SSA_{TC}$)
NPK from thermochemically treated sewage sludge ash ($NPK\text{-}SSA_{TC}$)

conditions which are very different from the field, where plant availability is dependent also on diffusion and mass flow. In this trial, a large number of summer rye seeds were cultivated in a small amount of quartz sand substrate fertilized with different levels of P from the fertilizers listed in Table 6. The design was chosen to make sure that "root soluble" and thus plant available nutrients were completely used up within a short period of time (43 days, 3 cuts). As can be seen in Table 7, water solubility of the tested fertilizers (except for the NPK) was close to zero. While the two precipitation products from sewage sludge as well as the NPK from SSA_{TC} showed high solubilities in NAC, (meat and) bone meal and the straight P were only fairly soluble in citric or even formic acid. Overall, the observed correlations between extractability and P uptake were only of moderate strength, however, they were significant in most cases (Table 8). The strongest correlations were found for net P uptake (shoots and roots) and extraction with citric acid, followed by neutral ammonium citrate with or without preceeding water extraction. This mirrors the low solubility of the tested recycling products. With regard to the products from thermochemically treated SSA, which were also subject of the investigation by Schick (2009), it also demonstrates quite clearly that at least at the time of our studies, these products were a quite heterogeneous group of varying chemical quality, so that no general statement about the best suited extraction procedure could be made. The study further confirms once more that citric acid extraction may be a viable procedure to characterize slowly available P forms in a fertilizer.

Table 7 P-content and relative solubility of test fertilizers in the Neubauer trial using different standard extractants (adapted from Kratz et al. 2010)

Type	P-content (%) (in MinA)	Relative solubility (in % of P in MinA)					
		FA	CA	NAC_{EU}	W + NAC_{FN}	AAC	W
Org NP_{BM}	6.5	76	34	27	33	8.5	0.93
Org NP_{MBM}	7.6	91	74	48	65	16	0.99
MBMA	11.3	48	50	15	15	7.0	0.04
$OrgNP_{SS}$	3.9	57	84	88	86	80	0.40
Struvite	12.0	97	100	92	51	6.3	0.73
$P\text{-}SSA_{TC}$	9.3	75	65	21	21	4.8	0.13
$NPK\text{-}SSA_{TC}$	3.2	101	103	92	92	70	73

Table 8 Pearson correlation coefficients (r) for the relationship between P solubility in different chemical extractants and P uptake in the Neubauer pot trial (from Kratz et al. 2010)

	Shoots cut 1	Shoots cut 2	Shoots cut 3	Roots	Shoots cut 1 − 3 + roots
n	84	81	76	84	76
MinA-P (%)	−0.550**	0.303**	0.222	−0.365**	−0.150
FA (%MinA)	0.414**	0.241*	0.469**	0.249*	0.395**
CA (%MinA)	0.590**	0.277*	0.488**	0.530**	0.722**
NAC_{EU} (% MinA)	0.479**	0.271*	0.480**	0.540**	0.625**
W + NAC_{FN} (% MinA)	0.588**	0.086	0.283*	0.518**	0.550**
AAC (%MinA)	0.611**	−0.170	−0.060	0.516**	0.385**
W (%MinA)	0.865**	−0.323**	−0.093	0.461**	0.387**

Significances *$p < 0.05$, **$p < 0.01$

In contrast to the two experiments described above, other researchers prefer alternative methods to estimate plant availability of fertilizer P from recycling products. For example, Cabeza et al. (2011) carried out a 2-year pot experiment with maize, testing various types of struvite (magnesium ammonium phosphate [MAP] precipitated from sewage sludge or waste water), a sinter Ca-P, a cupola furnace slag made from sewage sludge, a heavy metal depleted sewage sludge ash (of the same type as the products tested by Schick (2009) and Kratz et al. (2010), and a meat and bone meal ash. While they found poor correlations between plant P uptake and P solubility in water or citric acid, there was a strong and significant linear relationship between P uptake and isotopically exchangeable P (IEP), which gives a measure of the amount of P that dissolves in the soil solution and reacts with the soil (Cabeza et al. 2011). As explained by the authors, once the P dissolves in the soil solution, it may be adsorbed to soil particles or precipitate as a salt, but in both cases stays in equilibrium with the soil solution and remains plant available. In their experiment, Cabeza et al. (2011) kept one pot in each treatment without plants, but watered regularly through the whole experimental period to keep the soil moisture at the same level as in the planted pots. At the end of the 2-year period, samples from these unplanted pots were taken and spiked with $H_3^{33}PO_4$, shaken for 5 days and then the ratio $^{33}P/^{31}P$ was determined. From the results, relative increase in IEP as compared to the triplesuperphosphate (TSP) treatment was calculated for all treatments. The various types of struvite resulted in IEP increases (as well as plant uptake) in the same order of magnitude as TSP, both in a neutral loamy as well as in an acid sandy soil, while sinter Ca-P performed only well at low soil pH and Cupola slag at neutral soil conditions (the latter one particularly so in the second year, indicating the presence of slowly reacting P forms, which seems to be typical for thermically treated products). The two tested ashes showed only a low effectiveness, which is most probably the consequence of the presence of sparingly soluble P compounds such as chlorapatite, stanfieldite and of Al and Fe

(Adam et al. 2009; Nanzer et al. 2009). Cabeza et al. (2011) therefore suggested to use such ash products as raw materials for the fertilizer industry. A prerequisite for this, however, is that these products are improved insofar as they must display reliably homogenous and stable properties and quality.

Applicability for Organic P Fertilizers?

Organic P fertilizers constitute a wide range of different products including farmyard manure/slurry, composted or digested food waste, treated sewage sludge and different by-products from industry. In opposite to what often is assumed, the major part of P in most of these products is in an inorganic form (Table 9). Compared to mineral fertilizers, organic P fertilizers generally contain a lower share of soluble P and they have a large variability of accompanying elements and materials, which might influence P chemistry (Hedley and McLaughlin 2005). Further, because of the organic P fraction, P release is also dependent on conditions for microbial decomposition of organic matter.

Table 9 P forms in organic fertilizers

Fertilizer type	P forms	Reference
Sewage sludge (1–5 % P in dry substance)	60–90 % anorganic, depending on chemical or biological P precipitation and anaerobic or aerobic sludge treatment, thereoff <1–38 % water soluble Anorganic forms: Octo calcium phosphate ($Ca_8H_2(PO_4)_6.5H_2O$) Di calcium phosphate (-dihydrate) Fe-phosphate (Vivianite) Al-phosphate (Wavellite) Organic forms: Mono ester (inositol phosphate), diester, phospholipides	Frossard et al. (1994, 1996, 1997), Hedley and McLaughlin (2005), Mullins et al. (2005), Shober et al. (2006)
Farmyard manure (0.5 to >10 % P in dry substance)	Solid part: 60–90 % anorganic, mostly rather scarcely soluble: weakly crystalline Ca- and Mg phosphates Octo calcium phosphate Di calcium phosphate (-dihydrate) Organic forms (quickly mineralized): Phytates (pigs and poultry) Pyrophosphates, mono- and diester (ruminants)	Hedley and McLaughlin (2005), Mullins et al. (2005), Leinweber (1996), Leinweber et al. (1997a, b), Shober et al. (2006)
Biowaste compost (<1 % P in dry substance)	55–95 % anorganic, thereoff 3–20 % water soluble about 30–50 % of anorganic P as rather scarcely soluble, weakly crystalline Ca-P-compounds: Apatite, octo calcium phosphate Organic forms: Inositol phosphate, DNA	Frossard et al. (2002)

In line with the heterogenity of organic fertilizers, also within subgroups of organic fertilizers like farmyard manure or sewage sludge, the relative solubility of P in organic fertilizers shows a high variability (Kratz et al. 2010). This demonstrates the need for reliable analyses adapted to organic fertilizers for estimating their concentration of plant available P.

P Forms in Organic Fertilizers

Inorganic P forms found in farmyard manure and biowaste compost are mostly Ca-P compounds (Table 9). Both di- and octocalciumphosphate are found. These Ca-P compounds are also found in sewage sludge, but in addition, the composition of the inorganic P fraction depends on the waste-water treatment system used for reducing concentrations of P and organic matter in the waste water. For chemical P removal, Fe and/or Al-salts are often used to flocculate P, and the sewage sludge will then contain Fe/Al-phosphates in addition to P adsorbed to Fe/Al-(hydr)oxides (Frossard et al. 1994; Shober et al. 2006; Hedley and McLaughlin 2005). Shober et al. (2006) reported that P in Fe and Al treated sewage sludge were mainly adsorbed to Fe- and Al-hydroxides. Fe- and Al-phosphates were formed to a lesser extent. However, Frossard et al. (1997) found that 60–67 % of P was bound in vivianite (Fe-phosphate) in two $FeSO_4$-flocculated and anaerobically digested sewage sludges, showing that the P forms in sewage sludge varies, probably depending on the sewage water treatment systems used. Alternatively, P precipitation may be also achieved by using calcium oxide, usually as ($Ca(OH)_2$) which leads to the formation of $Ca_5(PO_4)_3(OH)$] [hydroxyapatite] in the precipitate (Baran 1985; Pinnekamp et al. 2007).

The exact composition of the organic P fraction in organic fertilizers is not well defined, but phosphomono- and diesters, phospholipids, nucleic acids, phytates, and pyrophosphates are found (Table 9). There are large variations between different organic P fertilizers in the composition of the organic P fraction. For instance, in pig and dairy cattle manure, 39 and 17 % of water extractable organic P were indentified as phytate and 43 and 15 % of NaOH extractable organic P were identified as phosphomonoesters, respectively (He and Honeycutt 2001).

Chemical Extractions for Estimating P Solubility of Organic Fertilizers

For the characterization of P availability in organic fertilizers, methods developed for characterization of mineral fertilizers, as well as methods originally developed for estimating available P in soils are used. For instance, for biowaste compost, the German society for quality control of compost (Bundesgütegemeinschaft Kompost e.V.) defines CAL (calcium acetate lactate)-extraction (according to Schueller 1969), which is also used for soils, as the official method to determine plant available P (Kehres and Pohle 1998). With this method, Timmermann et al. (1999) determined CAL-P in samples of 124 biowaste and 178 green forage composts and found between 8–18 % and 12–25 % CAL-P, respectively.

However, regarding the use of extraction methods developed for soils, most organic fertilizers are more chemically active than soils, and the buffer capacity and/or extraction capacity of the extraction solution may be exceeded when fertilizers are extracted. For instance, some extraction solutions are buffered to a certain pH, but if the buffer capacity is exceeded by extraction of a fertilizer, the extraction will be performed at different pH for different products. This was demonstrated by P-AL (ammonium lactate, Egnér et al. 1960) extraction of 11 different sewage sludges. The extraction solution is buffered to 3.75, but pH in the extracts of the different sewage sludges varied from 4.0 to 5.8 (Øgaard, unpubl. data). This may have implications on the relationship between extracted P and plant available P when diverse types of organic fertilizers are tested.

By chemical extractions of organic fertilizers both inorganic and organic P are extracted. However, if P in the extracts is measured by the molybdenum blue method (Murphy and Riley 1962), only inorganic P (orthophosphate) is recorded, whereas by measurements with Inductively Coupled Plasma Spectrometry (ICP) total P in the extracts is measured. Analyses of P-AL-extracts from 11 different sewage sludges revealed that ICP measurements gave values that were 7–53 % higher than the values from the molybdenum blue method (Øgaard, unpubl. data). Olsen P (Olsen et al. 1954) extraction of the same sewage sludges gave 20–300 % higher values with ICP measurement compared to the molybdenum blue method. Thereby, the choice of method for analyzing P in the extracts is crucial for the results. However, there is lack of knowledge about plant availability of extracted organic P, and it is not revealed if it is the total P or orthophosphate concentration in the extracts that best estimates plant available P.

By sequential fractionation procedures, P in the organic product may be separated into P fractions with different solubility. Many of the sequential fractionation procedures used are modifications of a procedure developed for soil by Hedley et al. (1982). One modified Hedley fractionation scheme, which has been used for organic fertilizers, includes extractions with H_2O, 0.5 M $NaHCO_3$, 0.1 M NaOH and 1 M HCl in the mentioned order (Sharpley and Moyer 2000). Phosphorus in all extracts is analyzed for both total P and inorganic P, and organic P is calculated by the difference between total and inorganic P. According to Hedley et al. (1982) the sum of H_2O- and $NaHCO_3$-extractable P constitutes the labile P fraction, 0.1 M NaOH extracts P bound more strongly to Fe and Al, and 1 M HCl extracts apatite-type minerals. Sharpley and Moyer (2000) extracted samples of dairy, poultry and swine manure and reported large differences in distribution of P on the P fractions between the different manure types. They found the distribution of inorganic P to be 21–51 % water extractable, 9–36 % $NaHCO_3$-extractable, 1–50 % NaOH-extractable and 1–33 % HCl-extractable, depending on manure type. Another modification of the Hedley fractionation procedure was proposed by Tiessen et al. (1983) and used by

Leinweber (1996) and Leinweber et al. (1997a, b) to characterize dry chicken manure and liquid swine manure. In this case, the first extractant was an anion exchange resin, with the extracted P representing P in solution and parts of labile P, followed by a 0.5 M $NaHCO_3$-extract comprising labile inorganic (remaining after resin extraction), as well as organic and microbial P, a 0.1 M NaOH-extract containing Al- and Fe-phosphates and finally digestion in 1 M H_2SO_4, representing relatively insoluble P such as apatite. The difference between total P in the manure (determined by microwave digestion with H_2O_2, HNO_3, HCl and HF) and the sum of these four fractions was termed residual P, comprising stable organic and insoluble inorganic P. The investigations by Leinweber et al. (1997a, b) demonstrated the large variability of animal manures even for the same animal species, yielding (in % of total P) between 26–31.5 % resin P, 10–31 % $NaHCO_3$-P, 3–13 % NaOH-P, 14–17 % H_2SO_4-P and 15–41 % residual P for dry chicken manure, and 11–41 % resin-P, 13–21 % $NaHCO_3$-P, 1.4–10 % NaOH-P, 18–27 % H_2SO_4-P and 17–39 % residual P for liquid swine manure. Water solubility in these samples was determined by Leinweber et al. as 27–30 % for dry chicken manure and 5–10 % for liquid swine manure. Ajiboye et al. (2004), comparing different types of organic amendments, found 20, 33 and 35–43 % of P being water extractable in dairy, beef and hog manure respectively. In contrast to this, they reported for anaerobically digested sewage sludge that only 10 % of total P were water extractable, whereas 14 % were $NaHCO_3$-extractable, 35 % NaOH-extractable and 30 % HCl-extractable. The authors explained the smaller proportion of labile P fractions in the sludge compared to animal manures by the combined effect of anaerobic digestion and higher Fe, Al und Ca contents in the biosolids.

Correlation of Extracted P with P Uptake by Plants

In different studies, a large number of different chemical extraction procedures have been tested for their ability to estimate plant available P in a wide range of organic products. The actual plant availability of P in the organic products has mainly been measured in pot experiments, but different studies have used different test crops and different length of the growth period. Further, the organic products have been mixed into quartz sand in some studies, whereas soils with different retention capacity for P have been used in other studies. However, put together, these studies give an insight into the complexity of this topic.

Alt et al. (1994) found that extraction with a mixture of $CaCl_2$ and DTPA (Diethylenetriaminepentaacetic acid) (Alt and Peters 1992) best estimated the P availability in composts and compost/peat mixtures. The composts included composts from plant residues and composts from biowaste. Phosphorus availability was measured as P uptake in chrysanthemum (*Dendranthema grandiflorum*). Extraction with 0.01 M $CaCl_2$ (Houba et al. 1990) also showed a good correlation with P uptake in this study, even though the extracted amount of P was small compared to the P uptake. Extractions with the CAL method, formate (Jungk 1961), or NH_4-

acetate (Gabriels and Verdonck 1991) overestimated available P in the composts and showed poor correlations with P uptake. However, the authors pointed out that the acid and well-buffered extraction solutions may perform better in other types of substrates where P is bound to Fe and Al rather than Ca, as was the case for these composts.

Ylivainio et al. (2008) demonstrated that the concentration of easily available P is of special importance at start of the growth. In a pot experiment with a slightly acid sandy soil, P application with dairy manure with a high share of water-soluble P resulted in slightly higher P uptake in the first cut of ryegrass compared to P applied as superphosphate, whereas P applied with meat bone meal, which had a low share of water-soluble P, resulted in significant lower yields. In the meat bone meal 90 % of P was in acid soluble form as determined by the Hedley et al. (1982) fractionation scheme. However, a significant contribution from this P fraction was observed from the second cut onwards. This shows that the fertilizer effect of different P fractions may be interpreted differently depending on the use of the organic fertilizer product.

Velthof et al. (1998) extracted a wide range of organic products with different P extraction methods and related extracted P to P uptake in a pot experiment where ryegrass was grown in a mixture of quartz sand and the tested organic product. The organic products included composted organic municipal solid waste, composted manure, digested pig slurry, anaerobic residue (organic municipal solid waste), earthworm compost and ground animal waste. Phosphorus was extracted with 0.05 M H_2SO_4, 0.025 M HCl + 0.03 M NH_4F (P-Bray I), 0.5 M $NaHCO_3$ (P-Olsen) and with iron oxide coated filter paper (Van der Zee et al. 1987). The iron oxide coated filter paper acts as a sink for P released from the substrate. Thereby, the P concentration in the extraction solution is kept low and P is released by equilibrium reactions. This mimics the effect of plant P uptake on P release from soil and fertilizers, where P concentration in the soil solution is reduced by plant P uptake. Extraction with iron oxide coated filter paper was the method that best estimated P availability in the organic products in this study. P extracted with this method explained 42 % of the variation in P uptake in ryegrass. The other tested extraction procedures showed a poor relationship with plant uptake. These results are comparable to results from a pot experiment with ryegrass grown in a sandy soil poor in nutrients and supplied with 11 different Fe and/or Al-precipitated sewage sludges (Øgaard 2013). Concentrations of P-AL and P-Olsen in the sludges were not related to plant P uptake, whereas P extracted with iron oxide coated filter paper was the P extraction method that best estimated P availability. Extraction with 0.0025 M $CaCl_2$ also showed a fairly good relationship with P uptake, despite low extracted amounts. However, the best parameters for estimating plant available P in these sewage sludges were the total concentations of Al and Fe, which were negatively correlated with plant P uptake. A regression analysis showed that 83 % of the variation in P uptake could be explained by the concentrations of Al and Fe in the sludges. A negative relationship between plant availability of P in sewage sludges and the total concentrations of Fe and Al was also demonstrated by Elliot et al. (2005). They tested plant availability of P in a pot experiment with 35 different sewage sludges. The test crop was Bahia grass (*Paspalum notatum* Flugge) and the soil was an acid, P deficient sandy soil with low content of organic matter. They found that P extracted with NAC was not correlated with plant availability of P in sewage sludge. The reason was that NAC dissolved Fe/Al hydoxides which are the components that adsorb P and reduce plant availability of P in Fe/Al-precipitated sewage sludges. The authors concluded that an extraction method for estimating plant availability of P in sewage sludge should extract a decreasing part of P as the concentration of Fe and Al in the sludge increases.

In line with this, García-Albacete et al. (2012) found that the short-term solubility of P in a wide range of biowastes of different origin and treatments including, among others, animal manure, sewage sludge and municipal solid waste, was better predicted from the total molar concentration of [Ca + Fe + Al] than from total P. Miller and O´Connor (2009) reported that the molar ratio of oxalate-extractable P to oxalate-extractable Fe and Al (P saturation index, PSI) in sewage sludges were well correlated with the plant availability of P. In soil it was found that the effect of sewage sludge application on PSI in soil was related to the concentration of P extracted from the sludge with iron oxide coated filter paper (Maguire et al. 2000).

Investigating the effect of application of organic fertilizers on soil P fractions gives indirect indications on the plant availability of applied P. Maguire et al. (2001) demonstrated that Mehlich-3 P and PSI in sewage sludge were related to the changes in water soluble P and easily desorbable P in the soil after application. However, the chemical properties of the soil influence the plant availability of P in sewage sludge. There was less change in easily releasable soil P fractions by sewage sludge application in soils with a high P retention capacity compared to soils with a lower P retention capacity (Maguire et al. 2001; Sarkar and O'Connor 2004).

Chemical processes such as precipitation-dissolution and adsorption-desorption are pH dependent, and therefore the availability of P in fertilizers is dependent on the pH in the soil receiving the fertilizer. For instance, Ca-bound P may be released in acid soils, but remain unavailable in calcareous soils. For Fe- and Al-bound P the opposite will be the case, P will be released at high pH and remain unavailable at low pH. Consequently, plant availability of P in fertilizers with Ca/Mg bound P must be evaluated differently than fertilizers with dominating Fe/Al bound P depending on the soil pH. Further, which chemical extraction best reflects plant availability of P in the fertilizer product is probably dependent on the soil type (calcareous or acid soil) that will receive the fertilizer.

Organic fertilizers may contain significant amounts of low molecular organic acids like citric acid, which increase the availability of P in soil by competition with P for the sorbing sites or by formation of complexes with Fe and Al (Lopez-Hernandez et al. 1979; Amberger and Amman 1984; Mnkeni and MacKenzie 1985). Different sources of organic fertilizers have different concentrations of low molecular organic acids. Øgaard (1996) found a lower concentration of organic acids in composted manure than in fresh manure. The difference in concentration of simple organic acids was reflected in different effects on P sorption in soil. Phosphorus added with fresh and composted manure was both adsorbed to a lesser degree in the soil than added inorganic P, but fresh manure reduced P sorption more than composted manure. Consequently, organic fertilizers may have an effect on P availability in soils, which may not be reflected in P extractions of the product.

Conclusions

Many different extraction procedures exist to estimate the plant availability of mineral P fertilizers. They differ from each other in their ability to solubilize the various P forms occurring in the different fertilizers. While each of them shows its advantages for the type of fertilizer it was originally designed for, it would be desirable to have procedures that make it possible to compare the potential of different types of fertilizer. Water and neutral ammonium citrate are shown to be the best to mirror the soil conditions predominating in temperate European climates (slightly acid to neutral), which makes them the preferable methods to estimate the amount of P from mineral fertilizers going into solution within short time (one vegetation period). The performance of these methods has been well established for mineral fertilizers in many vegetation trials.

Recycling products such as sewage sludge ashes or (meat and) bone meal derived products which are not additionally digested often contain considerable amounts of apatitic P-forms which are not soluble in water and if, at all, only slowly soluble in soil. Different investigations of recycled fertilizer products have shown that conventional extraction methods (such as neutral ammonium citrate or citric acid) may be also used to assess the agronomic potential of such P sources.

For characterization of P availability in organic fertilizers like farmyard manure, biowaste composts and treated sewage sludge, methods developed for characterization of mineral fertilizers, as well as methods originally developed for estimating available P in soils are used. Extraction of P with weak extraction solutions like 0.01 M $CaCl_2$ or with a P sink (iron coated filter paper) have seemed to be better related to plant availability of P in organic fertilizers than conventional acid extraction procedures or alkaline extraction with 0.05 M $NaHCO_3$. The availability of P in organic fertilizers seems to be controlled by their concentrations of Fe, Al and Ca. Consequently, the P availability may be predicted by the concentrations of these elements instead of P extractions.

Which extraction method is best at predicting the amount of plant available P in any type of fertilizer, however, also depends on the type of plant looked at. Some crops are more sensitive to low concentrations of easily releasable P than others, and the ability of the fertilizer to sustain the P concentration in soil solution by chemical equilibrium reactions is important for these crops. E.g. barley has a short period for nutrient uptake and needs easily available nutrients for producing high yields, whereas grass may utilize less soluble P sources when the plants are well established.

Furthermore, availability of P in all types of fertilizer products is dependent on soil pH, and the effect of soil pH is different for Ca-bound compared to Fe- and

Al-bound P. Ca-bound P is most available in acid soils, whereas Fe- and Al-bound P is most available at higher pH. Therefore, it should be investigated if the choice of chemical extraction should depend on the soil type (calcareous or acid soil) that will recieve the fertilizer.

Extraction methods are all static methods for determining a momentary solubility, which may not depict the continuous process of P dissolution taking place under field conditions. Because of a multitude of different fertilizers and growing conditions, chemical extractions will not always yield satisfactory results regarding the correlations between chemical solubility and plant availabilty. Therefore, some authors prefer alternative methods for estimating plant availability of recycled P fertilizers, particularly the isotopically exchangeable P in the fertilized soil, arguing that this parameter gives a direct measure of the amount of fertilizer P that dissolves in the soil solution or reacts with the soil but stays plant available.

References

Abrahamczik E, Niedermaier T (1968) Chemische Prüfverfahren. In: Scharrer K, Linser H (eds) Handbuch der Pflanzenernährung und Düngung, Band 2: Boden und Düngemittel (2. Hälfte), pp 1555–1614

Adam C, Vogel C, Wellendorf S, Schick J, Kratz S, Schnug E (2009) Phosphorus recovery by thermochemical treatment of sewage sludge ash—results of the European FP6-project SUSAN. In: Ashley K, Mavinic D, Koch F (eds) International conference on nutrient recovery from wastewater streams. IWA Publishing, London, pp 417–430

Ajiboye B, Akinremi OO, Racz GJ (2004) Laboratory characterization of phosphorus in fresh and oven-dried organic amendments. J Environ Qual 33:1062–1069

Alt D, Peters I (1992) Die CaCl2/DTPA-Methode zur Untersuchung gartnerischer Erden auf Mengen- und Spurenelemente. 1. Mitteilung: Phosphor, Kalium und Stickstoff. Agribiol Res 45:204–214

Alt D, Peters I, Fokken H (1994) Estimation of phosphorus availability in composts and compoat/peat mixtures by different extraction methods. Commun Soil Sci Plant Anal 25:2063–2080

Amberger A, Amann C (1984) Wirkungen organischer Substanzen auf Boden- und Düngerphosphate. Teil 2: Einfluss verschiedener organischer Stoffe auf die Mobilität von Dünger-P. Z. Pflanzenernaehr. Bodenkunde 147:60–70

Baran E (1985) Zusammensetzung und Duengewirkung von Phosphat—Faellungsprodukten aus der dritten Abwasserreinigungsstufe. Dissertation, Göttingen.

Bartos JM, Mullins GL, Sikora FJ, Copeland JO (1991) Availability of phosphorus in the water-insoluble fraction of monoammonium phosphate fertilizers. Soil Sci Soc Am J 55:539–543

Bingeman CW, Varner JE, Martin WP (1953) The effect of the addition of organic materials on the decomposition of an organic soil. Soil Sci Soc Proc 17:34–38

Brabson JA, Burch jr. WG (1964) How available is available? J Assoc Offic Agric Chem 47:439–442

Braithwaite AC (1987) The use of chemical solubility tests in comparing phosphate fertilizers. Fert Res 12:185–192

Cabeza R, Steingrobe B, Römer W, Claassen N (2011) Effectiveness of recycled P products as P fertilizers as evaluated in pot experiments. Nutr Cycl Agroecosys 91:173–184

Chien SH (1993) Solubility assessment for fertilizer containing phosphate rock. Fert Res 35:93–99

Chien SH, Hammond LL (1978) A comparison of various laboratory methods for predicting the agronomic potential of phosphate rocks for direct application. Soil Sci Soc Am J 42:935–939

Deeley DM, Yeates JS, Gilkes RJ (1987) Chemical testing procedures for poorly soluble P fertilizers used on acid leaching sands. Fert Res 14:101–111

Egnér H, Riehm H, Domingo WR (1960) Untersuchungen über die chemische Boden-Analyse als Grundlage für die Beurteilung des Nährstoffzustandes der Boden. Kungliga Lantbrukshögskolans Annaler 26:199–215

Elliot HA, Potter JM, Kang JH, Brandt RC (2005) Neutral ammonium citrate extraction of biosolids phosphorus. Commun Soil Sci Plant Anal 36:2447–2459

Finck A (1992) Dünger und Düngung. VCH Verlagsgesellschaft mbH, Weinheim. ISBN 3-527-28356-0

Fresenius R, Neubauer C, Luck E (1871) Gutachten über die besten Methoden der Analyse der künstlichen Dünger. Fres J Anal Chem 10:133–158

Frossard E, Tekely P, Grimal JY (1994) Characterization of phosphate species in urban sewage sludges by high-resolution solid-state ^{31}P NMR. Eur J Soil Sci 45:403–408

Frossard E, Sinaj S, Dufour P (1996) Phosphorus in urban sewage sludges as assessed by isotopic exchange. Soil Sci Soc Am J 60:179–182

Frossard E, Bauer JP, Lothe F (1997) Evidence of vivianite in $FeSO_4$-flocculated sludges. Water Res 31:2449–2454

Frossard E, Skrabal P, Sinaj S, Bangerter F, Traore O (2002) Forms and exchangeability of inorganic phosphate in composted solid organic waste. Nutr Cycl Agroecosys 62:103–113

Gabriels R, Verdonck O (1991) Physical and chemical characterization of plant substrates: towards a European standardization. Acta Hort 294:249–259

García-Albacete M, Martín A, Cartagena C (2012) Fractionation of phosphorus biowastes: characterisation and environmental risk. Waste Manag 32:1061–1068

Gericke S (1952) Untersuchungen über die Löslichkeitsverhältnisse der Phosphorsäure in Thomasschlacken. Stahl Eisen 72:1580–1583

Gericke S (1968) Thomasphophat. In: Scharrer K, Linser H (Hrsg.) Handbuch der Pflanzenernährung und Düngung, Band 2: Boden und Düngemittel (2. Hälfte), pp 1168–1202

Gilliam JW, Wright B (1964) Quality of phosphorus in mixed fertilizers as revealed by solubility in water, neutral ammonium citrate, and alkaline ammonium citrate solutions. J Agric Food Chem 12:205–207

Gisiger L, Pulver H (1959) Von den Eigenschaften der Rohphosphate und ihrer Wirkung als Dünger. Agrochimica 103:165–189

Hammond LL, Chien SH, Polo JR (1980) Phosphorus availability from partial acidulation of two phosphate rocks. Fert Res 1:37–49

Hammond LL, Chien SH, Roy AH, Mokwunye AU (1989) Solubility and agronomic effectiveness of partially acidulated phosphate rocks as influenced by their iron and aluminium oxide content. Fert Res 19:93–98

He Z, Honeycutt CV (2001) Enzymatic characterization of organic phosphorus in animal manure. J Environ Qual 30:1685–1692

Hedley MJ, Stewart JWB, Chauhan BS (1982) Changes in inorganic and organic soil phosphorus fractions induced by cultivation practices and by laboratory incubations. Soil Sci Soc Am J 46:970–976

Hedley M, McLaughlin M (2005) Reactions of phosphate fertilizers and by-products in soils. In: Sims TJ, Sharpley AN (eds) Phosphorus: agriculture and the environment. ASA-CSSA-SSSA Publishers, USA, pp 181–252

Higgins DJ (1982) Continuous solubility measurements on superphosphates. In: New Zealand Fertilizer Manufacturers' Research Association (Hrsg.) 7th research symposium: superphosphates and other phosphate fertilizers—a current appraisal. Auckland, 17–18 Nov 1982, pp 19–23

Hignett TP, Brabson JA (1961) Evaluation of water-insoluble phosphorus in fertilizers by extraction with alkaline ammonium citrate solutions. J Agric Food Chem 9:272–276

Hofmann E, Mager D (1951) Über Löslichkeit und Bewertung der Rohphosphate. Zeitschrift für Pflanzenernährung, Düngung und Bodenkunde 52:116–120

Houba VJ, Novozamsky I, Lexmond TM, van der Lee JJ (1990) Applicability of 0.01 M $CaCl_2$ as a single extraction solution for the assessment of the nutrient status of soils and other diagnosis purposes. Commun Soil Sci Plant Anal 21:2281–2290

Joulie MH (1873) Ètude sur le dosage de l'acide phosphorique dans tous les produits qui intéressent l'agriculture et la physiologie. Le Moniteur Scientifique de Quesneville 3:563–584

Junge A, Werner W (1989) Investigations on interactions of phosphorus compounds in partially acidulated phosphate rock and fertilizer effectiveness. Fert Res 20:129–134

Jungk A (1961) Über eine Schnellmethode zur Bestimmung des Nährstoffgehaltes gärtnerischer Erden. Zeitschrift Pflanzenernährung, Düngung, Bodenkunde 95:10–23

Kehres B, Pohle A (1998) Methodenbuch zur Analyse von Kompost. Bundesgütegemeinschaft Kompost e.V. Verlag Abfall Now e.V, Stuttgart. ISBN 3-928179-32-2

Kratz S, Haneklaus S, Schnug E (2010) Chemical solubility and agricultural performance of P-containing recycling fertilizers. Landbauforsch vTI Agric For Res 60:227–240

Leinweber P (1996) Phosphorus fractions in soils from an area with high density of livestock population. Zeitschrift für Pflanzenernährung und Bodenkunde 159:251–256

Leinweber P, Lünsmann F, Eckhardt KU (1997a) Phosphorus sorption capacities and saturation of soils in two regions with different livestock densities in Northwest Germany. Soil Use Manag 13:82–89

Leinweber P, Hintermaier L, BBB (1997b) Sequential extractions and 31P-NMR spectroscopy of phosphorus forms in animal manures, whole soils and particle-size separates from a densely populated livestock area in northwest Germany. Biol Fertil Soils 25(1):89–94

Léon LA, Fenster WE, Hammond LL (1986) Agronomic potential of eleven phosphate rocks from Brazil, Colombia, Perú and Venezuela. Soil Sci Soc Am J 50:798–802

Lopez-Hernandez D, Flores D, Siegert G, Rodriguez JV (1979) The effect of some organic anions on phosphate removal from acid and calcareous soils. Soil Sci 128:321–326

Maercker M (1895) Die Wertschätzung der Thomasmehle. Die landwirtschaftlichen Versuchsstationen 45:378–385

Maguire RO, Sims JT, Coale FJ (2000) Phosphorus solubility in biosolids-amended farm soils in the Mid-Atlantic region of the USA. J Environ Qual 29:1225–1233

Maguire RO, Sims JT, Dentel SK, Coale FJ, Mah JT (2001) Relationship between biosolids treatment process and soil phosphorus availability. J Environ Qual 30:1023–1033

Martens P (1943) Die 2 %ige, mit Calciumcitrat gesättigte Citronensäurelösung und ihre lösende Wirkung auf die Phosphatdüngemittel. Fresenius J Anal Chem 125:245–256

Mengel K, Kirkby EA (1987) Principles of plant nutrition, 4th edn. International Potash Institute, Switzerland 687 p

Miller M, O'Connor GA (2009) Longer-term phytoavailability of biosolids-phosphorus. Agron J 101:889–896

Mnkeni PNS, Mackenzie AF (1985) Retention of ortho- and polyphosphates in some Quebec soils as affected by added organic residues and calcium carbonate. Can J Soil Sci 65:575–585

Müller (1968) Feinvermahlenes Rohphosphat. In: Scharrer K, Linser H (Hrsg.) Handbuch der Pflanzenernährung und Düngung, Band 2: Boden und Düngemittel (2. Hälfte), pp 1129–1139

Mullins G, Joern B, Moore P (2005) By-product phosphorus: sources, characteristics and management. In: Sims TJ, Sharpley AN (eds) Phosphorus: agriculture and the environment. ASA-CSSA-SSSA Publishers, USA, pp 829–880

Mullins GL, Sikora FJ, Bartos JM, Bryant HH (1990) Plant availability of phosphorus in the water-insoluble fraction of commercial triple superphosphate fertilizers. Soil Sci Soc Am J 54:1469–1472

Murphy J, Riley JP (1962) A modified single solution method for the determination of phosphate in natural waters. Anal Chim Acta 26:31–36

Nanzer S, Janousch M, Huthwelker T, Eggenberger U, Hermann L, Oberson A, Frossard E (2009) Phosphorus speciation of sewage sludge ashes and potential for fertilizer production. In: Ashley K, Mavinic D, Koch F (eds) International conference on nutrient recovery from wastewater streams. IWA Publishing, London, pp 609–614

Neubauer H (1931) Die Keimpflanzenmethode nach Neubauer und Schneider. In: Honcamp F (Hrsg.) Düngemittel und Düngung. Handbuch der Pflanzenernährung und Düngerlehre, Band 2. Springer, Berlin, pp 882–902
Neubauer H, Schneider W (1923) Die Nährstoffaufnahme der Keimpflanzen und ihre Anwendung auf die Bestimmung des Nährstoffgehalts der Böden. Zeitschrift für Pflanzenernährung und Düngung A. Wissenschaftlicher Teil 2:329–362
NN (1966) Bericht über die Arbeitstagung der Fachgruppe Düngemitteluntersuchung am 12./13. Mai 1966 in Mannheim. LUFA-Mitteilungen 3:37–38
Olsen SR, Cole C, Watanabe FS, Dean LA (1954) Estimation of available phosphorus in soils by extraction with sodium bicarbonate. USDA Circular 939. U.S. Government Printing Office, Washington D.C
Peplinski B, Adam C, Michaelis M, Kley G, Emmerling F, Simon FG (2009) Reaction sequences in the thermochemical treatment of sewage sludge ashes revealed by X-ray powder diffraction—a contribution to the European project SUSAN. Z Kristallogr Suppl 30:459–464
Petermann A (1880) Ueber den landwirthschaftlichen Werth der sogenannten zurückgegangenen Phosphorsäure. Die landwirtsch Vers 24:310–350
Pinnekamp J, Montag D, Gethke, K, Goebel S, Herbst H (2007) Rückgewinnung eines schadstofffreien, mineralischen Kombinationsdüngers "Magensiumammoniumphosphat—MAP" aus Abwasser und Klärschlamm. UBA-Texte25/07. Forschungsbericht 202 33 308
Prochnow LI, Dillard EF, Austin ER, Chien SH, Calvo CG (2003) Modal analysis to estimate the composition of single superphosphates. Comm Soil Sci Plant Anal 34(15 and 16):2131–2147
Quin BF (1982) The use of citric-acid soluble and water-soluble phosphate to assess the agronomic value of fertilizers. In: New Zealand Fertilizer Manufacturers' Research Association (Hrsg.) 7th research symposium: superphosphates and other phosphate fertilizers—a current appraisal. Auckland, 17–18 Nov 1982, pp 40–54
Russell WE (1982) Measurement of phosphate availability and statutory requirements. In: New Zealand Fertilizer Manufacturers' Research Association (Hrsg.) 7th research symposium: Superphosphates and other phosphate fertilizers—a current appraisal. Auckland, 17–18 Nov 1982, pp 9–18
Sarkar D, O'Connor GA (2004) Plant and soil responses to biosolids-phosphorus in two Florida soils with high phosphorus content. Commun Soil Sci Plant Anal 35:1569–1589
Schachtschabel P (1960) Umwandlung der Düngerphosphate im Boden und Verfügbarkeit des Bodenphosphors. Landwirtsch Forsch SH 14:30–37
Schachtschabel P (1976) Kapitel XI: Nährstoffe. In: Scheffer/Schachtschabel (Hrsg.) Lehrbuch der Bodenkunde, pp 196–264
Scheel KC (1968a) Glühphosphate. In: Scharrer K, Linser H (eds) Handbuch der Pflanzenernährung und Düngung, Band 2: Boden und Düngemittel (2. Hälfte), pp 1203–1216
Scheel KC (1968b) Superphosphate. In: Scharrer K, Linser H (eds) Handbuch der Pflanzenernährung und Düngung, Band 2: Boden und Düngemittel (2. Hälfte), pp 1140–1167
Scheffer F (1956) Die "wirksame' Phosphorsäure bestimmt den Pflanzenertrag. Phosphorsäure 16:105–120
Schick J (2009) Untersuchungen zu P-Düngewirkung und Schwermetallgehalten thermochemisch behandelter Klärschlammaschen. Ph.D. dissertation, Technical University Braunschweig
Schilling G (2000) Pflanzenernährung und Düngung. Verlag Eugen Ulmer, Stuttgart
Schleede A, Meppen B, Jörgensen OB (1939) Zur Frage der Citronensäurelöslichkeit von Naturphosphaten (Apatiten). Zeitschrift für Angewandte Chemie 52:316–319
Schmitt L (1969a) Über die Beziehungen zwischen chemischer Löslichkeit und pflanzenphysiologischer Wirkung verschiedener Phosphatformen. 1. Mitteilung. Landwirtschaftliche Forschung 22:109–115
Schmitt L (1969b) Über die Beziehungen zwischen chemischer Löslichkeit und pflanzenphysiologischer Wirkung verschiedener Phosphatformen. 2. Mitteilung. Landwirtschaftliche Forschung 22:314–325
Schnug E, Rogasik J, Haneklaus S (2003a) Quantitative und qualitative Anforderungen der deutschen Landwirtschaft an Phosphor für die Düngung. In: Institut für

Siedlungswasserwirtschaft der RWTH Aachen, Umweltbundesamt (Hrsg.) Rückgewinnung von Phosphor in der Landwirtschaft und aus Abwasser und Abfall. Tagungsband zum Symposium am 06.-07. Februar 2003, Berlin, pp 1/1–1/14

Schnug E, Rogasik J, Haneklaus S (2003b) Die Ausnutzung von Phosphor aus Düngemitteln unter besonderer Berücksichtigung des ökologischen Landbaus. Landbauforsch Völkenrode 53:1–11

Schueller H (1969) Die CAL-Methode, eine neue Methode zur Bestimmung des pflanzenverfügbaren Phosphates in Boden. Z. Pflanzenernahr., Düngung, Bodenkde. 123:48–63

Sharpley A, Moyer B (2000) Phosphorus forms in manure and compost and their release during simulated rainfall. J Environ Qual 29(5):1462–1469

Shober AL, Hesterberg DL, Sims JT, Gardner S (2006) Characterization of phosphorus species in biosolids and manures using XANES spectroscopy. J Environ Qual 35:1983–1993

Sikora JF, Copeland JP, Mullins GL (1992) Apparent solubility products of phosphorus impurity compounds in commercial monoammonium phosphate fertilizers. Soil Sci Soc Am J 56:402–407

Sikora JF, Dillard EF, Copeland JP, Bartos JM (1989) Chemical characterization and bioavailability of phosphorus in water-insoluble fractions of three mono-ammonium phosphate fertilizers. J Assoc Off Anal Chem 72:852–856

Sikora JF, Mullins GL (1995) Bioavailability of citrate-insoluble phosphorus in monoammonium phosphate and triple superphosphate fertilizers. Soil Sci Soc Am J 59:1183–1188

Stephen RC (1982) The relationship between phosphorus solubility of single superphosphate in different extractants and the dry matter production of ryegrass in pots. In: New Zealand Fertilizer Manufacturers' Research Association (Hrsg.) 7th research symposium: superphosphates and other phosphate fertilizers—a current appraisal. Auckland, 17–18 Nov 1982, pp 24–31

Stephen RC (1985) The agronomic value of products resulting from the partial acidulation of North Carolina phosphate rock with phosphoric acid. Fert Res 8:67–73

Stephen RC, Condron LM (1986) An assessment of the agronomic efficiency of partially acidulated phosphate rock fertilizers. Fert Res 10:269–282

Terman GL, Hoffman GW, Wright BC (1964) Crop response to fertilizers in relation to content of "available" phosphorus. Adv Agron 16:59–100

Tiessen H, Stewart JWB, Moir JO (1983) Changes in organic and inorganic phosphorus composition of two grassland soils and their particle size fractions during 60–90 years of cultivation. J Soil Sci 34(4):815–823

Timmermann F, Kluge R, Stahr K, Zauner G (1999) Projekt Abfall Wasser Boden (PWAB), Forschungsvorhaben PW 95171 – Erarbeitung von Grundlagen für Anwendungsrichtlinien zur Verwertung geeigneter Rest- und Abfallstoffe im landwirtschaftlichen Pflanzenbau (Ackerbau). Abschlussbericht

Van der Zee SEATM, Fokkink LGJ, Van Riemsdijk WH (1987) A new technique for assessment of reversibly adsorbed phosphate. Soil Sci Soc Am J 51:599–604

Velthof GL, van Beusichem ML, Raijmakers WMF, Janssen BH (1998) Relationship between availability indices and plant uptake of nitrogen and phosphorus from organic products. Plant Soil 200:215–226

Werner W (1967) Die Rhenania-Dünger. Monographie über Herstellung, Eigenschaften und Wirkung der Rhenania-Dünger. Verlag M. and H. Schaper, Hannover

Ylivainio K, Uusitalo R, Turtola E (2008) Meat bone meal and fox manure as P sources for ryegrass (*Lolium multiflorum*) grown on a limed soil. Nutr Cycl Agroecosyst 81:267–278

Øgaard AF (1996) Effect of fresh and composted cattle manure on phosphate retention in soil. Acta Agric Scand Sect B Soil Plant Sci 46:98–105

Øgaard AF (2013) Plantetilgjengelig fosfor i avløpsslam—testing av analysemetodikk for tilgjengelig fosfor. Bioforsk Rapp 8(34):23 p (In Norwegian)

Variability of P Uptake by Plants

Kerstin Panten, Frauke Godlinski, Susanne Schroetter and Maximilian Hofmeier

Abstract Phosphorus (P) uptake affects the planning of P fertilisation of agricultural crops and subsequently the economic evaluation of farming practices and the quantification of agri-environmental burdens. The P uptake of plants is influenced by soil type, soil available P, availability of other essential nutrients, climate and weather conditions as well as by the nutrient use efficiency of the grown cultivar. Phosphorus coefficients are mainly derived from P fertiliser experiments, reflecting regional soil and weather conditions as well as farming practice. P coefficients commonly used for a single crop can vary substantially and hence rarely reflect the actual P uptake rates of crops. However, the main driving factor in P budgets is the yield level rather than the P concentration in plants. While it is difficult to determine a 'correct' P coefficient for all crops from which to calculate reliable agri-environmental indicators, it is important for farmers to be able to calculate their nutrient budgets as accurately as possible to avoid economic losses and environmental burdens.

Keywords Crop cultivar · Crop genotype · Spatial variability · Phosphorus coefficient · Phosphorus balance · Phosphorus uptake · Phosphorus availability

Introduction

Phosphorus concentrations of plants are used to estimate the plant uptake of P (P uptake = P concentration × dry matter). Based on such values, the P fertilisation of agricultural plants can be planned, economic efficiency of farming practices evaluated and agri-environmental burdens quantified. The accumulation and depletion of soil P is an indicator of the effectiveness of management practices to meet the plant growth needs in a cropping system (Bundy et al. 2005). Environmental impacts of fertilisation practices are frequently investigated through the calculation

K. Panten (✉) · F. Godlinski · S. Schroetter · M. Hofmeier
Federal Research Centre for Cultivated Plants, Institute for Crop and Soil Science, Julius Kühn-Institut (JKI), Bundesallee 50, 38116 Braunschweig, Germany
e-mail: kerstin.panten@julius-kuehn.de

E. Schnug and L.J. De Kok (eds.), *Phosphorus in Agriculture: 100 % Zero*,
DOI 10.1007/978-94-017-7612-7_8

of P budgets at farm, region or country level and require reliable P uptake coefficients. These P coefficients are mainly derived from individual experiments. Fertilisation practices on cultivated crops have a major influence on plant growth and hence on yield, but also on the environment as a result of nutrients leaching into ground water or being transported by runoff into surface waters (Klatt et al. 2003; Oenema et al. 2005; Withers and Haygarth 2007). Whilst over-fertilisation in some areas increases environmental P loads, large agricultural areas in the world, mainly in tropical and subtropical regions, are characterized by soil P deficiency constraining plant growth and yield (Stangel and von Uexküll 1990; Tan et al. 2005; Simpson et al. 2011). The 'minimum law' of Liebig describes the effect, that crop yield is limited by the nutrient least available (Mitscherlich 1909). This 'law' was extended by the fact that other factors, mainly soil physics, water availability, climate, and weather conditions also influence plant growth and limit crop yield (Mitscherlich 1909; Bates 1971; Fisher 1980). If the plant growth is restricted below its potential by one of these factors, there is an increased risk of a reduced uptake of plant available nutrients such as P and nitrogen. This may lead to a surplus that is subsequently at risk of leaching and runoff into ground and surface waters. Such nutrient losses are to a certain extent unavoidable particularly considering the unforeseeable weather conditions during the vegetation period at the time of fertilisation. On the other hand over-application of P can directly contribute to losses into water bodies and result in eutrophication. Maximising the efficiency of P utilisation will minimise the P fertiliser rate required for crop production (Helyar 1998) and therefore contribute to the reduction of nutrient losses.

Actual P fertilisation recommendations in developed countries are mainly based on the available P status of soils; hence soil analysis is widely used for soil fertility assessments (Withers et al. 2005; DÜV 2006; Weld and Sharpley 2006). In countries where the analysis of the available P status in soils is mandatory, such analyses are conducted mostly only every couple of years. For example, the German Fertiliser Application Ordinance (DüV 2006) requires for each field above one hectare in size one P analysis every six years. In addition to the available soil P, farmers use the knowledge about P concentrations of crops to calculate the P uptake of previous crops as well as the expected uptake of actual crops in their fertilisation plans. For farmers it is therefore important to know that P coefficients provided reflect the actual P uptake by crops. It is also important to know which crop parts are removed from the field and farm and which will remain because of variations in P concentrations in different plant parts of the aboveground biomass. Karlen et al. (1988) showed e.g. that aboveground P uptake of corn at physiological maturity in lower leaves, upper leaves, stem and tassel, and ear and shank was approximately 13, 10, 9, and 59 kg ha^{-1} P, respectively. Roots of most crops remain in the soil and can therefore be neglected when considering P removal from agricultural soils. Then again, the above ground biomass can take a variety of pathways into e.g. food and feed production, animal housing, or be used as a source of renewable energies. For the calculation of P removal from soils and for the subsequent fertilisation it makes a significant difference whether e.g. corn is harvested as a grain source or as silage corn for animal feeding or energy production.

This chapter will provide information about the variation in P concentrations of various major crops grown worldwide. According to the Food and Agriculture Organization of the United Nations (FAO) major crops produced worldwide in terms of tonnages were sugar cane, corn, paddy rice, wheat and potatoes in 2012 (FAO 2014a). In Europe, wheat was the major crop grown in 2012, followed by sugar beet, potatoes, corn and barley (FAO 2014b). The total area harvested worldwide was dominated by agricultural sites for wheat, corn and paddy rice production in 2012 (FAO 2014c). The focus of this chapter will therefore be on the P uptake of wheat, corn and rice.

Common P Coefficients in Use

Phosphorus coefficients are commonly used in agricultural statistics for the calculation of nutrient budgets at field, farm, region, and country level. These P coefficients are derived mainly from P fertiliser experiments, reflecting regional soil and weather conditions as well as farming practice. Therefore, they have an important influence on national nutrient budgets, for example the German gross P budget (Fig. 1), which is used as agri-environmental indicator (Panten et al. 2009). Farmers also use these coefficients to estimate the expected P uptake by crops at a certain yield level followed by the calculation of P fertiliser rates as well as for the planning of animal feedings. Whilst P coefficients for P uptake of plants are mostly given for fresh matter (FM) coefficients provided for animal feeding refer to dry matter (DM).

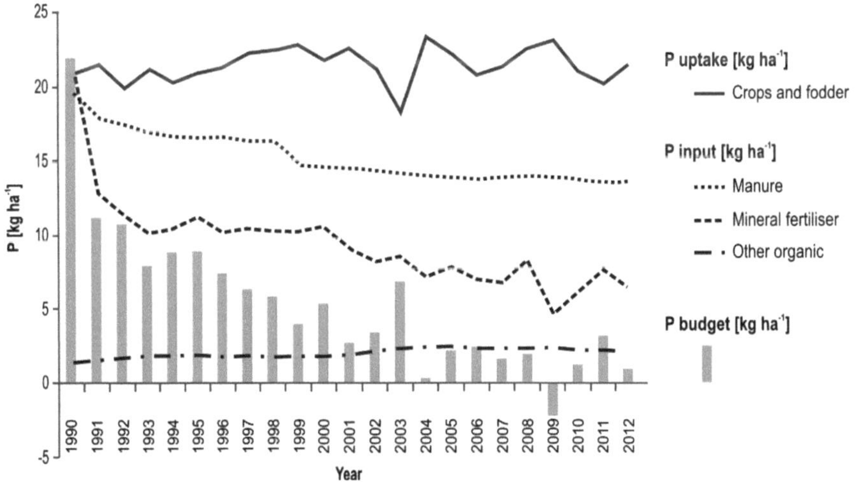

Fig. 1 German gross P budget (1990–2012) calculated according to the methodology of Eurostat/OECD (Eurostat 2013) using common P coefficients for Germany as well as some P input and output parameters

For better comparisons coefficients were sometimes recalculated. Table 1 provides an overview of commonly used P coefficients of some OECD countries for the calculation of gross nutrient budgets.

It can be seen that the P coefficients for a given crop vary substantially between countries as well as that the P budgets of countries calculated with such P coefficients show a wide range of P surpluses and deficits (Table 1). Some reported P budget surpluses for 2000–2004 were very high, like e.g. for Japan and Korea whilst many countries had nearly balanced P budgets. Most of the used P coefficients have been determined using data from field experiments, averaging the

Table 1 Phosphorus coefficients used for national gross P budgets in 2004 in selected OECD countries for wheat, rice and corn (FM) and the average P budgets between 2000 and 2004 in kg ha^{-1} calculated according to the OECD standards (OECD 2010)

Country	Wheat (grain)	Rice (grain)	Corn (grain)	Corn (silage)	Avg. P budget 2000–2004 (kg ha^{-1})
	P coefficients to convert harvested crops into P uptake (%)				
Australia	0.26	0.24	0.23	–	1
Austria	0.35	–	0.35	0.07	3
Belgium	0.34	–	0.29	0.05	25
Canada	0.37	–	0.33	0.25	1
Czech Republic	0.41	–	0.31	0.04	2
Denmark	0.30	–	–	0.07	11
France	0.40	0.22	0.31	0.24	5
Germany	0.35	–	0.35	0.07	5
Greece	0.34	0.30	0.32	0.09	4
Hungary	0.41	0.37	0.31	0.06	−1
Ireland	0.35	–	–	–	7
Japan	0.34	0.33	0.15	0.09	52
Korea	0.28	0.22	0.20	0.13	48
Mexico	0.35	0.28	0.30	0.11	1
Netherlands	0.34	–	0.33	0.20	22
Poland	0.35	–	0.63	0.06	2
Portugal	0.34	–	0.32	0.09	16
Slovak Republic	0.39	–	0.35	0.06	1
Switzerland	0.38	–	0.26	0.07	5
Turkey	0.38	0.30	0.32	–	6
United Kingdom	0.38	–	–	0.11	14
United States	0.36	0.28	0.27	0.11	3

Data was rounded to two decimal digits

known variation between years, climate, soil, and cultivars. Nonetheless, those values are questioned considering their importance for the farm management when calculating nutrient uptake followed by decisions about fertilisation practices. Interestingly, different coefficients are used for the calculation of the nutrient uptake of animals through fodder. For example, several regional offices for agriculture in Germany provide feed composition tables to farmers. These tables state for instance ranges from 0.28 to 0.34 % P in wheat grains (FM) and 0.24–0.29 % P in corn grains (FM; LLH 2009; LKSH 2011; LfL 2013). According to Flachowsky (2006) it is recommended that farmers have their feed analysed because of the known discrepancies between the feed composition tables and actual nutrient concentrations in feed. Unfortunately, the provided coefficients of feed composition tables are those generally used by farmers. Similar to Germany, the UK has published different P coefficients according to the latter use. Whilst for the OECD P budget 0.375 % P in winter wheat (FM) is used, Watson et al. (2009) recommend to farmers to use 0.29 % P concentration for the calculation of P budgets.

In terms of the P uptake of crops, Binford (2010) demonstrated on 668 corn grain samples of Delaware and Maryland (USA) that P removals from fields varied considerably between different fields but also within individual fields concluding that average P uptake should be used for the development of nutrient management plans. Furthermore, the results of these samples lead to the conclusion that the P removal concentrations officially published in the USA by universities and governmental agencies may be too high (Binford 2010). Winter wheat grain samples collected between 1995 and 2011 in northern Germany (50.35°–54.40°N; 9.48°–12.07°E) show a wide range of P concentrations (Fig. 2). Interestingly, the average P concentration of these 556 observations was only 0.23 % P (FM) which would result in a considerably lower P uptake by winter wheat crops than average values reported in

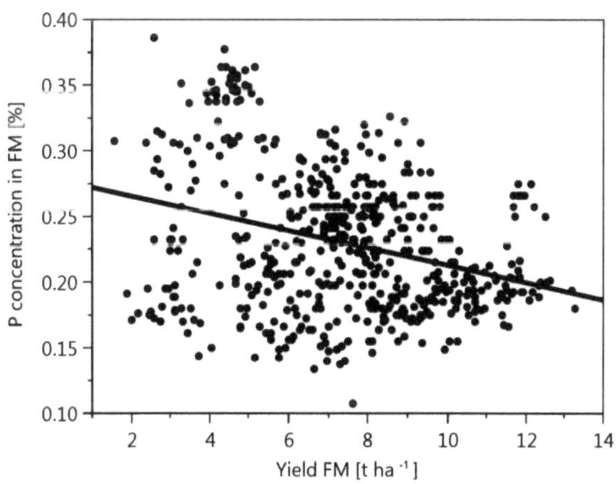

Fig. 2 Variation of P concentrations in winter wheat grain (FM = 86 % DS) sampled between 1995 and 2011 in northern Germany on various field experiments and on-farm trials (P concentration = 0.2794 − 0.0066 * FM yield; unpublished data)

the DüV and commonly used to calculate national P balances (0.35 %, Table 1), to plan fertilisation strategies (e.g. 0.35 %, LVLF 2008), and to calculate animal nutrition values (0.29 % LLH 2009; 0.34 %, LKSH 2011; 0.28 %, LfL 2013) in Germany. The following subchapters will provide an overview of some causes and ranges of such variations.

Variability of P Uptake

Unlike the values for P coefficients, references about P efficiency and plant uptake provide P concentrations related to the DM of crops. In the following provided values are therefore on a DM basis unless otherwise indicated.

Variability Between Cultivars and Genotypes

Plant breeding has two main targets. First, the creation of new products with advanced properties and second, the reduction of the economical and ecological production costs of a specific crop. Several types of improvement can decrease production costs. A new variety needs either to be able to produce the same amount of yield using less inputs e.g. fertilisers or provides a yield increase under the same production costs then another variety (Diepenbrock et al. 2009). Plant breeders aim therefore for particular objectives, e.g. yield increase, higher nutrient use efficiency, and better stress tolerance against insects, pathogens, frost, drought or heat. Batten (1992) discusses that there are economic reasons for seeking genotypes which achieve high yields of grain with a low concentration of P, removing less P from the soil. Increasing the nutrient use efficiency of crops has seldom been the main target of plant breeding focusing mainly on yield improvements (Hawkesford 2011). Meanwhile, it has been demonstrated for many plant species that under P limiting conditions root and shoot responses are genetically variable (Vance 2011). Early trials conducted by Gregory et al. (1928), Smith (1934), and Lyness (1936) demonstrate already the variation in P uptake and P concentrations between crop varieties in barley and corn. Such findings led to the attempts to breed varieties with higher P use efficiency (Sarić 1987; Jones et al. 1989; Manske et al. 2001; Manschadi et al. 2014) which is most important for regions where available soil P is not only limited by fertilisation but also by soil properties such as the low pH values of acidic soils. Such soils are limited by low levels of P, Ca and Mg as well as by toxic levels of Al and Mn. Furthermore, several studies concluded that low seed P concentrations impair seedling vigour in cereals and that nutrient reserves can affect the competitiveness of annual species (e.g. Zhang et al. 1990; Ros et al. 1997; Zhu and Smith 2001). These studies investigated seedling growth on infertile soils, and did not exclude seeds which might differ in more than the P concentration. Therefore, Rose et al. (2012) compared high P and low P containing rice seeds from several genotypes and detected genotype-specific responses but no reduction in seedling vigour, biomass and grain yields at maturity of plants. Nadeem et al. (2012) confirmed for corn, that P was not a limiting factor for seedling P nutrition during the first four weeks of early growth because of phytate hydrolysis and remobilisation of nonphytate P. Rose et al. (2012) suggested the need for research to define the minimum P levels in seeds required for adequate seedling growth across a range of environments.

Being grown in many different climatic regions and on a vast variety of soils the P use efficiency and the range of P concentrations of wheat, rice and corn cultivars has been investigated widely (Tables 2 and 3). Some studies investigated the P efficiency during early crop growth. Schnug and Strampe (1988) analysed P concentrations of 15 winter wheat varieties at shooting stage (0.37–0.42 %) and found differences in P concentrations of about 20 % assuming that such a difference may also represent a comparable difference in P uptake. Yaseen and Malhi (2011) explored 46 wheat genotypes for their tolerance to P deficiency stress and concluded that the P uptake was influenced by genotype and P fertilisation level during early growth stages. Rosa (1991) investigated the yield response of 10 wheat cultivars to P fertilisation on acidic soils with low P availability and toxic levels of Al and/or Mn in Brazil by growing wheat shoots in four different nutrient solutions. The shoots were harvested after 15 days. Phosphorus concentrations of plants varied between 0.48 and 0.65 % when grown in low P containing solutions (3.875 mg l^{-1} P) and between 0.83 and 1.11 % in high P containing solutions (15.5 mg l^{-1} P) leading to the conclusion that some cultivars were more efficient in P utilisation than others (Rosa 1991). The P use efficiencies of 22 corn genotypes with high efficiency ratios were investigated by Baligar et al. (1997) during a growth period of 28 days on P deficient acidic tropical soils containing high levels of Al and Mn. At the end of the experiment a wide range of P use efficiencies was recorded demonstrating that genotypes at least at the beginning of their growth varied considerably in their P uptake (Baligar et al. 1997).

For the calculation of nutrient budgets and to decide on fertilisation rates it is more important to investigate the variation in final P uptake between cultivars. Dikeman et al. (1982) evaluated the variation of P concentrations in different winter wheat cultivars across the Great Plains of the USA by merging the same quantity of grain samples of the same variety grown at 23 different locations. In 1973 the P concentration ranged from 0.415 to 0.470 % for 10 cultivars and in 1979 from 0.385 to 0.496 % for nine cultivars tested (Dikeman et al. 1982). A decade later, Batten (1994) analysed wheat grain samples collected from grain terminals representing different wheat grades (e.g. Prime Hard, Soft, Hard Red Winter, Soft Red Winter) from Australia, North America, and the UK in order to compare their nutrient concentrations. A wide range of P concentrations was found in these

Table 2 Variation of P concentrations between cultivars or genotypes during vegetative growth of major field crops reported in literature

Crop	Number of tested cultivars/genotypes	Growth stage	Range of P concentrations in DM (%)	Reference
Corn	9	19 days after sowing	0.097–0.455 Low: 0.097–0.118 Medium: 0.274–0.335 High: 0.383–0.455	Fageria and Baligar (1997a)/three fertilisation levels
Rice	20	28 days after sowing	0.261–0.328	Fageria and Baligar (1997b)
Rice	5	35 days after germination	0.344–0.692	Hafeez et al. (2010)
Wheat	10	15 days after planting	0.48–1.11 3.875 mg l^{-1} P: 0.48–0.65 15.5 mg l^{-1} P: 0.83–1.11	Rosa (1991)/two fertilisation levels
Wheat	8	14 days after planting	0.52–0.64	Gahoonia and Nielsen (1996)
Wheat	49	41 days after planting	0.12–1.19 Deficient: 0.12–0.18 Moderate: 0.18–0.28 Sufficient: 0.69–1.19	Osborne and Rengel (2002)/three fertilisation levels
Bread wheat	39	39 days after planting	0.157–0.449 20 mg kg^{-1} P: 0.157–0.229 80 mg kg^{-1} P: 0.280–0.449	Ozturk et al. (2005)/two fertilisation levels
Durum wheat	34	39 days after planting	0.153–0.474 20 mg kg^{-1} P: 0.153–0.223 80 mg kg^{-1} P: 0.330–0.474	Ozturk et al. (2005)/two fertilisation levels
Wheat	15	Data not available	0.310–0.385	Fageria et al. (2011)
Wheat	46	28 days after transplanting	Deficient: 0.142–0.526 Adequate: 0.651–1.319	Yaseen and Malhi (2011)/ two fertilisation levels

Table 3 Variation of grain P concentrations between cultivars or genotypes at harvest of major field crops reported in literature

Crop	Number of tested cultivars/genotypes	Range of P concentrations in DM (%)	Reference
Corn	4	0.40–0.57	Lyness (1936)
Corn	6	0.25–0.30	Vyn and Tollenaar (1998)
Corn	10	0.22–0.54	Heckman et al. (2003)
Corn	(668 samples)	0.190–0.516	Binford (2010)
Corn	10	0.28–0.35 (2006) 0.34–0.51 (2007)	Ferreira et al. (2012)
Corn	20	0.20–0.43 Low: 0.24–0.43 High: 0.20–0.38	Bayuelo-Jiménez and Ochoa-Cadavid (2014)/two fertilisation levels
White rice	8	0.096–0.121	McCall et al. (1953)
Rice	100	0.04–0.36	Hussain et al. (1987)
Rice	15 (1991) 17 (1992)	*1991*: 0.051–0.299 Low: 0.051–0.299 Medium: 0.057–0.282 High: 0.139–0.280 *1992*: 0.146–0.388 Low: 0.179–0.388 Medium: 0.205–0.322 High: 0.146–0.363	Fageria et al. (1995)/two years; three fertilisation levels
Brown rice	–	0.24–0.31	Marr et al. (1995)/90 samples of Amaroo brown rice
Milled rice	8	0.32–0.43	Ren et al. (2006)
Wheat	38	0.283–0.524	Lolas et al. (1976)
Wheat	38	0.13–0.41	Erdman and Moul (1982)
Wheat	27	Flour: 0.065–0.110 Bran: 0.571–0.840	Peterson et al. (1983)/flour and bran
Wheat	9	0.189–0.600 2 kg P: 0.201–0.600 10 kg P: 0.189–0.353 40 kg P: 0.344–0.465 80 kg P: 0.393–0.538	Jones et al. (1992)/four fertilisation levels
Wheat	7	0.32–0.46	Calderini et al. (1995)
Wheat	4	0.349–0.368	Barrier-Guillot et al. (1996)
Wheat	8	0.29–0.40	Schulthess et al. (1997)
Wheat	30	0.119–0.333	Gill et al. (2004)
Wheat	6	0.35–0.44	Steiner et al. (2007)

samples. For example, Australian Soft Wheat collected from seven ports ranged from 0.199 to 0.339 % P in grains (Batten 1994). For calcareous Indian soils, Gill et al. (2004) demonstrated that the P concentration varies substantially between 30 wheat cultivars ranging from 0.119 to 0.333 % which is even wider than the range observed by Batten (1994) in Australian Soft Wheat but narrower than the established range across all tested samples of 0.199–0.454 % P. In rice similar ranges were found. For rough, brown, and milled rice total P concentrations of 0.17–0.39, 0.17–0.43, and 0.08–0.15 %, respectively, are provided in an overview by Juliano and Bechtel (1985). Grain samples of Amaroo brown rice collected from individual farms in three rice growing regions in Victoria and New South Wales (Australia) contained a range of 0.24–0.31 % P (Marr et al. 1995). Wider ranges of P contents were found in corn grain samples by Heckman et al. (2003) who reported P concentrations of 0.22–0.54 % in ten different hybrids grown in five different US states. In a different study conducted by Ferreira et al. (2012) ten tested corn cultivars of five different technological levels (landrace variety, commercial variety, and double, triple, and single cross-hybrids) showed genetic differences in grain P concentrations. However, these results were superimposed by weather effects during the growth period of two consecutive years in a subtropical environment. Earlier experiments by Baker et al. (1970) on corn hybrids with very low, low, medium, and high P accumulation capacity already showed different efficiencies to utilise soil P. These findings were confirmed by Chen et al. (2012) on two inbred lines known for their difference in P efficiency. Corn has gained interest not only as a staple food for humans but also as animal feed as well as a biomass source for energy plants leading subsequently to an increase in corn cultivation. Therefore, not only the variation in grain P concentrations are of interest, but also the total above ground P uptake in cases of silage maize or biomass production. The nutrient uptake of different corn plant parts (lower leaves, upper leaves, stalk and tassel, ear and shank) was measured by Karlen et al. (1988) in a study with very high (336 kg ha^{-1} y^{-1} P) fertiliser application rates. They showed that at harvest in contrast to other plant nutrients about 85 % of the assimilated P had accumulated in the ear and shanks of the examined variety (Karlen et al. 1988). Consequently, the difference in terms of P removal between corn grain and whole plant harvest is small. A study from Ning et al. (2013) compared the P accumulation and post-silking shifts from vegetative organs into grains of six old and new corn varieties differing in leaf longevity. In this experiment, new varieties took up more P after silking than old ones, but all varieties used large amounts of pre-silking P reserves of the vegetative organs for the grain formation (Ning et al. 2013). Leaf longevity increases yield due to a longer period of high photosynthetic activity, unfortunately, maximal yields seem associated with low nutrient use efficiency (Gregersen 2011). Conclusively it can be stated that the P use efficiency and uptake by crops is not only influenced by environmental factors but also by the chosen cultivar.

Because factors influencing the P uptake are not easily separated, most studies on the P uptake ability of crops investigated not only the differences between

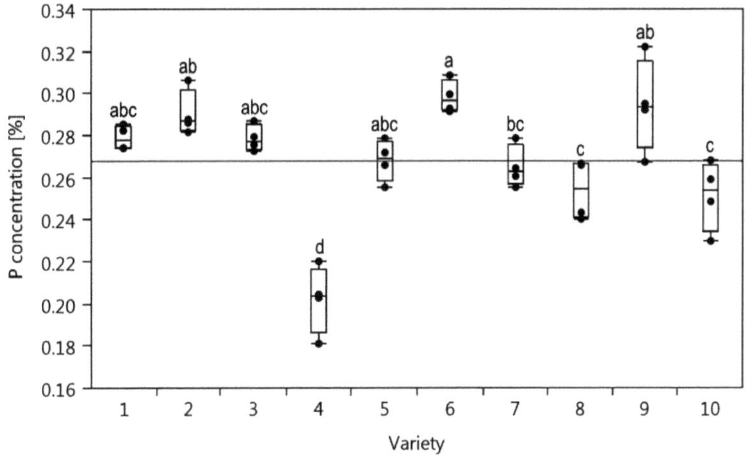

Fig. 3 Variation of P concentrations in soft winter wheat grain samples of 10 varieties tested across two soil types (unpublished data). Levels not connected by the same letter are significantly different (Tukey-Kramer t-test). The mean value of all 10 varieties is indicated by the grey line. Samples provided by the Lower Saxony Agricultural Chamber from a winter wheat variety trial conducted in northern Germany in 2013

cultivars but also at the same time their behaviour under a variety of soil P conditions. Jessop et al. (1984) evaluated the P uptake of 22 wheat varieties including older lines, conventional height and semi-dwarf varieties and material from breeding programmes grown under low (2 kg ha^{-1} P) and high (40 kg ha^{-1} P) P conditions. Results of this experiment showed that the grain P concentration of two wheat cultivars varied between 0.295 % (low P) and 0.359 % (high P) for the variety Charter, and between 0.175 % (low P) and 0.294 % (high P) for Narrabri Sun 66M (Jessop et al. 1984). In 2013, a variety trial of the Lower Saxony Agricultural Chamber (Germany) was evaluated. Ten high to very high yielding soft winter wheat cultivars showed significant differences in P concentrations across two different soil types (sandy and loess loamy soil), both soils well supplied with P. The variation of the P concentration in grain samples ranged from 0.202 to 0.298 % P (Fig. 3), although these 10 varieties reflected only a minority of the licensed varieties in Germany. All of these analysed grain samples were well below the coefficient of 0.41 % P (0.35 % P in FM) used for the calculation of nutrient budgets in Germany. Soil effects could not be observed in this experiment which is not surprising given the sufficient availability of soil P.

To test the effect of soil (calcareous versus acid), fertilisation (0 versus 35 kg ha^{-1} P) and genotypes (30 spring wheat varieties), two field trials were set up by Manske et al. (2001) between 1993 and 1995 in Mexico. Results of these experiments showed that the genotypes had significant interactions by year, soil, and P level (Manske et al. 2001). The effect of an increasing P level on 25 rice cultivars was tested by Fageria et al. (1988) finding significant differences in P use efficiencies. Wissuwa and Ae (2001) analysed genotypic variation of 30 rice varieties grown on P deficient and annually P fertilised (60 kg ha^{-1}) soils and identified a high variation for P uptake ranging from 0.6 to 12.9 mg P per plant on the P deficient soil and from 17.3 to 66.7 mg P per plant on the fertilised soil. The P uptake per plant was strongly dependent on the available soil P resulting in a 4 to 40 fold P uptake from fertilised soils. Considering the variation of P concentrations of wheat, rice and corn cultivars and genotypes (Tables 2 and 3) it can be expected that such differences will also exist for other agricultural crops.

It is not only the total P concentration that considerably impacts on the effective use of P in human and animal nutrition but also the form of P. Humans and nonruminant animals excrete most of the phytic acid P they consume which impacts on the nutrition as well as on the environment (Bryant et al. 2005). Barrier-Guillot et al. (1996) reported for wheat crops receiving higher doses of P than required that these seemed to store the excessive P in the phytic form. To test if cultivars can also have genetic differences in their phytic acid P content, Bryant et al. (2005) compared two rice genotypes, a control and a low phytic acid mutant of it. The total P concentration ranged from 0.31 to 0.36 % P for the control and from 0.32 to 0.36 % P for the low phytic acid mutant in the years 2000–2002. Phytic acid P was in all three years reduced by 42–45 % in the low phytic acid genotype (Bryant et al. 2005). The reduction in phytic acid P was accompanied by an almost equivalent increase in inorganic P. Ren et al. (2007) confirmed these findings using the low phytic acid mutant of a different rice variety. Bryant et al. (2005) conclude that the nutritional value of white rice and rice bran as human food or animal feed can be enhanced by the use of low phytic acid varieties.

Spatial Variability of P Uptake by Crops

With the introduction of variable rate (VR) technology as a management tool for precision agriculture in the early 1990s, variable rate P fertilisation was discussed (Mallarino and Schepers 2005). Common fertiliser recommendations based on yield data and soil P values can easily be transferred into spatially VR maps for P application. Given the environmental burdens caused by over fertilisation of P, spatially variable P application rates according to soil P levels and crop needs would reduce these effects (Sharpley 1995; Pote et al. 1999; Syers et al. 2008). Conventional farming practices with uniform fertilisation rates can increase the naturally occurring soil heterogeneity throughout the years if other factors like e.g. water availability lead to reduced P uptake in parts of the field. Furthermore, remaining soil P can be transported from some parts of the field and accumulated in others. Even though acquired knowledge about spatial heterogeneity led to a better understanding of sampling strategies (Pocknee et al. 1996; Marchant and Lark 2007; Lark 2009; Oliver 2010; Vašát et al. 2010), most fields are sampled, if at all, following simple sampling schemes as e.g. provided by Finck (1979) leading to average soil P values neglecting spatial variability. Guided sampling based on spatially varying yield data will be much more beneficial to farmers allowing them to adjust their fertilisation maps (Johnston 2005). Since soil heterogeneity of e.g. available soil P across agricultural fields is well known (Table 4) (e.g. Schnug et al. 1985; Cahn et al. 1994; Kristensen et al. 1995; Nolin et al. 1996; Mallarino et al. 1998; Schepers et al. 2000; Mallarino and Schepers 2005), the question arises to what extend the spatial variation influences the P uptake of plants.

Bullock and Lowenberg-DeBoer (2007) stated correctly that 'little is known about how yields respond to inputs', even though yield response functions have been calculated for a long time, site-specific yield response functions are still in their early development. From the findings of variety trials on different soils and on

Table 4 Top soil available P ranges of single fields of various sizes as reported in literature

Soil type[a]	Field size (ha)	Top soil[b] P minimum ($mg\ kg^{-1}$)	Top soil[b] P maximum ($mg\ kg^{-1}$)	Reference
Silty clay loam—silt loam	3.3	25	226	Cahn et al. (1994)
Sandy loam—sandy clay loam	10	20	90	Kristensen et al. (1995)
Sandy loam	10	20	90	Kristensen et al. (1995)
Sod-gleic sandy loam	5–10 (5 fields)	48 26 13 74 34	423 148 113 523 158	Buciene and Svedas (1997)
Sandy	14	24	55	Yang et al. (1999)
Sandy loam	14 14	14 6	100 45	Panten (2002)
Sand—loamy sand	7.9 12.2 9.1 14.1	87 77 92 81	165 196 309 325	Panten (2002)
Loamy sand	7.3 6.6 10.9 11.6 5.7	79 50 87 59 94	127 190 239 168 222	Panten (2002)

(continued)

Table 4 (continued)

Soil type[a]	Field size (ha)	Top soil[b] P minimum (mg kg^{-1})	Top soil[b] P maximum (mg kg^{-1})	Reference
Sandy loam—loam	9.3	48	221	Panten (2002)
Loamy sand—sandy loam	11.2 13.6 13.5 13.5 23.1	106 58 29 32 62	395 199 222 159 340	Panten (2002)
Sandy loam—loamy clay	11.3	39	110	Panten (2002)
Loamy sand—loam	10.5	34	401	Panten (2002)

[a]Soil type description depends on reference
[b]Depth of the top soil samples varies between studies

soils with different P levels, it is known that available soil P can also influence the P uptake of plants (Jessop et al. 1984; Manske et al. 2001; Wissuwa and Ae 2001). To demonstrate possible ranges of spatial variation in P uptake, two case studies conducted on winter wheat and winter barley in 2002 have been evaluated. Both experiments took place in northern Germany (9.47°E, 52.58°N) on a loamy sand with a mean annual precipitation of 699 mm. Table 5 provides an overview about some soil parameters and the P uptake by grain and straw, sampled in a 30 m raster at harvest.

The experimental sites have a considerable variation in available top soil P. Nonetheless, even low soil P areas supplied sufficient P for plant growth. It can be seen that P concentrations in grain and straw also vary substantially across both experimental sites (Table 5). However, the variation of P uptake is mainly driven by the yield variation (Figs. 4 and 5). P concentrations at shooting stage are of little use for the prediction of the final P uptake and are hardly related to the final P concentration in grains (Figs. 4 and 5). Therefore, under the tested soil P conditions yield is a good predictor for the spatially varying P uptake, provided that the actual mean P concentration of the crop is known and used instead of P coefficients from nutrient tables. This confirms Wittry and Mallarino (2004) conducting an experiment with variable fertilisation rates observing yield responses in maize-soybean rotations only in areas with soil P below 20 mg kg^{-1}. Results of long term field experiments conducted by Johnston et al. (2001) showed likewise that at soil levels below 20 mg kg^{-1} available P, yield was small and P budgets were therefore positive. With yield mapping data and the mean P concentration of the harvested crop, farmers will be able to calculate spatially varying P removals from fields and can create VR application maps to avoid excessive P fertilisation. This practice will be inexpensive in comparison to intensive soil sampling campaigns to capture the spatial heterogeneity of available soil P across fields. Nevertheless, it will not

Table 5 Variation of soil and plant properties of two case studies conducted in northern Germany in 2002 on a winter wheat (WW) field of 7.3 ha, and on a winter barley (WB) field of 11.6 ha (unpublished data)

Experimental site	Top soil chemistry (0–30 cm) $n = 72$ (*WW*); $n = 117$ (*WB*)									Top soil texture (0–30 cm) $n = 17$ (*WW*); $n = 15$ (*WB*)								
	Corg (%)			pH			Available P (mg kg^{-1})			Sand [63–2000 μm] (%)			Silt [2–63 μm] (%)			Clay [<2 μm] (%)		
	Min	Max	Mean	Min	Max	Mean	Min	Max	Mean	Min	Max	Mean	Min	Max	Mean	Min	Max	Mean
WW	1.29	3.10	2.17	5.4	6.3	5.9	44	138	75	36	66	56	29	59	40	4	7	5
WB	1.04	3.27	1.46	5.9	7.1	6.2	37	100	58	27	52	35	43	67	59	5	7	6
	P concentrations (%) $n = 72$ (*WW*); $n = 117$ (*WB*)									P uptake (kg ha^{-1}) $n = 72$ (*WW*); $n = 117$ (*WB*)								
	Crop at shooting			Grain			Straw			Grain			Straw			Total		
	Min	Max	Mean	Min	Max	Mean	Min	Max	Mean	Min	Max	Mean	Min	Max	Mean	Min	Max	Mean
WW	0.31	0.50	0.41	0.17	0.25	0.21	0.07	0.13	0.09	8.4	25.1	17.7	3.4	6.8	5.1	12.5	30.6	22.8
WB	0.28	0.47	0.36	0.26	0.37	0.31	0.09	0.19	0.13	7.0	38.4	25.2	2.8	10.5	5.8	9.8	46.4	30.9

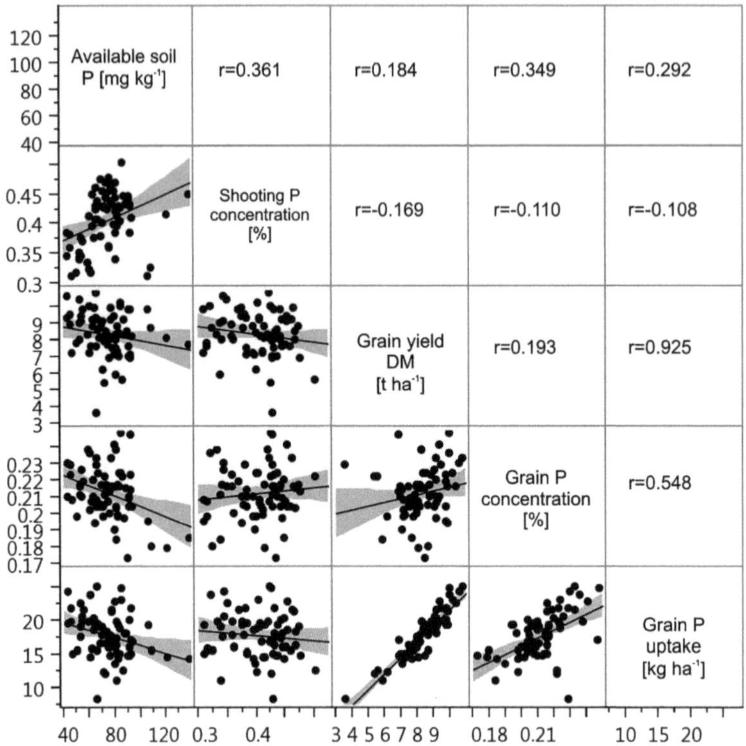

Fig. 4 Correlations matrix between available soil P, P concentration during shooting, yield, grain P concentrations and grain P uptake recorded 2002 in a winter wheat field (unpublished data). Shades next to the regression line represent the confidence interval

account for the spatial variation of available soil P and therefore does not reduce already existing heterogeneity or decrease soil P levels in areas oversupplied with P. Mandatory standard soil sampling, if possible guided by yield maps instead of sampling according to fixed sampling schemes, can then be used every few years to detect changes in soil P and help farmers to adjust the VR application map manually for areas with initially high available soil P and by this make use of residual soil P.

Studies about the spatial variability of P uptake carried out on P deficient soils are not known. But drawn from the results of field and pot experiments conducted under P deficit conditions, it can be assumed that crops grown on fields with variations from deficient to sufficient available soil P will be influenced in their P uptake. Therefore, yield response functions of these fields will also need to be spatially varying and simple P uptake calculations will not be possible.

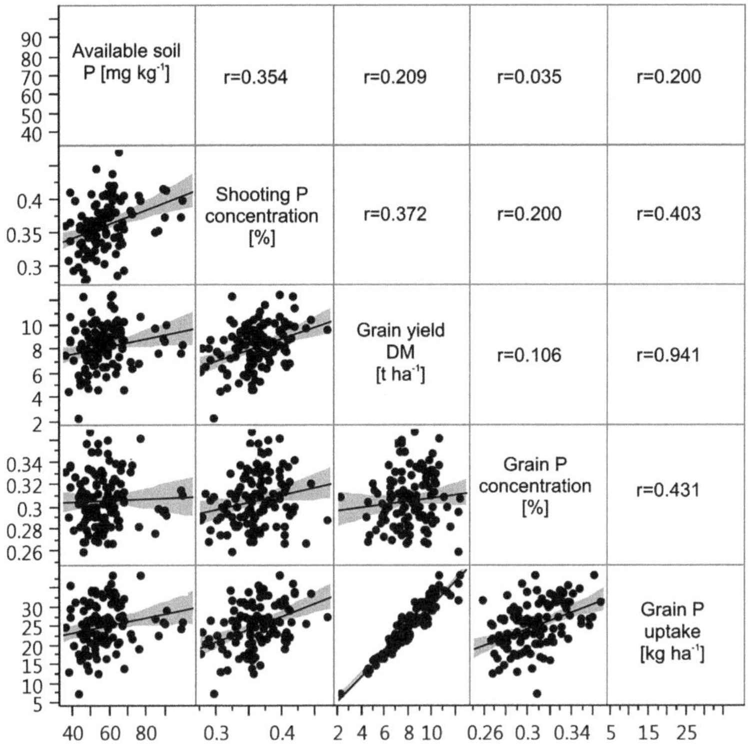

Fig. 5 Correlations matrix between available soil P, P concentration during shooting, yield, grain P concentrations and grain P uptake recorded 2002 in a winter barley field (unpublished data). Shades next to the regression line represent the confidence interval

Conclusions

The P uptake of plants is not only affected by soil type, soil available P, availability of other essential nutrients as well as climate and weather conditions (Syers et al. 2008) but also by the nutrient use efficiency of the grown cultivar. Given all these influencing factors it is not surprising that P uptake by crops can vary substantially between agricultural fields and it will be difficult to define a singular or average P coefficient for nutrient budgeting and human and animal nutrition. Even though most regional offices for agriculture in Germany conduct trials to test new varieties and provide farmers with recommendations for suitable cultivars according to regional specifics, P concentrations of crops are not frequently analysed. Commercial plant breeders might have knowledge about the P efficiency of their cultivars. Unfortunately, in regions where P deficient soils are rare, this knowledge is not passed on to farmers and therefore P efficiency of cultivars cannot be used as a selection criterion when choosing crop varieties. Hence, a new evaluation of P coefficients seems to be recommended to gain insights into P uptakes of modern varieties grown under actual conditions. The known discrepancies between published P coefficients and actual measured P concentrations of crops will lead to wrong conclusions when it comes to nutrient budgeting on higher administrative levels and consequently to misleading agri-environmental indicators for agricultural policy planning. Just the change of the P coefficient for P uptake of cereals in Germany from 0.35 to 0.30 % would result in an average increase of the P budget surplus by 1.3 kg ha^{-1} y^{-1} P, estimated on the basis of the 2002–2012 P budget data. Based on the actual market price for triple-superphosphate fertilisers in Germany this accounts for annual economical losses of about 2.70 € ha^{-1} arable land for farmers and an additional surplus of 22,000 t P based on the total area of arable land, that are either absorbed into soils or discharged into the environment. While it is difficult to define a 'correct' P coefficient for all crops to calculate reliable agri-environmental indicators, it is important for farmers to be able to calculate their nutrient budgets as accurately as possible to avoid economic losses.

In regions with intensive livestock farming, animal manure provides significant amounts of P for agricultural use. Like e.g. in Germany where 60 % of the annually applied P on arable land originates from manure and only 40 % from mineral and other organic fertilisers than manure (Fig. 1). Unfortunately, intensive livestock farming is not equally spread across the agriculturally used area but is instead concentrated in some regions. The Ministry for Environment, Energy and Climate Protection of Lower Saxony, Germany provides e.g. regional P budgets, showing a wide range of 0.0–40 kg ha^{-1} P surplus (NMUEK 2014). Therefore, in some regions farmers struggle to meet the legislative requirements when calculating their farm P budgets. In this case a reduction of P uptake coefficients of crops will result in even higher calculated surpluses. In contrast, farmers applying only mineral P risk economic losses if oversupplying P as well as when P is deficient. Spiess (2011) reported nearly closed P cycles for Switzerland similar to the P budget of Germany but also stated that nutrient budgets at regional levels are required to identify areas with high P surpluses. Regional P budgets for France showed that livestock farm regions are accumulating P whilst crop farming regions relying on mineral fertilisers have balanced P budgets (Senthilkumar et al. 2012). Continued miscalculations of P budgets at the farm level will be followed by a nutrient deficiency or nutrient surplus with the risk of economic losses in the first and P losses into the environment in the latter case. It is therefore highly recommended to use measured P concentrations of crops instead of values from nutrient tables for fertilisation planning.

References

Baker DE, Jarrell AE, Marshall LE, Thomas WI (1970) Phosphorus uptake from soils by corn hybrids selected for high and low phosphorus accumulation. Agron J 62:103–106

Baligar VC, Pitta GVE, Gama EEE, Schaffert RE, de Bahia Filho AFC, Clark RB (1997) Soil acidity effects on nutrient use efficiency in exotic maize genotypes. Plant Soil 192:9–13

Barrier-Guillot B, Casado P, Maupetit P, Jondreville C, Gatel F, Larbier M (1996) Wheat phosphorus availability: 1-In vitro study; factors affecting endogenous phytasic activity and phytic phosphorus content. J Sci Food Agric 70:62–68

Bates TE (1971) Factors affecting critical nutrient concentrations in plants and their evaluation: a review. Soil Sci 112:116–130

Batten GD (1992) A review of phosphorus efficiency in wheat. Plant Soil 146:163–168

Batten GD (1994) Concentrations of elements in wheat grains grown in Australia, North America, and the United Kingdom. Aust J Exp Agric 34:51–56

Bayuelo-Jiménez JS, Ochoa-Cadavid I (2014) Phosphorus acquisition and internal utilization efficiency among maize landraces from the central Mexican highlands. Field Crop Res 156:123–134

Binford GD (2010) Amounts of nutrients removed in corn grain at harvest in Delaware. In: Proceedings of the 19th world congress of soil science, soil solution for a changing world, Brisbane, Australia, 1–6 Aug 2010, pp 26–29

Bryant RJ, Dorsch JA, Peterson KL, Rutger JN, Raboy V (2005) Phosphorus and mineral concentrations in whole grain and milled low phytic acid (*lpa*) 1-1 rice. Cereal Chem 82:517–522

Buciene A, Svedas A (1997) Spatial variability of soil agrochemical properties and crop yield in Lithuania. In: Proceedings of the 1st European conference on precision agriculture, Warwick, UK, 7–10 Sept 1997, vol 1, pp 71–78

Bullock DS, Lowenberg-DeBoer J (2007) Using spatial analysis to study the values of variable rate technology and information. J Agric Econ 58:517–535

Bundy LG, Tunney H, Halvorson AD (2005) Agronomic aspects of phosphorus management. In: Sims JT, Sharpley AN (eds) Phosphorus: agriculture and the environment. ASA-CSSA-SSSA, Madison, Wisconsin, USA, pp 685–727

Cahn MD, Hummel JW, Brouer BH (1994) Spatial analysis of soil fertility for site-specific crop management. Soil Sci Soc Am J 58:1240–1248

Calderini DF, Torres-León S, Slafer GA (1995) Consequences of wheat breeding on nitrogen and phosphorus yield, grain nitrogen and phosphorus concentration and associated traits. Ann Bot 76:315–322

Chen F-J, Liu X-S, Mi G-H (2012) Varietal differences in plant growth, phosphorus uptake and yield formation in two maize inbred lines grown under field conditions. J Integr Agric 11 (10):1738–1743

Diepenbrock W, Ellmer F, Léon J (2009) Ackerbau, Pflanzenbau und Pflanzenzüchtung. Eugen Ulmer KG, Stuttgart, pp 264–349

Dikeman E, Pomeranz Y, Lai FS (1982) Mineral and protein contents in hard red winter wheat. Cereal Chem 59(2):139–142

DüV (2006) Verordnung über die Anwendung von Düngemitteln, Bodenhilfsstoffen, Kultursubstraten und Pflanzenhilfsmitteln nach den Grundsätzen der guten fachlichen Praxis beim Düngen (Düngeverordnung – DüV). http://www.gesetze-im-internet.de/bundesrecht/d_v/gesamt.pdf. Accessed 18 July 2012

Erdman JA, Moul RC (1982) Mineral composition of small-grain cultivars from a uniform test plot in South Dakota. J Agr Food Chem 30(1):169–174

Eurostat (2013) Nutrient budgets—methodology and handbook, Version 1.02. Eurostat and OECD, Luxembourg, pp 1–112

Fageria NK, Baligar VC (1997a) Phosphorus-use efficiency by corn genotypes. J Plant Nutr 20:1267–1277

Fageria NK, Baligar VC (1997b) Upland rice genotypes evaluation for phosphorus use efficiency. J Plant Nutr 20:499–509

Fageria N, Wright R, Baligar V (1988) Rice cultivar evaluation for phosphorus use efficiency. Plant Soil 111(1):105–109

Fageria NK, Santana EP, de Castro EM, Moraes OP (1995) Resposta diferencial de genótipos de arroz de sequeiro à fertilidade do solo. Rev Bras Cienc Solo 19:261–267

Fageria N, Baligar VC, Jones CA (2011) Growth and mineral nutrition of field crops, 3rd edn. CRC Press, Boca Raton, FL, USA, pp 1–560

FAO (2014a) Top production—World (total)—2012. http://faostat.fao.org/site/339/default.aspx. Accessed 21 Aug 2014

FAO (2014b) Top production—Europe (total)—2012. http://faostat.fao.org/site/339/default.aspx. Accessed 21 Aug 2014

FAO (2014c) FAOSTAT. http://faostat.fao.org/site/567/default.aspx. Accessed 21 Aug 2014

Ferreira CF, Vargas Motta AC, Prior SA, Reissman CB, dos Santos NZ, Gabardo J (2012) Influence of corn (*Zea mays* L.) cultivar development on grain nutrient concentration. Int J Agron 2012:ID842582

Finck A (1979) Dünger und Düngung – Grundlagen und Anleitung zur Düngung der Kulturpflanzen. Verlag Chemie, Weinheim, New York, pp 1–442

Fisher MJ (1980) The influence of water stress on nitrogen and phosphorus uptake and concentration in Towensville Stylo (*Stylosanthes humilis*). Aust J Exp Agr 20:175–180

Flachowsky G (2006) Futtermittelkundliche Aspekte. Landbauforsch Volk, Sonderheft 289:101–111

Gahoonia TS, Nielsen NE (1996) Variation in acquisition of soil phosphorus among wheat and barley genotypes. Plant Soil 178:223–230

Gill HS, Singh A, Sethi SK, Behl RK (2004) Phosphorus uptake and use efficiency in different varieties of bread wheat (*Triticum aestivum* L.). Arch Acker Pfl Boden 50:563–572

Gregersen PL (2011) Senescence and nutrient remobilization in crop plants. In: Hawkesford MJ, Barraclough P (eds) (2011) The molecular and physiological basis of nutrient use efficiency in crops. Wiley-Blackwell, Chichester, UK, pp 83–102

Gregory FG, Crowther F, Beaven ES (1928) A physical study of varietal differences in plants. Part I. A study of the comparative yields of barley varieties with different manurings. Ann Bot-London 42:757–770

Hafeez F, Aziz T, Maqsood MA, Ahmed M, Farooq M (2010) Differences in rice cultivars for growth and phosphorus acquisition from rock phosphate and mono-ammonium phosphate sources. Int J Agric Biol 12:907–910

Hawkesford MJ (2011) An overview of nutrient use efficiency and strategies for crop improvement. In: Hawkesford MJ, Barraclough P (eds) The molecular and physiological basis of nutrient use efficiency in crops. Wiley-Blackwell, Chichester, UK, pp 5–19

Heckman JR, Sims JT, Beegle DB, Coale FJ, Herbert SJ, Bruulsema TW, Bamka WJ (2003) Nutrient removal by corn grain harvest. Agron J 95:587–591

Helyar KR (1998) Efficiency of nutrient utilization and sustaining soil fertility with particular reference to phosphorus. Field Crop Res 56:187–195

Hussain AA, Maurya DM, Vaish CP (1987) Studies on quality of indigenous upland rice (*Oryza sativa*). Indian J Genet Plant Breed 47:145–152

Jessop RS, Jones GP, Blair GJ (1984) Performance of 22 wheat varieties under low and high phosphorus conditions and implications for the selection of P-efficient varieties. In: Proceedings of the 3rd international congress on phosphorus compounds, Brussels, 4–6 Oct 1983, pp 445–454

Johnston AE (2005) Phosphorus nutrition of arable crops. In: Sims JT, Sharpley AN (eds) Phosphorus: agriculture and the environment. ASA-CSSA-SSSA, Madison, Wisconsin, USA, pp 495–519

Johnston AE, Poulton PR, Syers JK (2001) Phosphorus, potassium and sulphur cycles in agricultural soils. The International Fertiliser Society, Proceedings No: 465

Jones GPD, Blair GJ, Jessop RS (1989) Phosphorus efficiency in wheat—a useful selection criterion? Field Crop Res 21:257–264

Jones GPD, Jessop RS, Blair GJ (1992) Alternative methods for the selection of phosphorus efficiency in wheat. Field Crop Res 30:29–40

Juliano BO, Bechtel DB (1985) The rice grain and its gross composition. Rice—chemistry and technology. American Association of Cereal Chemistry, St. Paul, Minnesota, USA, pp 17–57

Karlen DL, Flannery RL, Sadler EJ (1988) Aerial accumulation and partitioning of nutrients by corn. Agron J 80:232–242

Klatt JG, Mallarino AP, Downing JA, Kopaska JA, Wittry DJ (2003) Soil phosphorus, management practice, and their relationship to phosphorus delivery in the Iowa Clear Lake agricultural watershed. J Environ Qual 32(6):2140–2149

Kristensen K, Simmelsgaard SE, Djurhuus J, Olesen SE (1995) Spatial variability of soil physical and chemical parameters. In: Seminar on site specific farming, Koldkægaard, Århus, Denmark, 20–21 Mar 1995, pp 39–55

Lark RM (2009) Estimating the regional mean status and change of soil properties: two distinct objectives for soil survey. Eur J Soil Sci 60:748–756

LfL (2013) Gruber Tabellen zur Fütterung der Milchkühe, Zuchtrinder, Schafe, Ziegen. Bayerische Landesanstalt für Landwirtschaft (LfL). pp 1–94. http://www.lfl.bayern.de/mam/cms07/publikationen/daten/informationen/p_36967.pdf. Accessed 16 July 2014

LKSH (2011) Futtermitteltabellen. Landwirtschaftskammer Schleswig-Holstein (LKSH). pp 1–9. http://www.lksh.de/fileadmin/dokumente/Landwirtschaft/Tier/Rinder/2011/Futterwerte-wichtigste-Futtermittel.pdf. Accessed 16 July 2014

LLH (2009) Empfehlungen zur Energie- und Nährstoffversorgung für Milchkühe und Jungrinder sowie Futterwerte der in Hessen gebräuchlichen Futtermittel. Landesbetrieb Landwirtschaft Hessen (LLH). pp 1–6. http://www.llh.hessen.de/tierproduktion/rinder.html. Accessed 16 July 2014

Lolas GM, Palamidis N, Markakis P (1976) The phytic acid-total phosphorus relationship in barley, oats, soybean, and wheat. Cereal Chem 53(6):867–871

LVLF (2008) Richtwerte für die Untersuchung und Beratung sowie zur fachlichen Umsetzung der Düngeverordnung (DüV). http://www.mil.brandenburg.de/cms/detail.php/bb1.c.246175.de?highlight. Accessed 31 Jan 2013

Lyness AS (1936) Varietal differences in the phosphorous feeding capacity of plants. Plant Physiol 11:665–688

Mallarino AP, Schepers JS (2005) Role of precision agriculture in phosphorus management practices. In: Sims JT, Sharpley AN (eds) Phosphorus: agriculture and the environment. ASA-CSSA-SSSA, Madison, Wisconsin, USA, pp 881–908

Mallarino AP, Wittry DJ, Dousa D, Hinz PN (1998) Variable-rate phosphorus fertilization: on-farm research methods and evaluation for corn and soybean. In: Robert PC et al (eds) Precision agriculture. Proceedings of the 4th international conference, Minneapolis, 19–22 July 1998, Part A, pp 687–696

Manschadi AM, Kaul H-P, Vollmann J, Eitzinger J, Wenzel W (2014) Developing phosphorus-efficient crop varieties—an interdisciplinary research framework. Field Crop Res 162:87–98

Manske GGB, Ortiz-Monasterio JI, van Ginkel M, Gonzáles RM, Fischer RA, Rajaram S, Vlek PLG (2001) Importance of P uptake efficiency versus P utilization for wheat yield in acid and calcareous soils in Mexico. Eur J Agron 14:261–274

Marchant BP, Lark RM (2007) Optimized sample schemes for geostatistical surveys. Math Geol 39:113–134

Marr KM, Batten GD, Blakeney AB (1995) Relationships between minerals in Australian brown rice. J Sci Food Agr 68:285–291

McCall ER, Jurgens JF, Hoffpauir CL, Pons WA, Stark SM, Cucullu AF, Heinzelman DC, Cirino VO, Murray MD (1953) Composition of rice: Influence of variety and environment on physical and chemical composition. J Agr Food Chem 1:988–993

Mitscherlich EA (1909) Das Gesetz des Minimums und das Gesetz des abnehmenden Bodenertrages. Landwirtschaftliche Jahrbücher 38:537–552

Nadeem M, Mollier A, Morel C, Vives A, Prud'homme L, Pellerin S (2012) Seed phosphorus remobilisation is not a major limiting step for phosphorus nutrition during early growth of maize. J Plant Nutr Soil Sc 175:805–809

Ning P, Li S, Yu P, Zhang Y, Li C (2013) Post-silking accumulation and partitioning of dry matter, nitrogen, phosphorus and potassium in maize varieties differing in leaf longevity. Field Crop Res 144:19–27

NMUEK (2014) Nährstoffeintrag durch Düngung. Quellen und Einträge. http://www.umwelt.niedersachsen.de/umweltbericht/herausforderungen/eutrophierung/naehrstoffeintrag_durch_duengung/quellen_und_eintraege/quellen-und-eintraege-89000.html. Accessed 21 Aug 2014

Nolin MC, Guertin SP, Wang C (1996) Within-field spatial variability of soil nutrients and corn yield in a Montreal Lowlands clay soil. In: Robert PC et al (eds) Precision agriculture. Proceedings of the 3rd international conference, Minneapolis, 23–26 June 1996, pp 257–270

OECD (2010) Environmental performance of agriculture. OECD agricultural statistics (database). doi:10.1787/data-00173-en. Accessed 08 Jan 2013

Oenema O, van Liere L, Schoumans O (2005) Effects of lowering nitrogen and phosphorus surpluses in agriculture on the quality of groundwater and surface water in the Netherlands. J Hydrol 304:289–301

Oliver MA (2010) Geostatistical applications for precision agriculture. Springer, Dordrecht, Heidelberg, London, New York, pp 1–331

Osborne LD, Rengel Z (2002) Screening cereals for genotypic variation in efficiency of phosphorus uptake and utilisation. Aust J Agr Res 53:295–303

Ozturk L, Eker S, Torun B, Cakmak I (2005) Variation in phosphorus efficiency among 73 bread and durum wheat genotypes grown in a phosphorus-deficient calcareous soil. Plant Soil 269:69–80

Panten K (2002) Ein Beitrag zur Fernerkundung der räumlichen Variabilität von Boden- und Bestandesmerkmalen. Landbauforsch Volk, Sonderheft 232:1–135

Panten K, Rogasik J, Godlinski F, Funder U, Greef J-M, Schnug E (2009) Gross soil surface nutrient balances: the OECD approach implemented under German conditions. Landbauforsch Volk 59:19–28

Peterson CJ, Johnson VA, Mattern PJ (1983) Evaluation of variation in mineral element concentrations in wheat flour and bran of different cultivars. Cereal Chem 60:450–455

Pocknee S, Boydell BC, Green HM, Waters DJ, Kvien CK (1996) Directed soil sampling. In: Robert PC, Rust RH, Larson WE (eds) Proceedings of the 3rd international conference on precision agriculture, Minneapolis, 23–26 June 1996. ASA; CSSA; SSSA, Madison, Wisconsin, pp 159–168

Pote DH, Daniel TC, Nichols DJ, Sharpley AN, Moore PA, Miller DM, Edwards DR (1999) Relationship between phosphorus levels in three Ultisols and phosphorus concentrations in runoff. J Environ Qual 28(1):170–175

Ren X-L, Liu Q-L, Wu D-X, Shu Q-Y (2006) Variations in concentration and distribution of health-related elements affected by environmental and genotypic differences in rice grains. Rice Sci 13:170–178

Ren X-L, Liu Q-L, Fu H-W, Wu D-X, Shu Q-Y (2007) Density alteration of nutrient elements in rice grains of a low phytate mutant. Food Chem 102:1400–1406

Ros C, Bell RW, White PF (1997) Effects of seed phosphorus and soil phosphorus applications on early growth of rice (*Oryza sativa* L.) cv. IR66. Soil Sci Plant Nutr 43:499–509

Rosa OS (1991) Wheat breeding for better efficiency in phosphorus use. In: Saunders DA (ed) Wheat for the non-traditional warm areas. Proceedings of an international conference, 29 June–3 Aug 1990, Foz do Iguacu, Brazil, pp 333–351

Rose TJ, Pariasca-Tanaka J, Rose MT, Mori A, Wissuwa M (2012) Seeds of doubt: re-assessing the impact of grain P concentrations on seedling vigor. J Plant Nutr Soil Sci 175:799–804

Sarić MR (1987) Progress since the first international symposium: 'genetic aspects of plant mineral nutrition', Beograd, 1982, and perspectives of future research. Plant Soil 99:197–209

Schepers JS, Schlemmer MR, Ferguson RB (2000) Site-specific considerations for managing phosphorus. J Environ Qual 29(1):125–130

Schnug E, Strampe U (1988) Sortentypische Unterschiede der Nährelementkonzentrationen bei Winterweizen. J Agron Crop Sci 160:163–172

Schnug E, Lamp J, Knoop F (1985) Regionale Variabilität von Merkmalen der Bodenfruchtbarkeit II. Beispiele und praktische Bedeutung. Mitt Deut Bodenkdl G 43 (2):679–684
Schulthess U, Feil B, Jutzi SC (1997) Yield-independent variation in grain nitrogen and phosphorus concentration among Ethiopian wheats. Agron J 89:497–506
Senthilkumar K, Nesme T, Mollier A, Pellerin S (2012) Regional-scale phosphorus flows and budgets within France: the importance of agricultural production systems. Nutr Cycl Agroecosyst 92:145–159
Sharpley AN (1995) Dependence of runoff phosphorus on extractable soil phosphorus. J Environ Qual 24(5):920–926
Simpson RJ, Oberson A, Culvenor RA, Ryan MH, Veneklaas EJ, Lambers H, Lynch JP, Ryan PR, Delhaize E, Smith FA, Smith SE, Harvey PR, Richardson AE (2011) Strategies and agronomic interventions to improve the phosphorus-use efficiency of farming systems. Plant Soil 349:89–120
Smith SN (1934) Response of inbred lines and crosses in maize to variations of nitrogen and phosphorus supplied as nutrients. J Am Soc Agron 26(9):785–804
Spiess E (2011) Nitrogen, phosphorus and potassium balances and cycles of Swiss agriculture from 1975–2008. Nutr Cycl Agroecosyst 91:351–365
Stangel PJ, von Uexküll HR (1990) Regional food security; demographic and geographic implications. Phosphorus requirements for sustainable agriculture in Asia and Oceania. IRRI, Manilla, Philippines, pp 21–43
Steiner T, Mosenthin R, Zimmermann B, Greiner R, Roth S (2007) Distribution of phytase, total phosphorus and phytate phosphorus in legume seeds, cereals and cereal by-products as influenced by harvest year and cultivar. Anim Feed Sci Tech 133:320–334
Syers JK, Johnston AE, Curtin D (2008) Efficiency of soil and fertilizer phosphorus use. Reconciling changing concepts of soil phosphorus behaviour with agronomic information. FAO Fertilizer Plant Nutr Bull 18:1–123
Tan ZX, Lal R, Wiebe KD (2005) Global soil nutrient depletion and yield reduction. J Sustain Agr 26(1):123–146
Vance CP (2011) Phosphorus as a critical macronutrient. In: Hawkesford MJ, Barraclough P (eds) The molecular and physiological basis of nutrient use efficiency in crops. Wiley-Blackwell, Chichester, UK, pp 229–264
Vašát R, Heuvelink GBM, Borůvka L (2010) Sampling design optimization for multivariate soil sampling. Geoderma 155:147–153
Vyn TJ, Tollenaar M (1998) Changes in chemical and physical quality parameters of maize grain during three decades of yield improvement. Field Crop Res 59(2):135–140
Watson C, Topp K, Stockdale L (2009) A guide to nutrient budgeting on organic farms. http://www.organicresearchcentre.com/manage/authincludes/article_uploads/iota/technical-leaflets/a-guide-to-nutrient-budgeting-on-farms.pdf. Accessed 1 Aug 2014
Weld J, Sharpley AN (2006) Phosphorus indices. In: Radcliff DE, Cabrera ML (eds) Modeling phosphorus in the environment. CRC Press, New York, USA, pp 301–332
Wissuwa M, Ae N (2001) Genotypic variation for tolerance to phosphorus deficiency in rice and the potential for its exploitation in rice improvement. Plant Breed 120:43–48
Withers PJA, Haygarth PM (2007) Agriculture, phosphorus and eutrophication: a European perspective. Soil Use Manage 23:1–4
Withers PJA, Nash DA, Laboski CAM (2005) Environmental management of phosphorus fertilisers. In: Sims JT, Sharpley AN (eds) Phosphorus: agriculture and the environment. ASA-CSSA-SSSA, Madison, Wisconsin, USA, pp 781–827
Wittry DJ, Mallarino AP (2004) Comparison of uniform- and variable-rate phosphorus fertilization for corn-soybean rotations. Agron J 96:26–33
Yang C, Anderson GL, King JH, Chandler EK (1999) Comparison of uniform and variable rate fertilization strategies using grid soil sampling, variable rate technology, and yield monitoring. In: Proceedings of the 4th international conference on precision agriculture, St. Paul, USA, 19–20 July 1998, pp 675–686

Yaseen M, Malhi SS (2011) Exploitation of genetic variability among wheat genotypes for tolerance to phosphorus deficiency. J Plant Nutr 34(5):665–699
Zhang M, Nyborg M, McGill WB (1990) Phosphorus concentration in barley (*Hordeum vulgare* L.) seed: influence on seedling growth and dry matter production. Plant Soil 122:79–83
Zhu Y-G, Smith SE (2001) Seed phosphorus (P) content affects growth, and P uptake of wheat plants and their association with arbuscular mycorrhizal (AM) fungi. Plant Soil 231:105–112

Management Options for an Efficient Utilization of Phosphorus in Agroecosystems

Bettina Eichler-Löbermann, Silvia Bachmann, Stefanie Busch, Katja Schiemenz, Thomas Krey, Verena Pfahler and Ralf Uptmoor

Abstract This chapter focuses on management options to increase the utilization of phosphorus (P) in agroecosystems like (i) catch cropping and mixed cropping, by (ii) application of beneficial microorganisms, and by (iii) P recycling with wastes and residues.

Keywords Cropping system · Microorganism · Digestate · Biomass ash

Catch Cropping and Mixed Cropping

The better utilization of P in agroecosystems by catch cropping and mixed cropping based on plant specific P mobilization and transformation as well as on the complementary use of P in space and time. However, also other aspects regarding soil fertility play a role in this sense, e.g. the reduction of erosion.

To improve P acquisition from soil plants can either change the root morphology and/or excrete compounds, which enhance the availability of P to plants (Richardson et al. 2011; Eichler-Löbermann et al. 2008; Wang et al. 2007). Plant root exudates mainly consist of a complex mixture of amino acids, organic acids, different ions, sugars, vitamins, nucleosides, enzymes and root border cells which have major direct or indirect effects on the acquisition of nutrients required (Nurruzzaman et al. 2006; Schilling et al. 1998). Usually, significant interactions between these exudates and the soil microflora occur which modify the efficiency of P acquisition (Bachmann et al. 2011; Richardson et al. 2009).

B. Eichler-Löbermann (✉) · S. Bachmann · S. Busch · K. Schiemenz · T. Krey · V. Pfahler · R. Uptmoor
Faculty of Agricultural and Environmental Science, University of Rostock, Justus-von-Liebig-Weg 6, 18059 Rostock, Germany
e-mail: bettina.eichler@uni-rostock.de

E. Schnug and L.J. De Kok (eds.), *Phosphorus in Agriculture: 100 % Zero*,
DOI 10.1007/978-94-017-7612-7_9

Catch Cropping

Catch cropping is the integration of a crop plant with short vegetation time into a crop rotation between two main crops. Catch crops are mainly used as green manure and cover the soil in temperate climates from late autumn until the next spring. Using catch crops as green fertilizer can increase cropping system sustainability and soil fertility by enriching organic matter in the soil through plant residues, reducing erosion by soil covering, and holding back nutrients by biological fixation (Li et al. 2014; Eichler-Löbermann et al. 2008; Nziguheba and Bünemann 2005; Sørensen 1991).

Catch cropping may enhance the P availability (i) during their growth by modifying of soil properties and mobilizing P; and (ii) after decomposition, when the organic P bound in the catch crop may provide an easily accessible form of P to succeeding crops (Eichler-Löbermann et al. 2008; Eichler-Löbermann and Schnug 2006). The P which was mobilized by a catch crop during its growing time in excess to its own requirements can also improve the P availability for a main crop. The integration of legume catch crops in the crop rotation seems to be especially suitable in this sense, since, beside of the P supply, they may enhance the nitrogen (N) supply of the main crops.

Catch crops which have a high P uptake efficiency seem to be suitable for improving the P nutrition status of soils, since they can accumulate relatively high amounts of P, even when the P content in soil is low. This P accumulated in the plant tissue is released upon decomposition. A high P uptake efficiency requires biochemical P mobilizing strategies or larger root systems which can explore a great soil volume (see above).

High P efficiencies were found for phacelia (*Phacelia tanacetifolia*) on loamy soils and buckwheat (*Fagopyrum esculentum*) and serradella (*Ornithopus sativus*) on sandy soils (Eichler-Löbermann and Schnug 2006; Table 1). In a pot experiment incorporation of phacelia plants increased the P uptake of the following maize (*Zea mays*) more than an application of Triple-Super P or cattle manure (Eichler 2004). Outstanding high P uptakes were found for amaranth (*Amaranthus cruentus*) (Bachmann et al. 2011; Brandt et al. 2011; Ojo et al. 2010), which was related to a considerable decrease of soil pH values—mainly due to the excretion of organic acids (Li et al. 2006). The prediction of Prelease from plant tissue is difficult, since the decomposition of the green manure material is a complex process depending on P content in soil, amount and N/P ratio of the green material, P mineralization rates, and P cycling (Oberson and Joner 2005).

However, apart from these positive effects, catch cropping cannot replace P fertilizers, but it should be seen as an opportunity to improve the utilization of P given (Table 1).

Nutrients mainly get lost from agricultural soils due to erosion and leaching especially when the soil is not covered by plants. These losses also occur when crops are cultivated which have a slow-growing juvenile phase. Catch cropping may help to reduce nutrient losses, since erosion is significantly reduced when the

Table 1 Features of different crop plants (cultivated as catch crops in three different trials) and their influence on P content in soil and P uptake of the following main crop

Crop	Root length* (m pot^{-1})	Root/shoot ratio* (m g^{-1} DM)	Activity of acid phosphatase* (p-nitrophenol, µg g^{-1} h^{-1})	Cation/anion ratio of nutrient uptake**	P content in soil solution*** (PO_4^- P, mg l^{-1})	P content in soil *** (P_{DL}, mg 100 g^{-1} soil)	P uptake main crops**** (kg ha^{-1})
Oilseed rape	107.5 a	4.9 a	227.3 c	5.3 ab			
Maize	578.5 c	13.8 c	276.7 d	8.4 c			
Phacelia	88.1 a	9.3 b	122.6 b	13.7 d	0.35 d	4.91 b	24.7 b
Buckwheat	274.3 b	10.2 b	262.1 d	9.4 c	0.30 c	4.41 a	23.3 a
Comm. ryegrass	205.7 b	14.4 b	73.0 a	7.7 bc	0.22 a	4.40 a	23.0 a
Lupine	49.5 a	9.3 b	217.1 c	9.5 c			
Pea	65.9 a	5.1 a	231.8 c	6.9 b			
Serradella	68.7 a	9.7 b	302.6 e	8.1 bc	0.32 c	4.80 b	23.9 ab
Amaranth	42.1 a	7.6 ab	301.5 e				
Oil narved				6.3 b			
Oil radish				4.0 a	0.25 b	4.79 b	23.5 a
Yellow mustard				4,2 a			

Adopted from Eichler-Löbermann and Schnug (2006)

Different letters (a–e) label significant differences between the variants (ANOVA, $\alpha = 0.05$)

* Pot trial A [6 kg of loamy sand pot^{-1} (low in plant available P)] after 7 weeks of vegetation time under greenhouse conditions

** Pot trial B [6 kg of sandy loam pot^{-1} [extreme low in plant available P)] in a two-year pot trial under greenhouse conditions. Average of two years

*** Field experiment (loamy sand), Average of 6 sampling dates, from spring 1999 to autumn 2001

****Field experiment (loamy sand), Average of P uptake of main crops (Summer oilseed rape, Summer barley, Summer wheat) from 1999 to 2001

soil is thoroughly covered by crops, and nutrients can be temporarily fixed biologically by catch crops. On the other hand, an experiment with shallow soil lysimeters (0.25 m deep) showed, that after freezing–thawing cycles, P concentrations in the leachate in the lysimeters with ryegrass, oilseed radish and red clover were higher than in the control without a catch crop (Li et al. 2014). Probably, these high P concentrations in the upper soil layer were related to the high P uptake (P mobilization) by these crops and following high P release. In a deeper soil layer (0.9 m deep), which is relevant for losses into the ground water, however, catch crops can significantly reduce the P concentrations in soil solution. This is probably related to lower velocity of water percolation when the soil is covered with catch crops (Eichler 2004).

Mixed Cropping

Mixed cropping is the cultivation of two or more crops together in rows or no distinct row arrangement. The agricultural practice to cultivated different cultivars or species simultaneously is a centuries-old practice and mainly applied in small-scale farming and organic farming (Ehrmann and Ritz 2014; Malézieux et al. 2009). The multispecies agroecosystems can result in an enhanced productivity compared with that of monoculture. The main advantage of mixed cropping is the efficient utilization of resources, such as light, water, and nutrients. This can be explained by two major processes: complementarities and facilitation (Hinsinger et al. 2011; Fridley 2001).

The resource complementarity reduces the niche overlap and the success of mixed cropping is based on the lack of interference in the spatial and temporal dimension between species. The spatial dimension is mainly based on the differences in rooting pattern. The temporal dimension may occur when the growth patterns of component crops differ in time so that the crops make their major demands on resources at different times (Devi et al. 2014). A complementary root system might also decrease the risk of erosions (Pekrun 2014).

Facilitation occurs when one species (or genotype) enhances the growth of another species. Species can modify the biotic/abiotic environment of its roots (rhizosphere), ultimately benefitting the intercropped species by increasing P availability. This can happen directly or indirectly. A direct facilitation comprises mechanisms as favorable mobilization of a P compound in soil. The indirect mechanisms comprise e.g. beneficial changes in soil mycorrhizal or microbial communities (Hinsinger et al. 2011; Callaway 2007).

Usually there are primary and secondary components in mixed cropping systems (Morris and Garrity 1993), whereas the primary crop components planted at almost optimum plant density and intended to yield near its full potential. The secondary component, however, is planted at a density often lower than its optimum and provides the primary crops with resources, mainly plant nutrients.

Table 2 Plant P uptake, soil P pools and enzyme activities in soil for sole and mixed cropping under different soil water contents in a pot experiment in Rostock in 2012

	WHC 80 %	WHC 30 %				
	P uptake (mg pot^{-1})	P uptake (mg pot^{-1})	Pdl (mg kg^{-1})	Pw (mg kg^{-1})	Microbial activity*	Alk. pase**
Maize	103 e	32.9 bc	40.1 ab	8.17 ab	49.7 ab	23.1 abc
Sorghum	87.5 de	37.3 bc	39.2 ab	9.54 abc	34.1 ab	15.9 a
Faba bean	53.2 a	19.8 a	42.5 bc	11.3 c	36.0 ab	22.7 abc
Soybean	63.4 b	31.9 b	40.5 ab	11.5 c	25.1 a	17.1 ab
Ma + Fb	87.0 de	43.5 c	44.1 c	10.6 bc	56.9 b	25.2 bc
Ma + Sb	91.7 d	34.9 bc	38.7 a	6.93 a	35.4 ab	27.2 c
So + Fb	81.5 cd	38.8 bc	41.8 abc	9.81 abc	53.2 ab	20.8 abc
So + Sb	75.6 c	37.0 bc	39.6 ab	9.11 ab	51.9 ab	25.8 bc

Lower-case letters indicate significant different means between the bacterial treatments; Duncan-test $p \leq 0.05$

*activity of dehydrogenase DHA, TPF, µg/g/h, **alkaline phosphatase in soil; p-nitrophenol, µg/g/h

WHC = water holding capacity, Ma = maize, So = sorghum, Fb = Faba bean, Sb = soybean

Combined cultivation of cereals and legumes is very often practiced because of their complementarity in the use of N resources (Tosti and Guiducci 2010), which may positively affect the yield and protein content of cereals (Mikic et al. 2015; Pekrun 2014).

The interactions among crops are context dependent and shift from facilitation to competition as environmental stress decreases (stress-gradient-hypothesis; Betencourt et al. 2012; Daleo and Iribarne 2009). Thus, usually positive effects of mixed cropping are found under unfavorable conditions. Accordingly, in a pot experiment with different water supply positive yield effects of mixed cropping [maize + runner bean (*Phaseolus coccineus*) or sorghum (*Sorghum bicolor*) + faba bean (*Vicia faba*)] were mainly found under drought stress (Table 2). In a field experiment under P deficiency the yields of mixed cropped maize + runner bean were not different to maize in sole cropping, although the N supply in the mixed cropping treatments was only 50 % of that of the sole cropping. The same was true for Sorghum + Andean lupin (*Lupinus mutabilis*). Under optimal P supply however, the sole cropping was superior to the mixed cropping treatments (Table 3).

Thus, it can be concluded that the combination of several crops in an agroecosystem have positive effects on the biodiversity, soil fertility and plant health, and may also help to improve the P utilization in soils and save P fertilizers.

Table 3 Crop yield (whole plant, dry matter) of sole and mixed cropping in a field experiment in dependence of long-term P supply (Rostock)

	Without	TSP	Cattle manure	Biowaste compost	Compost + TSP
2013 DM, t ha^{-1}					
Maize	11.9	13.9	13.8	16.5*	15.2*
Maize + Bean**	14.0	13.6	13.0	14.3	15.2
2014 DM, t ha^{-1}					
Maize	17.0	18.7	19.0	20.2*	22.1*
Maize + Bean**	15.4	17.4	16.9	16.3	18.4*
Sorghum	9.1	9.8	11.7	13.3*	12.4*
Sorghum + Lupin***	9.9	8.8	8.8	8.1	11.9

* significantly different from the control without P supply (Duncan-Test, $p \leq 0.05$)
** *Phaseolus coccineus*, *** *Lupinus mutabilis*
TSP = Triple-Super-P

Beneficial Microorganisms and Organic Matter Management

Like plants microorganisms (MOs) are also capable to improve the availability of P in soil and are of great importance regarding the soil P cycle. The active mobilization of inorganic P sources and the hydrolysis of organic P compounds by MOs is an elemental factor to counteract the loss of P plant availability and thus a source of energy for high P utilization.

Mycorrhizal fungi, especially arbuscular mycorrhizal fungi (AMF), are considered to be the most important microbial symbionts for the majority of crops and have a special importance regarding the mobilization and the transport of P in soils (Read and Perez-Moreno 2003). Fungi are highly adept at obtaining P, they respond to critically low P levels by increasing the activity of their P uptake systems i.e. release of phosphatases enzymes to cleave P from organic sources (Deacon 2006). Fungi also tend to accumulate and store P in excess of their immediate requirements; therefore it is not surprising that plants have mycorrhizal associations, as a mechanism to improve P uptake. P deficiency promotes mycorrhizal colonization, but high levels of plant P inhibit infection by AM fungi (Mohammad et al. 2003). When P deficiency occurs, plants release large amounts of sugars and organic acids into the rhizosphere which are utilized by AMF for growth. Hyphae of AMF greatly increase the absorbing surface area of the roots, bridge the depletion zone, and grow into areas with an adequate supply of P.

Bacterial strains capable of dissolving P have been found, for example in the genera Enterobacter (Vassilev et al. 1997) and Pseudomonas (Egamberdiyeva and Höflich 2003, Table 4).

Beside the activation of native MOs in the soil the application of microbial inoculants as so called biofertilizers is an option to utilize the P mobilization effects of MOs. The main efforts in this field have focused on fungi, the application of

Table 4 Water soluble soil P content, alkaline phosphatase and root colonization by arbuscular mycorrhizal fungi as affected by and bacterial treatments in a field experiment under P deficiency (Rostock), test crop maize

PGPR	Autumn 2007			Autumn 2008		
	P_w (mg kg^{-1})	AIP p-np (μg g^{-1} h^{-1})	AMF (%)	P_w (mg kg^{-1})	AIP p-np (μg g^{-1} h^{-1})	AMF (%)
B−	10.3 a	11.1 a	18.8 a	8.88 a	21.5 a	19.2 a
BE+	9.68 a	11.6 a	27.0 b	8.83 a	21.6 a	20.7 a
BP+	12.6 b	9.81 a	30.0 b	10.7 b	26.4 a	31.8 b
Mean	10.7	10.8	25.2	9.46	23.2	24.2

Adopted from Krey et al. (2013)
B−, without microbial application; BE+, *Enterobacter radicincitans*; BP+, *Pseudomonas fluorescens*, (n = 4), p-np = p-nitrophenol
Lower-case letters indicate significant different means between the bacterial treatments; Duncan-test $p \leq 0.05$

plant-growth promoting rhizobacteria (PGPR) however has also been proposed as a component of sustainable nutrient management (Richardson and Simson 2011). PGPR can enhance plant growth in various ways, not only by the increase of nutrient availability in soil, but also by release of growth stimulating hormones or protection against soil-borne pathogens (Vassilev et al. 2006). Bacterial inoculants added to promote the plant growth also affect the AMF. Usually positive synergistic effects were observed (Martinez et al. 2004).

Organic fertilizers, like compost and manure, serve as valuable plant P sources but they can also enhance soil microbial activity and maintain soil enzymes in their active forms (Krey et al. 2011, 2013; Oberson et al. 2010). The organic matter supply plays also an important role for the successful inoculation of soil with MOs. Bünemann et al. (2004) emphasized the important role of organic substances in soil for the conversion of P trough soil MOs and found that the microbial biomass is limited by C and N rather than P availability. In a laboratory experiment the microbial P increased substantially when soil was supplied with C and N. The linkage between soil organic C and microbial P is a major reason why cropped soils usually have lower microbial P contents than forest soils or grassland (Selles et al. 1995). However, application of organic fertilizers can mask bacterial effects on plant growth by the generally increased nutrient supply. In a field experiment the application of PGPR increased plant growth of maize and soil P pools primarily in the P-deficient treatment (Krey et al. 2013). Similar results were found with other bacterial strains by Dobbelaere et al. (2002) and Egamberdiyeva (2007), where the bacterial plant growth promotion was stronger in soils with a low nutrient supply. Accordingly, the application of biofertilizers should be rather done on P poor soils.

Recycling Products (Biogas Digestates and Biomass Ashes)

Biogas Digestates

In the last decade the amount of biogas produced in Europe was increasing from 135,014 TJ in 2003 to 566,114 TJ in 2013 (EU-28; Eurostat 2015). As substrates for the anaerobic digestion animal manures, energy crops, wastes from food industry, crop residues and municipal wastes are mainly used (Möller and Müller 2012). Only about 20 % of the input substrate is converted into biogas, the rest remains in the digester. It was estimated that in Germany approximately 65.5×10^6 m^3 biogas residues (hereafter referred to as digestates) are produced annually (Möller and Müller 2012). The amount of nutrients in the digestates depends thereby on the input substrate (Nkoa 2014; Möller et al. 2008; Roschke 2003). Results regarding the P availability in the digestates varied. Studies of Field et al. (1984) and Umetsu et al. (2001) showed that the availability of P decreases due to the anaerobic digestion. This was explained by the precipitation of less readily available P forms like hydroxylapatite (Güngör et al. 2007). If and how much P precipitates as less available mineral phosphates depends on several characteristics of the digestate, like input material and biogas process. However, applications of digestates can increase crop P uptake and readily available P pools in soil to a comparable extend as high soluble P fertilizers like Triple-Super-P (TSP) (Loria and Sawyer 2005). Comparable P fertilizer effects between unfermented dairy slurry and the digestate were found in a field experiment carried out by Bachmann et al. (2014).

During anaerobic digestion the easily degradable organic compounds become mineralized, and it can be expected that less but more stable organic matter is returned to field with the digestate which is less available for microorganisms compared to unfermented materials. Following, the mineralization of organic C when applied as digestate was often found to be lower than that from undigested materials (Thomsen et al. 2012). In this context differences in the activity of soil enzymes (dehydrogenase and alkaline phosphatase) in dependence of the fertilizer treatments were found in a field experiment with biogas digestates. For both enzymes the activity was significantly higher when the unfermented digestate was applied in comparison to the biogas digestate and control treatment (Table 5).

Due to the high water content of digestates, a separation of solid and liquid parts is an important tool to achieve a better handling of these residues. Mechanical separation with screw presses or decanter centrifuges are efficient and simple techniques and are widely used (Møller et al. 2002). Only a few studies investigated so far the fertilizer effects of the solid and a liquid phase (e.g. Pagliari and Laboski 2013; Hjorth et al. 2010). The data from two biogas plants (plant A: dairy slurry and maize silage; plant B: maize silage, grass silage, whole plant silage), which separate the digestate in a solid and a liquid phase showed that P is mainly found in the solid fraction (unpublished data; Table 6). The mineral P predominates in the digestate,

Table 5 Water soluble P, dehydrogenase activity and alkaline phosphatase activity in soil of a 5-year field trial amended with unfermented (ingestate) and fermented (digestate) cattle slurry compared to a control without fertilizer application (NK)

	Year	Ingestate	Digestate	NK
P_w (mg kg^{-1})	2009	12.5 a	14.0 b	12.1 a
	2010	14.7 a	14.6 a	13.3 a
	2011	14.3 b	14.3 b	12.0 a
	2012	13.1 b	14.1 c	11.4 a
	2013	13.9 b	14.9 b	12.0 a
DH-activity (µg TPF g^{-1})	2009	n.d	n.d	n.d
	2010	24.0 b	11.2 a	12.8 a
	2011	26.5 b	13.2 a	12.9 a
	2012	22.4 b	10.6 a	10.6 a
	2013	33.1 b	19.0 a	19.8 a
Alk. pase (µg p-np g^{-1})	2009	n.d	n.d	n.d
	2010	43.2 b	24.5 a	26.1 a
	2011	43.5 b	23.0 a	23.6 a
	2012	48.8 b	25.3 a	30.8 a
	2013	57.6 b	32.0 a	38.4 a

Lower-case letters indicate significant different means between fertilizer treatments within one season; Duncan-test $p \leq 0.05$

*n.d.: not determined; TPF: triphenylformazan; p-np: p-nitrophenolphosphate

In the seasons from 2009 to 2012 maize (*Zea mays* L.) was grown and in 2013 beet root (*Beta vulgaris* ssp. vulgaris)

Table 6 Percentage of total content of dry matter (DM), phosphorus (P), nitrogen (N), ammonium N (NH_4-N), potassium (K), and magnesium (Mg) in the digestate before and after liquid–solid separation

	DM (% FM)	Total P (g kg^{-1} FW)	Total N (g kg^{-1} FW)	NH_4^-N (% total N)	Total K (g kg^{-1} FW)	Total Mg (g kg^{-1} FW)
Digestate A	6.64	0.75	4.22	49	3.00	0.83
Liquid	4.80	0.55	3.89	54	3.51	0.63
Solid	**32.7**	**3.22**	**7.54**	**41**	**2.90**	**3.11**
Digestate B	6.41	0.64	4.60	33	3.24	0.41
Liquid	4.92	0.44	4.20	38	3.40	0.31
Solid	**28.3**	**2.79**	**8.63**	**19**	**3.42**	**1.82**

whereas the organic P represents only about 5–20 % of the total P (Bachmann et al. 2016; Dou et al. 2000).

One can conclude that biogas digestates are equally suitable P fertilizers compared to unfermented slurries or high soluble mineral P fertilizers. However, the continuous application could have negative effects on soil microorganisms as indicated by the decreased activity of the dehydrogenase and alkaline phosphatase. Future research should thus investigate more the long-term effects of different digestates on the availability of P in the soil including the effect on microbial P cycling.

Biomass Ashes

Ashes from combustion of biomass are the oldest mineral fertilizer in the world. Biomass ashes are nearly free of N but contain P and other nutrients needed for plant nutrition (Schiemenz and Eichler-Löbermann 2010; Patterson et al. 2004). Furthermore, biomass ashes can also be used as liming agents (Mandre 2006), and can stimulate the microbial activities in the soil (Demeyer et al. 2001).

The nutrient compositions of ashes are affected by different factors. Particularly, the kind of biomass combusted influences the quality and the nutrient values of ashes. Beside the raw material used, the combustion process itself affects the chemical composition of the ashes. The P concentration in biomass ashes may vary between less than 1 up to more than 10 % (see Table 7).

Results regarding the availability of the P in the biomass ashes are also different. According to Ohno and Erich (1990), only a small part of P added with wood ash is available for plants. A poor solubility of P in biomass ashes was also described by Pels et al. (2005). In contrast, positive effects of alfalfa stem ash on Olsen P in loamy soil were detected by Mozaffari et al. (2002). Increased available P contents in soil and better growth of maize due to fertilization with cocoa husk ash were reported by Onwuka et al. (2007). Phongpan and Mosier (2003) found positive effects of rice hull ash on rice yield and Ikpe and Powell (2002) reported higher millet yields after application of millet ash.

The P uptake efficiency and P mobilization mechanisms of crops are also important for a high utilization of the applied P with the ashes. In a study combining three different biomass ashes with eight different crops a relatively high

Table 7 P contents in biomass ashes

Type of ash	P content %
Bagasse ash (Jamil et al. 2004)	0.01
Alfalfa stem ash (Mozaffari et al. 2002)	0.90
Horticulture ashes (Zhang et al. 2002)	0.04–1.00
Wood ashes (Erich and Ohno 1992; Saarsalmi et al. 2001; Hytönen 2003)	0.90–1.70
Wheat straw ash (Hytönen 2003)	1.30
Rape straw ash (Hytönen 2003)	2.10
Poultry litter ash (Yusiharni 2001; Codling 2006)	5.00
Cereal ash (Eichler et al. 2008b)	10.4

Table 8 Effect of biomass ashes on P uptake of main crops and catch crops (P uptake of the shoots in mg pot^{-1}) in a pot experiment on sandy loam

Fertilizer treatment	Maize	Blue lupin	Summer barley	Oilseed rape	Oil radish	Phacelia	Italian ryegrass	Buck-wheat
CON	44.3 a	53.6 ab	67.1 a	47.2 a	57.2 a	110.9 a	51.5 ab	110.5 bc
	(100)	(100)	(100)	(100)	(100)	(100)	(100)	(100)
TSP	69.2 c	61.6 bc	92.7 b	78.4 b	94.1 c	144.4 c	80.4	137.1 d
	(156)	(115)	(138)	(166)	(165)	(130)	(156)	(124)
RMA	71.7 cd	64.9 c	94.6 b	73.2 b	89.3 c	150.3 c	73.5 c	120.8 cd
	(162)	(121)	(141)	(155)	(156)	(136)	(143)	(109)
SA	56.7 b	56.3 bc	70.6 a	67.9 b	71.4 b	142.0 c	55.9 b	78.2 a
	(128)	(105)	(105)	(144)	(125)	(128)	(109)	(71)
CA	77.0 d	74.0 d	88.6 b	73.0 b	86.7 c	150.0 c	80.1 c	127.5 cd
	(174)	(138)	(132)	(155)	(152)	(135)	(156)	(115)
KCl	45.4	46.4 a	62.7 a	51.9 a	64.6 ab	123.5 b	47.8 a	86.6 ab
	(102)	(87)	(93)	(110)	(113)	(111)	(93)	(78)
Mean	60.7	59.5	79.4	65.2	77.2	136.8	64.9	110.1

Adopted from Schiemenz and Eichler-Löbermann (2010)
CON = control, TSP = TripleSuperP, RMA = rape meal ash, SA = straw ash, CA = cereal ash, KCl = potassium chloride; *** $p \leq 0.001$; different characters indicate significant different means at $p \leq 0.05$ within a column
Relative values in brackets; control = 100

bio-availability of P in the biomass ashes was found by Schiemenz and Eichler-Löbermann (2010) (see also Table 8). However, difference occurred between the crops x biomass ash treatments. Thus, the interactive effects of biomass ashes and cultivated crops may have an additional effect on P utilization from ashes. Another interesting approach is the ash application to green fertilizer catch crops. High P uptakes of these catch crops can guarantee high P release after decomposition. Since catch crops additionally supply the soil with organic material, an application of biomass ashes to catch crops seems very promising.

References

Bachmann S, Uptmoor R, Eichler-Löbermann B (2016) Phosphorus distribution and availability in untreated and mechanically separated biogas digestates. Sci Agric 73:9–17

Bachmann S, Wenzel S, Eichler-Löbermann B (2011) Co-digested dairy slurry as a phosphorus and nitrogen source for *Zea mays* L. and *Amaranthus cruentus* L. J Plant Nutr Soil Sci 174:908–915

Bachmann S, Gropp M, Eichler-Löbermann B (2014) Phosphorus availability and soil microbial activity in a 3 year field experiment amended with digested dairy slurry. Biomass Bioenergy 70:429–439

Betencourt E, Duputel M, Colomb B, Desclaux D, Hinsinger P (2012) Intercropping promotes the ability of durum wheat and chickpea to increase rhizosphere phosphorus availability in a low P soil. Soil Biol Biochem 46:181–190

Brandt C, Balko C, Eichler-Löbermann B (2011) Interactive effects of soil water content and phytin supply on phosphorus nutrition of different crops species. Agric For Res 61:317–326

Bünemann E, Smithson PC, Jama B, Frossard E, Oberson A (2004) Maize productivity and nutrient dynamics in maize-fallow rotations in western Kenya. Plant Soil 264:195–208

Callaway RM (2007) Positive interactions and interdependence in plant communities. Springer, Dordrecht

Codling EE (2006) Laboratory characterization of extractable phosphorus in poultry litter and poultry litter ash. Soil Sci 171:858–864

Daleo P, Iribarne O (2009) Beyond competition: the stress-gradient hypothesis tested in plant-herbivore interactions. Ecology 90:2368–2374

Deacon JW (2006) Fungal biology. Wiley-Blackwell, Oxford

Demeyer A, Nkana JCV, Verloo MG (2001) Characteristics of wood ash and influence on soil properties and nutrient uptake: an overview. Bioresour Technol 77:287–295

Devi KN, Shamurailatpam D, Singh TB, Athokpam H, Singh NB, Singh NG, Singh LN, Singh AD, Chanu OP, Singh SR, Devi KP, Devi LS (2014) Performance of lentil (*Lens culinaris* M.) and mustard (*Brassica juncea* L.) intercropping under rainfed conditions. AJCS 8 (2):284–289

Dobbelaere S, Croonenborghs A, Thys A, Ptacek D, Okon Y, Vanderleyden J (2002) Effect of inoculation with wild type *Azospirillumbrasilense* and *A. irakense* strains on development and nitrogen uptake of spring wheat and grain maize. Biol Fertil Soils 36:284–297

Dou Z, Toth JD, Galligan DT, Ramberg CF, Ferguson JD (2000) Laboratory procedures for characterizing manure phosphorus. J Environ Qual 29:508–514

Egamberdiyeva D (2007) The effect of plant growth promoting bacteria on growth and nutrient uptake of maize in two different soils. Appl Soil Ecol 36:184–189

Egamberdiyeva D, Höflich G (2003) Influence of growth-promoting bacteria on the growth of wheat in different soils and temperatures. Soil Biol Biochem 35:973–978

Ehrmann J, Ritz K (2014) Plant: soil interactions in temperate multi-cropping systems. Plant Soil 376:1–29

Eichler B (2004) Möglichkeiten zur Einflussnahme auf Phosphorkreisläufe für die Gestaltung nachhaltiger Bodennutzungssysteme. Habilitation thesis, University of Rostock, Germany

Eichler-Löbermann B, Schnug E (2006) Crop plants and the availability of phosphorus. In Dekker M (eds) Encyclopedia of soil science, pp 348–350. doi:10.1081/E-ESS-120025131

Eichler-Löbermann B, Köhne S, Kowalski B, Schnug E (2008) Effect of catch cropping on phosphorus bioavailability in comparison to organic and inorganic fertilization. J Plant Nutr 31:659–676

Eichler-Löbermann B, Schiemenz K, Makadi M, Vago I, Köppen D (2008b) Nutrient cycling by using residues of bioenergy production—II Effects of biomass ashes on plant and soil parameters. Cereal Res Commun 36:1259–1262

Erich MS, Ohno T (1992) Phosphorus availability to corn from wood ash amended soils. Water Air Soil Poll 64:475–485

Eurostat (2015) Primary production of renewable energy by type. European Commission

Field JA, Caldwell JS, Jeyanayagam S, Reneau RB, Kroontje W, Collins ER (1984) Fertilizer recovery from anaerobic digesters. Trans ASAE 27:1871–1876

Fridley JD (2001) The influence of species diversity on ecosystem productivity: how, where, and why? Oikos 93:514–526

Güngör K, Jürgensen A, Karthikeyan KG (2007) Determination of phosphorus speciation in dairy manure using XRD and XANES Spectroscopy. J Environ Qual 36:1856–1863

Hinsinger P, Betencourt E, Bernard L, Brauman A, Plassard C, Shen J, Tang X, Zhang F (2011) P for two, sharing a scarce resource: soil phosphorus acquisition in the rhizosphere of intercropped species. Plant Physiol 156:1078–1086

Hjorth M, Christensen KV, Christensen ML, Sommer SG (2010) Soli-liquid separation of animal slurry in theory and practice. A review. Agron Sustain Dev 30:153–180

Hytönen J (2003) Effects of wood, peat and coal ash fertilization on Scots Pine foliar nutrient concentrations and growth on afforested former agricultural peat soils. Silva Fennica 37:219–234

Ikpe FN, Powell JM (2002) Nutrient cycling practices and changes in soil properties in the crop-livestock farming systems of western Niger Republic of West Africa. Nutr Cycl Agroecosyst 62:37–45

Jamil M, Qasim M, Umar M, Subhan A (2004) Impact of organic wastes (bagasse ash) on the yield of wheat (*Triticum aestivum* L.) in a calcareous soil. Int J Agr Biol 6:468–470

Krey T, Caus M, Baum C, Ruppel S, Eichler-Löbermann B (2011) Interactive effects of plant growth promoting rhizobacteria and manuring on the phosphorus supply of *Zea mays* L. and *Brassica napus* L. J Plant Nutr Soil Sci 174:602–613

Krey T, Vassilev N, Baum C, Eichler-Löbermann B (2013) Effects of long-term phosphorus application and plant growth promoting rhizobacteria on maize phosphorus nutrition under field conditions. Euro J Soil Biol 55:124–130

Li T, Ma G, Zhang X, Wang C (2006) Change characteristics of organic acid und amino acid in root exudates in different grain amaranth genotypes. Plant Nutr Fert Sci 11:647–653

Li L, Ulén B, Bergkvist G, Aronsson H (2014) Freezing-thawing effects on phosphorus leaching from catch crops. Nutr Cycl Agroecosyst 99:17–30

Loria ER, Sawyer JE (2005) Extractable soil phosphorus and inorganic nitrogen following application of raw and anaerobically digested swine manure. Agron J 97:879–885

Malézieux E, Crozat Y, Dupraz C, Laurans M, Makowski D, Ozier-Lafontaine H, Rapidel B, De Tourdonnet S, Valantin-Morison M (2009) Mixing plant species in cropping systems: concepts, tools and models. A review. Agron Sustain Dev 29:43–62

Mandre M (2006) Influence of wood ash on soil chemical composition and biochemical parameters of young Scots pine. Proc Estonian Acad Sci Biol Ecol 55:91–107

Martinez A, Obertello M, Pardo A, Ocampo JA, Godeas A (2004) Interactions between *Trichoderma pseudokoningii* strains and the arbuscular mycorrhizal fungi *Glomus mosseae* and *Gigaspora rosea*. Mycorrhiza 14:79–84

Mikic A, Cupina B, Rubiales D, Mihailovic V, Sarunait L, Fustec J, Antanasovic S, Krstic Đ, Bedoussac L, Zoric L, Đorđevic V, Peric V, Srebri M (2015) Models, developments, and perspectives of mutual legume intercropping. Adv Agron 130:337–419

Mohammad MJ, Hamad SR, Malkawi HI (2003) Population of arbuscular mycorrhizal fungi in semi-arid environment of Jordan as influenced by biotic and abiotic factors. J Arid Environ 53:409–417

Möller K, Müller T (2012) Effects of anaerobic digestion on digestate nutrient availability and crop growth: a review. Eng Life Sci 12:242–257

Möller K, Stinner W, Deuker A, Leithold G (2008) Effects of different manuring systems with and without biogas digestion on nitrogen cycle and crop yield in mixed organic dairy farming systems. Nutri Cycl Agroecosyst 82:209–232

Møller HB, Sommer SG, Ahring BK (2002) Separation efficiency and particle size distribution in relation to manure type and storage conditions. Bioresour Technol 85:189–196

Morris RA, Garrity DP (1993) Resource capture and utilization in intercropping; non-nitrogen nutrients. Field Crops Res 34:319–334

Mozaffari M, Russelle MP, Rosen CJ, Natar EA (2002) Nutrient supply and neutralizing value of alfalfa stem gasification ash. Soil Sci Soc Am J 66:171–178

Nkoa R (2014) Agricultural benefits and environmental risks of soil fertilization with anaerobic digestates: a review. Agron Sustain Dev 34:473–492

Nurruzzaman M, Lamers H, Bolland MDA (2006) Distribution of carboxylates and acid phosphatase and depletion of different phosphorus fractions in the rhizosphere of a cereal and three grain legumes. Plant Soil 281:109–120

Nziguheba G, Bünemann EK (2005) Organic phosphorus dynamic in tropical agroecosystems. In: Turner BL, Frossard E, Baldwin DS (eds) Organic phosphorus in the environment. CABI Publishing, Wallingford, pp 243–268

Oberson A, Joner EJ (2005) Microbial turnover of phosphorus in soil. In: Turner BL, Frossard E, Baldwin DS (eds) Organic phosphorus in the environment. CABI Publishing, Wallingford, pp 133–164
Oberson A, Tagmann HU, Langmeier M, Dubois D, Mäder P, Frossard E (2010) Fresh and residual phosphorus uptake by ryegrass from soils with different fertilization histories. Plant Soil 334:391–407
Ohno T, Erich MS (1990) Effect of wood ash application on soil pH and soil test nutrient levels. Agric Ecosyst Environ 32:223–239
Ojo OD, Akinrinde EA, Akoroda MO (2010) Residual effects of phosphorus sources in grain amaranth production. J Plant Nutr 33:770–783
Onwuka MI, Osodeke VE, Okolo NA (2007) Amelioration of soil acidity using cocoa husk ash for maize production in Umudike area of south east Nigeria. Trop Subtrop Agroecosyst 7:41–45
Pagliari PH, Laboski CAM (2013) Dairy manure treatment effects on manure phosphorus fractionation and changes in soil test phosphorus. Biol Fert Soils 49:987–999
Patterson SJ, Acharya SN, Thomas JE et al (2004) Integrated soil and crop management: Barley biomass and grain yield and canola seed yield response to land application of wood ash. Agron J 96:971–977
Pekrun C (2014) Was verträgt sich mit Mais. DLG-Mitteilungen 2:72–73
Pels JR, De Nie DS, Kiel JHA (2005) Utilization of ashes from biomass combustion and gasification. In: 14th European biomass conference and exhibition, Paris, France. ECN-RX-05-182, 17–21 Oct 2005
Phongpan S, Mosier AR (2003) Impact of organic residue management in nitrogen use efficiency in an annual rice cropping sequence of lowland Central Thailand. Nutr Cycl Agroecosyst 66:233–240
Read DJ, Perez-Moreno J (2003) Mycorrhizas and nutrient cycling in ecosystems—a journey towards relevance? New Phytol 157:475–492
Richardson AE, Simpson RJ (2011) Soil microorganisms mediating phosphorus availability. Plant Physiol 156:989–996
Richardson AE, Barea JM, McNeill AM, Combaret CP (2009) Acquisition of phosphorus and nitrogen in the rhizosphere and plant growth promotion by microorganisms. Plant Soil 321:305–339
Richardson AE, Lynch JP, Ryan PR, Delhaize E, Smith FA, Smith SE, Harvey PR, Ryan MH, Veeneklaas EJ, Lambers H, Oberson A, Culvenor RA, Simpson RJ (2011) Plant and microbial strategies to improve the phosphorus efficiency of agriculture. Plant Soil 349:121–156
Roschke M (2003) Verwertung der Gärrückstände. In Ministerium für Landwirtschaft Umweltschutz und Raumordnung des Landes B (eds) Leitfaden Biogas, pp 29–33
Saarsalmi A, Mälkönen E, Piirainen S (2001) Effects of wood ash fertilization on forest soil chemical properties. Silva Fennica 35:355–368
Schiemenz K, Eichler-Löbermann B (2010) Biomass ashes and their phosphorus fertilizing effect on different crops. Nutr Cycl Agroecosyst 87:471–482
Schilling G, Gransee A, Deubel A, Lezovic GS, Ruppel S (1998) Phosphorus availability, root exudates, and microbial activity in the rhizosphere. J Plant Nutr Soil Sci 161:465–478
Selles F, Campbell CA, Zentner RP (1995) Effect of cropping and fertilization on plant and soil phosphorus. Soil Sci Soc Am J 59:140–144
Sørensen JN (1991) Effect of catch crops on the content of soil mineral nitrogen before and after winter leaching. Z Pflanzenernähr Bodenkd 155:61–66
Thomsen IK, Olesen JE, Møller HB, Sørensen P, Christensen BT (2012) Carbon dynamics and retention in soil after anaerobic digestion of dairy cattle feed and faeces. Soil Biol Biochem 58:82–87
Tosti G, Guiducci M (2010) Durum wheat-faba bean temporary intercropping: effects on nitrogen supply and wheat quality. Eur J Agron 33:157–165
Umetsu K, Kondo R, Tani M, Hayashi T (2001) Fertilizer value of anaerobically co-digested dairy manure and food processing wastes. In: Tkahashi J, Young BA (eds) Greenhouse gases and animal agriculture, proceedings. Elsevier, Amsterdam, pp 331–342

Vassilev N, Toro M, Vassileva M, Azcon R, Barea JM (1997) Rock phosphate solubilization by immobilized cells of Enterobacter sp. in fermentation and soil conditions. Bioresour Technol 61:29–32

Vassilev N, Vassileva M, Nikolaeva I (2006) Simultaneous P-solubilizing and biocontrol activity of microorganisms: potentials and future trends. App Microbiol Biotechnol 71:137–144

Wang DM, Marschner P, Solaiman Z, Rengel Z (2007) Below ground interactions between intercropped wheat and Brassicas in acidic and alkaline soils. Soil Biol Biochem 39:961–971

Yusiharni BE (2001) An evaluation of chicken litter ash, wood ash and slag for use as lime and phosphate soil amendments. Master Thesis. University of Mataram, Indonesia

Zhang F-S, Yamasaki S, Nanzyo M (2002) Waste ashes for use in agricultural production: I. Liming effect, contents of plant nutrients and chemical characteristics of some metals. Sci Total Environ 284:215–225

Phosphorus—The Predicament of Organic Farming

Hans Marten Paulsen, Ulrich Köpke, Astrid Oberson and Gerold Rahmann

Abstract Today on many sites organic farming might neglect phosphorus (P) fertilization due to soil reserves build up by P fertilization of former farming systems. Additionally organic farming has restricted itself to the use of non-solubilised rock phosphate as mineral fertilizer source that only has limited plant availability on agricultural soils with adequate pH. Also recycling of P from the food chain back to organic agriculture is not consequently realised. These predicaments of organic farming endanger its future sustainability. The article gives an overview on the state of knowledge of the use of topsoil and subsoil P reserves by plants in organic production, on possibilities to activate them by biological measures for direct use by plants and for their redistribution on farms. As in legislation phosphate rock is defined as prevailing mineral fertilizer source for organic farming, a detailed view is given on the use of its P in dependence of plant variety. Also attempts to improve its fertilizer value by mixtures with organic matter are reviewed. It is obvious that plant roots and soil organisms can activate P from the soil and keep it in biological turnover. This is a promising concept to use soil reserves, but quantification of time-spans offering sufficient P supply just by this approach is not possible. Therefore, while promoting root density, root penetration and biological activity in soils, site specific analyses of plant available P in soil, analyses of field and farm P-budgets as well as of plants P supply are necessary to decide on P-fertilization. Soluble P sources, preferably from recycling processes, should be developed for future fertilization demands of organic farming.

Keywords Soil P reserves · Biological P mobilization · P redistribution · Phosphate rock · Recycling P

H.M. Paulsen (✉) · G. Rahmann
Thünen-Institute of Organic Farming, Trenthorst, Germany
e-mail: hans.paulsen@thuenen.de

U. Köpke
Institute of Organic Agriculture (IOL), University of Bonn, Bonn, Germany

A. Oberson
Institute of Agricultural Sciences, Group of Plant Nutrition,
Swiss Federal Institute of Technology (ETH), Zurich, Switzerland

E. Schnug and L.J. De Kok (eds.), *Phosphorus in Agriculture: 100 % Zero*,
DOI 10.1007/978-94-017-7612-7_10

Introduction

Phosphorus (P) is an essential nutrient for all living organisms and is therefore a key element of agricultural production. Declining P resources along with widespread occurrence of P deficient soils threaten agricultural production worldwide (Vance 2001; Schröder et al. 2011). Organic farming has restricted itself to the recycling of P by livestock manure, to the use of hardly soluble rock phosphate as mineral fertilizer and to the use of P-containing organic materials (Commission Regulation (EC) No 889/2008). In farms located on P-rich soils, P exported from fields by agricultural products may exceed P input by fertilizers for quite some time without yield reductions, while in other situations low P availability may limit crop and animal production. In this chapter we will address: (i) the importance of accounting for the soil P status; (ii) the establishment of P input-output balances at the plot and farm scale to identify in-farm P transfer; (iii) ways to mobilise soil-borne and fertilizer P by biological measures; and (iv) the introduction of additional P fertilizers such as P recycled from the food industry and food consumption chain and waste water in order to sustain future organic production.

Organic Farming on Sites with Different Fertilization History

Reserves and availability of soil bound P are dependent on specific site conditions, i.e., soil type, clay mineralogy, soil organic matter, pH, Al and Fe-oxides and former fertilizer inputs. Additionally, limited fertilizer access and different fertilizing practices lead to high variation in the magnitudes of P imbalances worldwide (Vitousek et al. 2009) and inside continents (Cobo et al. 2010; McDonald et al. 2011). This volume focuses on agricultural production in countries, which discharge P to the Baltic Sea. In this region and in central Europe organic farming benefits from P reserves that accumulated in the soil due to over-fertilization during past decades (Barberis et al. 1996). In Germany, P input into agriculture with fertilizers and feedstuff imports exceeded the outputs during the period 1950–2000 by about 800 kg ha^{-1} P in the new and 1100 kg ha^{-1} P in the old federal states (Köster and Nieder 2007). Also in many other European countries (Schröder et al. 2011), decades of P fertilization higher than the crop removal increased the soil P levels. Several long-term experiments in Europe have shown only slightly lower crop yields after decades of omitted P fertilization compared with P-fertilized plots (Jungk et al. 1993; Gransee and Merbach 2000; Gallet et al. 2003), and the great potential of residual soil P in supplying crops has been highlighted (Sattari et al. 2012). Nevertheless, with negative long-term budgets, available P in the soil will decrease (Oehl et al. 2002; Frossard et al. this volume) and organic farming will increasingly depend on P amendments in the future (Loes and Ogaard 2001;

Newman 1997). The question is when P availability in organically cropped soils will limit yields and which measures and sources can be used to replenish the plant available P pool in the soil.

P Budgets of Organic Farming Systems Are Often Negative

Due to their feedstuff imports and manure management, organic livestock farms are often described with balanced or even positive P farm budgets (Watson et al. 2002; Berry et al. 2003) although their P export via milk products and bones of slaughtered animals is higher compared with arable farms. Compilation of published information for mixed crop livestock farms showed P balance ranges at farm gate from −25 to +23 kg P ha^{-1} a^{-1} (Oberson and Frossard 2005). However, if farms were exclusively using feed from farm-own production, and if they had no other P imports, they would inevitably reduce their P reserves by product export in the long run. Internal P flows, like P transfer from grassland via the stable to the fields in ruminant keeping systems might prolong the phase of sufficient P supply to arable crops but will exhaust the P reserves of grassland. The relevance of in-farm P flows was determined on the basis of trial results on farming systems with grassland, livestock and/or biogas manure, and with and without imports of organic matter (Möller 2009, Table 1).

Here without P-import, negative yearly P budgets between −2 and −3 kg ha^{-1} occurred in grassland and arable land. A biomass import with an N-flow of 40 kg ha^{-1} N via a biogas plant changed the balance to positive values of 1.8 and 3.6 kg ha^{-1} P on grassland and arable land, respectively. When calculating P-balances, soil P losses via leaching can normally be neglected, but may play a role on light sandy soils with high ground water table where P losses of 0.04–0.2 kg ha^{-1} a^{-1} P have been reported (Sorensen and Rubaek 2012; Liu et al. 2012). Loss of P with soil particles and plant residues by erosion can be significantly influenced by cultivation method and its timing. Early cultivation and quick establishment of vegetation covers before the rainfall season decreased losses as well as contour instead of downhill cultivation (Withers et al. 2007; Barbosa et al.

Table 1 P balances (kg ha^{-1} P) of grassland and arable land of two organic dairy farming systems (29.8 % grassland) fertilizing with livestock slurry without P import (P circle) or biogas manure with P import by external substrates (digestion of all field residues, manure and external plant silage: P import of 5.3 kg ha^{-1})

	Grassland, 1 ha			Arable land, 1 ha		
	Export	Return	Balance	Export	Return	Balance
P circle[a]	−13.8	+10.9	−2.9	−17.0	+14.8	−2.3
P import[a]	−16.2	+17.9	+1.8	−26.0	+29.6	+3.6

Adopted from (Möller 2009)

[a]P import via mineral feed 0.26 kg ha^{-1}

2009). Whereas concentrations of P in runoff water are related to the P content of the soil surface and are more constant over the year (Puustinen et al. 2007). In-field P-transfer and P-losses from theses sources are highly dependent from landscape and rainfall. In an experimental approach on loamy sites in England P losses with erosion and runoff were quantified. Between 0.03 and 0.58 kg ha^{-1} P were transferred under different cultivation and drilling regimes. At the same time 32–787 kg ha^{-1} sediment were moved (Withers et al. 2007). Whereas Schröder et al. (2011) summarized the variability of P losses associated with all forms of erosion (wind and water) from other worldwide studies and suggests values between 3 and 20 kg ha^{-1} P that might be transferred in the landscape according to the different site conditions and usages.

Modin-Edman et al. (2007) found negative balances for ley fields between 10 and 20 kg ha^{-1} a^{-1} P according to the management of an organic farm over 36 years in Sweden. But a considerable amount of this P remains in the farm cycle and can be transferred to other fields. Data from an organic dairy farm in North Germany without any external P-fertilizer input showed an average P flow from pasture to the stable of 4 kg ha^{-1} a^{-1} P via forage cuts. Here, on the sites that were mown according to differing biomass yields between the plots, 3.5–12.5 kg ha^{-1} a^{-1} P were exported in the first cut prior to grazing. P flow with slurry back to the grassland was only 1.4 kg ha a^{-1} (Ohm et al. 2015). Nevertheless, small P amounts like this mobilised by plants from soils in other farm parts may help to reduce balance gaps in arable land. In any case, when soil reserves are depleted, those buffers have to be refilled sometime. These in-farm P-flows are of special interest because they are acquired by plants from the soils and bound in biomass keeping P in an active biological turn-over. Arable fields of organic farms specialised on cash crops cannot benefit from P transfer from grassland within the farm. Here the introduction of an in-farm transfer by biogas manure generated from green manure crops could induce N flows that are more evenly (Stinner et al. 2008) and also the distribution of P acquired by plants from the soil might be improved. Ohm et al. (2015) quantified the P-off-take of aboveground red-clover grass on an organic farm with 17–32 kg ha^{-1} a^{-1} with two or more harvests and depending of the yield produced.

Withdrawal balances are not always closely related to chemical analyses in soil P changes. This shows the importance of unaccounted soil reserves, e.g., P from the subsoil (Garz et al. 2000), and its buffering capacity in practice. Thus, sooner or later all systems need P import by external sources, like organic vegetable production, which often relies on high amounts of N and P purchased with imported organic manures or green waste composts (Schmutz et al. 2007). Nevertheless, sometimes P surpluses occur in those systems (Nelson and Janke 2007). Consequently, budgets of P at the farm and field level should be used to get an overview of P supply. They should be accompanied by soil tests and leaf tissue analyses if deficits are expected. An on-farm improvement of P supply might include a targeted use and distribution of P mobilised from the soils by different plant covers.

Soil P Tests to Detect the Need for P Inputs

Proper P management requires that the availability of P soil reserves can be described quantitatively. The common practice is to orientate P fertilization based on chemical soil tests recommended, e.g., by advisory centres. The interpretation of extracted inorganic P contents is based on yield response curves from fertilized conventional fields and green house trials (Cai et al. 1997). As an example, data evaluation of 4737 recent P-fertilization trials in Germany and Austria resulted in the conclusion that insufficient P availability requiring external P fertilizers is reached when P contents determined by soil test (CAL/DL methods) are lower than 4.4 mg 100 g^{-1} P in soil. It was found that on sites with a soil concentration below this threshold yield increases over 4 % by P-fertilization were highly probable. The limit of 4 % yield increase was set by the median yield effect by P fertilization found in the data set (Kuchenbuch and Buczko 2011). Generally it was shown in conventional systems that even if nutrient contents in the top soil classified by CAL and DL method are 'very low' (A) (0–3.5 mg 100 g^{-1} P) or 'low' (B) (3.5–7.4 mg 100 g^{-1} P), there is only little probability to increase yields by additional P fertilization, even when highly soluble P-fertilizers were applied (Table 2). Since N is often the first limiting nutrient in organic farming and yield levels are lower, the common practice for fertilizer recommendations in organic farming is to indicate the soil test level "low" (B) as sufficient (Kolbe and Schuster 2011).

But the P needs of crops for the reduced yield levels apparent under organic growth conditions are not well defined. Compared with conventional agriculture, soils under long-term organic farming have lower available P contents (Loes and Ogaard 2001; Oehl et al. 2002). Limited P availability may for a while sustain organic farming yields. However, with continued depletion, available P will become a limiting factor. Thus organic farming cannot avoid maintaining a sufficient available P status through P fertilizer inputs. The evolution of the available soil P status can be followed by P budgets combined with a regular monitoring of the soil available-P status by soil testing.

Table 2 Average yields of P fertilization trials in Germany, P uptake and available P concentrations in soil as upper yield response level to P-fertilization with superphosphate (Schachtschabel and Köster 1985)

	n	Yield		P uptake (kg t^{-1})	P response
		Seed, root, tuber (t ha^{-1})	Straw, leaves (t ha^{-1})		P (CAL) (mg 100 g^{-1})
Wheat	35	6.1	3.5	1.7	<3.5
Barley	32	5.2	3.5	1.7	<3.5
Sugar beet	16	65	4.4	3.5	<3.5
Maize (kernel use)	7	6.6	3.5	2.2	<4
Potato	6	36	7.4	1.3	<4

The use of different soil extraction methods in different areas of the world tends more to be dependent on history, experiences and existing data backgrounds than scientific reasons. However, no chemical method extracts only available P, but usually also comprises a fraction of non-available P and may not extract all available P (Demaria et al. 2005). Soils under organic and conventional farming tend rather to differ in inorganic than organic P fractions, except microbial P, which is often higher under organic than conventional farming conditions (Oehl et al. 2001). This is illustrated by Keller et al. (2012) where 30 years of organic farming decreased inorganic P fractions, while content and hydrolysability of organic P in soils of a long term field experiment were not significantly affected. These results also show that organic P forms contained in farmyard manure were mineralized and may explain why organic P mineralization measured in organically cropped soils from this same field experiment was greater than in soil receiving exclusively mineral fertilizers (Oehl et al. 2004). However, in organic farming the mineralization of organic P sources from green manure crops, plant residues and organic fertilizers is of utmost importance in supporting the aim of basing organic production on nourishing plants primarily through the soil ecosystem (Commission Regulation (EC) No 889/2008).

As common soil analysis thresholds are limited in the prediction of yield effects, special soil tests for organic farming have been developed (Haneklaus et al. 2005). The method of Balzer and Balzer-Graf (1984), for instance, suggested the ratio of P soluble by different acids as predictor of P supply to plants through soil biological activity. However, the usability of this approach for agricultural practice has not been consequently verified and described.

In organic P related research, process-oriented approaches have been applied to study microbial functions in the soil P dynamics of both organic and conventionally cropped soils. Examples are microbial biomass turnover and organic P mineralization studies (Oehl et al. 2001, 2004) as well as enzyme assays (Keller et al. 2012). The use of isotope studies has much advanced the quantification of the biotic processes in P dynamics (e.g. Frossard et al. 2011). However, also in soils under organic cropping, abiotic processes largely govern P availability (Oehl et al. 2004; see Frossard et al. same volume).

Mobilising Soil P Reserves by Biological Measures

Using Suitable Root Systems and Adapted Crop Rotations

Root excretions can lower the pH (H^+, organic anions), directly transform P in a plant available form (acid phosphatase) or indirectly enhance P solubilization by enhancing microbial growth (Jungk and Claassen 1989; Tarafdar and Claassen 2003; Hinsinger 2001). Moreover, combined mechanisms occur according to the plants species. But diffusion of orthophosphate ions often limits P uptake of plants from soil (Kraus et al. 1987; Bhat and Nye 1974; Bieleski 1973). Therefore root architecture, root geometry, root-length density and active root surface area, increased by mycorrhiza in most crops, are important factors that govern the ability of plants to acquire nutrients from the solid phase of the soil (Vance et al. 2002; Gahoonia et al. 1997; Chapin 1980; Claassen et al. 2002; Gosling et al. 2006). So, P uptake efficiency can be increased by the use of cultivars with increased root hair development and might help plants to satisfy their high P demands especially in early growth stages (Gahoonia et al. 1992; Gahoonia and Nielsen 2004; Römer and Schilling 1986). Thus, the enhancement of active root surface in soil to assure intensive P exploration of plants should be an inevitable goal of agricultural production. Organic farming should entail cropping patterns that display different root systems and strategies for efficient P uptake, either in diversified crop rotations and/or by using intercrops.

In a degraded prairie soil it was shown that continuous alfalfa growth could restore lost P reserves by P mobilisation from stable soil reserves. This was determined by a sequential extraction showing an increase to more instable P and a decrease of residual P in the root zone after 7 years (Daroub et al. 2001). Also, white lupin is described to efficiently acquire sparingly soluble phosphate from soils, whereas faba bean is more efficient in the uptake of P from soluble P-fractions (Mat Hassan et al. 2010). Under P-deficient conditions white lupin develops dense cluster roots releasing P mobilising root exudates (Gilbert et al. 1999). Also buckwheat is described to have an effective P-uptake from sparingly soluble phosphates (Eichler-Löbermann et al. 2008; Schiemenz and Eichler-Löbermann 2011). This effect could be related to high Ca uptake and release of H^+-Ions in the root zones solubilising Ca-phosphates and to high root-surface to root-weight ratio in its long and fine roots, as well as to high phosphatase activity in the rhizosphere (Amann and Amberger 1989). Depending on the chemical soil characteristics the P forms are dominated by either Fe- and Al-bound P, or by Ca-bound P forms. They determine whether a pH-increase or decrease is required for P mobilisation (Gahoonia et al. 1992). Thus, the most important process of P mobilisation is soil and fertilizer specific.

Acidification related to N_2-fixation enables root systems of legumes to increase utilization of P derived from rock phosphate (Marschner 2012). From a literature overview, it appears that dicots displaying taproot systems combined with intensive rhizosphere activity can be introduced in organic crop rotations to strengthen P availability by the biochemical rhizosphere strategy, whereas monocots make P available by intensively exploring the soil by using the high rooting density strategy.

When grown in mixed cropping, P-efficient crops can directly enhance the P nutrition of the cropping partner (Horst and Waschkies 1987; Li et al. 2003; El Dessougi et al. 2003). But the effective P uptake, e.g., by white lupin, might also compete for P in the following crops (Cavigelli and Thien 2003). Moreover, the cultivation, mulching or harvest and redistribution of more common cover crops like grass-clover may enhance the biological P cycling on farm. Direct on-farm cycling of P with plant material as mulch in vegetables was analysed by Riley et al.

(2003) with an apparent P recovery of 14 % of the transferred P in the first year. In this connection the nutrient (and P) value and nutrient relations of the plant tissue can be influenced by crop type and plant age of the cover crop and might be designed by targeted harvest times for special fertilization purposes (Sorensen and Thorup-Kristensen 2011).

Enhancing Soil Microbial Activity

Soil microbes colonise the rhizosphere, enhanced by root exudates. Phosphorus mobilisation from insoluble inorganic and organic soil reserves can take place by soil acidification or the exudation of, e.g., phosphatases by microbes (Richardson and Simpson 2011). Organic inputs can significantly stimulate soil microbial activity (Elfstrand et al. 2007) and soil P mobilisation (Cavigelli and Thien 2003). It was found that several indicators on biological P cycling are enhanced under organic farming (see compilation in Oberson et al. 2011). But microbes and plants are also competing for the orthophosphate available in the soil solution after solubilisation of inorganic P or after the decomposition of microbial and plant materials. Thus, the amount of available P that may be derived from these biological processes is difficult to predict. So P release to plants from microorganisms is highly dependent on management factors as well as substrate attributes, soil environment and organic matter quality, e.g., the C/P ratio of organic materials (Richardson and Simpson 2011).

Organic farming aims at fostering biological P mobilisation by plants, mycorrhiza and other soil microorganisms. Processes of P-mobilisation hold promise for increasing the use efficiency of P added with mineral or organic sources (Simpson et al. 2011). Nevertheless, Keller et al. (2012), found no indication that the greater microbial activity of organically cropped soils resulted in the use of stable P forms (see also Frossard et al. this volume). In any case, keeping P in organic matter and microorganisms which have high P-turnover rates will save P from entering recalcitrant inorganic forms in soil. C supply to the root zone (Helal and Sauerbeck 1986) enhances the turnover of C-limited microorganisms and the resulting P turnover, and might improve P supply to plants (Richardson et al. 2009).

P Acquisition from the Subsoil

Compared with the topsoil—the layer is usually defined by the cultivation depth of in arable land, e.g. 20–30 cm—, the spatial accessibility of the less mobile P in the subsoil is generally lower. But efficient P utilization in organic agriculture should entail P resources in subsoil horizons. The proportion of P in the deeper layers is considered to range from 25 to 70 % of the total amount of P that is determined in the profile (Schwertmann and Huith 1975; Godlinski et al. 2004). Crops have been identified carrying P into the subsoil due to the fact that high amounts of organic P are translocated downwards by the roots, especially under a high supply of organic fertilizers (Oehl et al. 2002; Franchini et al. 2004). Decomposition of these roots directly influences the P availability in the subsoil (Campbell et al. 1993).

The size and morphology of the root system as well as the extension of the rhizosphere determine the ability of crops to access P from deeper soil layers. Here fast root growth to deeper soil layers takes place in macropores (Böhm and Köpke 1977; Watt et al. 2006). Macropores are either formed by physical processes such as swelling and shrinking or biogenically (old root channels and earthworm borrows). Especially taproot systems are able to open access to the subsoil by forming round shaped larger sized biopores that can be used by subsequent crops to more rapidly develop roots in deeper soil layers. Crop-species that can enhance the earthworm population, especially anaecic earthworms like *Lumbricus terrestris*, should be preferred in organic agriculture. According the results of Dreesmann (1993), a positive feeding habitat that supplies high amounts of nutrients to the worms in the form of non-decomposed plant residues can affect abundance of earthworms and density of earthworm burrows. This effect is further enhanced by soil rest, i.e., given by ley crops such as grass clover or lucerne grass mixtures or fallow crops established for longer than one year.

Compared with the bulk soil, the drilosphere that is influenced by earthworm excreta is characterized by a higher bacteria density (Devliegher and Verstraete 1997), higher population of nitrifying bacteria and higher dehydrogenase activity, and also by higher contents of microbial P and alkaline phosphatase activity meeting higher P contents in the tapestries of the walls coated with worm cast (Pankhurst et al. 2002). Furthermore, these earthworm-influenced soil spaces show higher basal respiration and microbial biomass (Pierret et al. 1999), as well as higher contents of microbial P (Tiunov and Scheu 1999).

By means of ^{32}P tracer methods, Murdock and Engelbert (1958) showed that the subsoil substantially contributed to P nutrition of maize before topsoils were extensively fertilized with P. The P delivery from the subsoil is enhanced when P content in the topsoil is low, as demonstrated with long-term field trials (Wechsung and Pagel 1993; Gransee and Merbach 2000; Stumpe et al. 1994). Investigations of Fleige et al. (1981) resulted in a subsoil contribution of 30 % to spring wheat. When available P decreased under long-term organic management, measurable amounts of P were transferred from the subsoil to the topsoil (Oehl et al. 2002). In contrast, redelivery of P to the subsoil by plant roots was determined in a situation of a P-rich topsoil (Franchini et al. 2004). Due to the increased use of P from the subsoil and recalcitrant P forms with depleting plant available soil P, subsoil P might be of key importance for a plant's supply in organic farming.

Screening of various soil samples from several lysimeter trials with 15 years running time at moderate P fertilization showed a range of 372–620 and 153–305 mg kg^{-1} total P in topsoil and subsoil layers, respectively. At average bulk densities of 1.3 and 1.6 g cm^3 in loamy and sandy soils respectively, between 1450 and 2418 kg ha^{-1} total P were bound in the topsoils (0–30 cm), and between 1591 and 3172 kg ha^{-1} total P bound in the subsoils (30–65 cm, Godlinski et al. 2004).

From another German silty site total P-reserves in the topsoil (0–30 cm) were reported with 2600 kg ha^{-1} and in subsoil (30–105 cm) with 5600 kg ha^{-1} (Barej et al. 2014). Long lasting fertilization trials in Great Britain (157 years, loam), Germany (98 years, loam) and Poland (77 years, sand) ended up with 480, 370 and 206 mg kg^{-1} total P in the topsoils without P fertilization and up to 1200, 942 and 580 mg kg^{-1} in fertilized plots. That is 1248, 1480 and 773 kg^{-1} ha^{-1} P total without P input and up to 3120, 3768 and 2175 kg^{-1} ha^{-1} in P fertilized plots, respectively (Blake et al. 2000). Looking at the binding and distribution of P in subsoils, it is apparent that total P contents in topsoils and subsoils were significantly correlated with organic C contents (Godlinski et al. 2004). Also bulk soil's available inorganic and organic P reserves are much greater (187 kg ha^{-1} in 45–75 cm depth) than in the soil biopores (4 kg ha P) on the silty soil. Generally the rhizosphere soil directly around the roots is specially enriched with organic P and C (Barej et al. 2014). In any case enhancing root access is essential to require subsoil P (Kautz et al. 2013).

Suitable Fertilizers Must Be Discussed

In organic mixed farms, animal manure is the main P source, permitting farm internal P recycling. P from this source is fully accounted for in fertilization recommendations and P budgets (Flisch et al. 2009), thus suggesting that this P becomes fully available to crops. In the short term, i.e. in a pot experiment lasting a few months, plants took up less P from ruminant faeces than from water-soluble mineral fertilizer P (Oberson et al. 2010). The lower recovery of manure P than from mineral P was explained by the relatively large proportion of P in faeces being either in an organic form or in an inorganic form with low solubility. Over the longer term manure P may become available, as suggested by the absence of a significant increase in organic and recalcitrant inorganic P forms in soils that were organically managed for 30 years (Keller et al. 2012). In bio-dynamic farming, composting of farmyard manure is a common practice. Composting of suitable organic municipal and agricultural wastes was shown to reduce the P availability by microbial immobilisation in organic P forms (Frossard et al. 2002; Mahimairaja et al. 1995). This may explain why residual P fertilizer value is lower for manure composts compared with mineral P-fertilizer (Oberson et al. 2010). Fertilizer P use efficiency generally might be improved by banding of P in the root zone, sufficient soil moisture and fertilizer forms adapted to soil properties (McLaughlin et al. 2011).

The use of mineral P fertilizers—which is mainly restricted to rock phosphate in organic farming—is potentially linked to soil contamination with Cd (Kratz and Schnug 2005) as well as U (Kratz and Schnug 2006). Furthermore, the use of rock phosphate is often inadequate because it has to be fully accounted for in the P budget while it is in a more or less insoluble form. At soil pH higher than 6, P release from this P source is doubtful and unpredictable (Arcand and Schneider

2006; Römer 2006; Steffens 1987; McLaughlin et al. 2011). Also the frequently discussed use of rock powders, e.g., of basaltic or volcanic origin as fertilizer in organic farming is ineffective due to low nutrient (and also P) contents and low plant availability (Blum et al. 1989).

P mobilisation from rock phosphate by plants known to have different P uptake strategies has been assessed in organic farming (Steffens et al. 2003). Indirect measurements of P uptake of wheat, white lupin and faba bean fertilized with rock phosphate and of succeeding cover crops were performed in pot trials. Spinach showed an increased P mobilisation from rock phosphate with white lupin pre-crops. Since rapeseed uses the same pools of P elucidated by root exudates from soils like white lupin, reduced yields occurred in rape when following white lupin. Whereas, when grown after pre-crops wheat or faba bean, its yields were not affected. Compared with the use of rock phosphate, easily soluble P-fertilizer (basic slag) that was applied in parallel treatments significantly increased P uptake of grains and straw of all main and cover crops explored. So, P mobilisation from rock phosphate is crop specific and can be used supplementary. But with high P demands of crops and/or with further declining soil P reserves, it does not seem adequate to rely solely on the strength of plants in P mobilisation from rock phosphates for sufficient supply to organic crops.

Several attempts to improve solubility of phosphate rock by mixture with organic wastes are reported from India. Rock phosphate dissolution in manure is dependent on H^+ concentrations from organic acids and nitrification processes and from the continuous immobilisation of $Ca^{2+}$and PO_4^{2-} in the zone of dissolution (Mahimairaja et al. 1995). But when judged by plant P uptake, effects of activation of rock phosphate P by manures or composts are difficult to explain due to mobilising and immobilising side effects of the applied substrate. However with the 32P radiotracer technique, increased rates of P derived from ground rock phosphates (0.15 mm sieve opening) added to biogas manures from kitchen wastes (pH 6.65) of more than 20 % were found. These could be further enhanced by inoculation with phosphate solubilizing fungi (*Aspergillus niger*, *Penicillium oxalicum*; 2 g mycelia per kg biowaste on dry weight basis). The inoculation with fungi decreased pH in the substrate to 5.78 and 5.82 (Shrivastava et al. Shrivastava et al. 2011). The P release of different rock phosphates (0.16 mm sieve opening) during composting with poultry manure (that is rich in Ca^{2+} and has a pH of 8.9) was quantified by the increase of NaOH extractable P between 8 and 27 %. Lowering pH by addition of S^0 significantly increased that process (Mahimairaja et al. 1995). Biswas and Narayanasamy (2006) compared P uptake from rice straw composts enriched with different ground rock phosphates (100 kg straw, 2 and 4 kg P as rock phosphate (0.15 mm sieve opening), 5 kg cow dung, urea solution, *Trichoderma viride*). For different rock phosphate sources they reported 67–89 % of the P uptake of mung bean pods and 56–98 % P uptake of the straw compared with soluble P fertilizer (diammon-phosphate, DAP). The addition of a P-mobilising fungus (*Aspergillus awamori*) to the different sources increased the P uptake further. There on average from 72 to 91.2 % P uptake of pods and from 70.3 to 91.2 % P uptake of straw were realized from the different fertilizers and relative to the uptake from

DAP. Olsen-P contents in soil after application of these enriched composts increased slowly compared with the use of soluble mineral fertilizer. This hints at a slow release of inorganic P from previously immobilised P in organic forms from the enriched composts.

But the possible release of soluble P to plants during the decomposition of organic matter is in competition to the demand of microorganisms and inorganic P immobilisation. Also, after application in fields, phosphate release is strongly dependent on soil and environmental conditions. So, high variability in results has to be expected. Therefore, improving the release of plant available P from rock phosphate by mixtures with organic materials is uncertain and must be further developed for practical application in organic farming. For this, trials with standardized materials, ground and sieved to different sizes, are necessary. P transformations and P release in different organic materials apparent in or sold to organic farms must be explored. Solid and liquid manures as well as certified composts with different pH and grade of decomposition should be used for the identification of suitable application forms of rock phosphate. Up to now no consistent results on mixtures releasing considerable amounts of plant available P are on hand for practical application in organic farming.

Besides mixtures of rock phosphate with organic matter, elemental sulphur and the inoculation with sulphur-oxidizing bacteria, such as *Thiobacillus thiooxidans*, are considered as effective to increase the availability of P (Aria et al. 2010). Those specially developed rock phosphate or bone meal fertilizers generating acid phases to dissolve P (Evans and Price 2009; Schnug et al. 2003) or microbial amendments to increase P-availability (Schnug and Haneklaus 2006) have not yet been introduced in practice.

In the discussion whether acid treatment of rock phosphate to increase plant availability of P can be used in organic farming, the reported limits of mobilisation of different P sources in biological systems must be clearly addressed. Since contradictory results have been found, at least for special cases, i.e. basic soil conditions, calcareous soils with high pH (Cornish 2009; Hamid et al. 2007), the use of soluble P fertilizers such as superphosphate should be allowed.

To close future P gaps P-containing residues from food industry and from human waste and wastewater streams should be recycled also in organic farming. But xenobiotica and heavy metal impacts to agricultural soils must be strictly avoided (Rahmann et al. 2009; Arthurson 2009). New technologies have been developed to produce P-rich sewage sludge or ashes containing low amounts of heavy metals (Adam et al. 2008; Nowak et al. 2010). Nevertheless, due to their different origins and processing, their P availability differs widely (Römer 2013) and agriculture has to care not to be used as depository of industrial waste with no or only low P availability. If in the future markets are able to deliver these new fertilizers, fulfilling the concept of recycling and not impacting the soil with undesired contaminants, organic agriculture might accept these materials to balance P budgets on the long-term. Whether NP-fertilizers from recycling processes containing mineral nitrogen forms, like struvite (magnesium ammonium phosphate) from waste water cleaning, might be integrated in organic fertilization strategies (Dawson et al. 2008) requires further discussion. Also ethical aspects concerning the use of conventional sources containing P, e.g., residues from animal production with low animal welfare standards must be discussed. Organic farmers using bone-, leather-, horn-, hair- or feather-meal, and also farm collaborations shifting livestock and biogas manures and also their P from conventional to organic systems, should be prepared to face public debates on the consistency of nutrient cycling in organic farming before rules on fertilization in organic farming might have to be changed.

Conclusion for Research Needs and Perspectives

The role of a dense and deep plant rooting to increase P uptake by physical measures is undoubted and the positive use of chemical and biological interactions in the root zone has been recognized. But the potential of a consequent increase of biological P mobilisation from soils and fertilizers by management has not sufficiently been quantified. The extent to which plants and soil- and root-microbes can mobilise the P reserves and feed them into the active soil-plant farm P cycle is still open (Guppy and McLaughlin 2009). These biological measures to enhance P mobilisation from soils and fertilizers are distant from confirmed knowledge that can be used in management advices to secure sustainable P supply in organic farming. It is not clear which yield level can be sustained by soil P reserves and their biological and chemical mobilisation. The common expectation is that P has to be imported in the system according to the off-take if the soil reserves are depleted. A hidden depletion of available P in soils under long-term organic management by the use of P sources with low availability in the soil/plant P cycle should be avoided. But despite the knowledge that high soil reserves can be available, measures efficiently prolonging the use phase of this resource are still unknown.

Due to the complex factors determining mobility of P in soil each component of the P cycle in organic farms should be analysed to decide on the necessity and the right form and place for P imports. The decision on adequate P sources to satisfy the typical high P demand in early growth stages of plants (Römer and Schilling 1986) and the biological-chemical P availability in soils should be considered at the same time. Of genuine importance for organic systems might be the role of P for the rhizobial N fixation with legumes that are hampered by insufficient P supply. Deeper knowledge of an adequate P supply for legumes would be beneficial for a systematic improvement and understanding of the nutrient interactions in organic agricultural systems (Oberson et al. 2013; Römer and Lehne 2004; Hogh-Jensen et al. 2002).

P fertilizing strategies for organic farming should clearly address site-specific soil reserves. And in the future identification of suitable fertilizers from P deposits and human waste streams general principles must be revised to overcome the legal predicament of organic farming on the right fertilizer source. An efficient P use from P applied in the fields should be aspired by agricultural measures, by bio-chemical enhancement of P availability in new P sources and by the use of known P fertilizers with plant available P. These aspects should be a driver of future development of organic farming process chains, while P is a limited resource worldwide.

References

Adam C, Pepelinski B, Kley G, Kratz S, Schick J, Schnug E (2008) Phosphorrückgewinnung aus Klärschlammaschen - Ergebnisse aus dem EU-Projekt SUSAN. Österreichische Wasser- und Abfallwirtschaft 3–4 (Special Issue "IFAT 2008"):55–64

Amann C, Amberger A (1989) Phosphorus efficiency of buckwheat (Fagopyrum esculentum). Z Pflanz Bodenkunde 152:181–189

Arcand MM, Schneider KD (2006) Plant- and microbial-based mechanisms to improve the agronomic effectiveness of phosphate rock: a review. An Acad Bras Cienc 78:791–807

Aria MM, Lakzian A, Haghnia GH, Berenji AR, Besharati H, Fotovat A (2010) Effect of Thiobacillus, sulphur, and vermicompost on the water-soluble phosphorus of hard rock phosphate. Bioresour Technol 101:551–554

Arthurson V (2009) Closing the global energy and nutrient cycles through application of biogas residue to agricultural land—potential benefits and drawbacks. Energies 2:226–242

Balzer FM, Balzer-Graf UR (1984) Bodenanalyse system Dr. Balzer. Lebendige Erde 1:13–18, 66–71; 4:151–156

Barberis E, Marsan FA, Scalenghe R, Lammers A, Schwertmann U, Edwards AC, Maguire R, Wilson MJ, Delgado A, Torrent J (1996) European soils overfertilized with phosphorus. 1. Basic properties. Fert Res 45:199–207

Barbosa FT, Bertol I, Luciano RV, Gonzalez AP (2009) Phosphorus losses in water and sediments in runoff of the water erosion in oat and vetch crops seed in contour and downhill. Soil Till Res 106:22–28

Barej JAM, Pätzold S, Perkons U, Amelung W (2014) Phosphorus fractions in bulk subsoil and its biopore systems. Eur J Soil Sci 65:553–561

Berry PM, Stockdale EA, Sylvester-Bradley R, Philipps L, Smith KA, Lord EI, Watson CA, Fortune S (2003) N, P and K budgets for crop rotations on nine organic farms in the UK. Soil Use Manag 19:112–118

Bhat KKS, Nye PH (1974) Diffusion of phosphate to plant roots in soil. Plant Soil 41:383–394

Bieleski RL (1973) Phosphate pools, phosphate transport, and phosphate availability. Annu Rev Plant Physiol 24:255

Biswas DR, Narayanasamy G (2006) Rock phosphate enriched compost: an approach to improve low-grade Indian rock phosphate. Bioresour Technol 97:2243–2251

Blake L, Mercik S, Koerschens M, Moskal S, Poulton PR, Goulding KWT, Weigel A, Powlson DS (2000) Phosphorus content in soil, uptake by plants and balance in three European long-term field experiments. Nutr Cycl Agroecosys 56:263–275

Blum WEH, Herbinger B, Mentler A, Ottner F, Pollok M, Unger E, Wenzel WW (1989): Zur Verwendung von Gesteinsmehlen in der Landwirtschaft. I. Chemisch-mineralogische Zusammensetzung und Eignung von Gesteinsmehlen als Düngemittel. Z Pflanz Bodenk 15:421–425

Böhm W, Köpke U (1977) Comparative root investigations with two profile wall methods. Z Acker Pflanzenb 144:297–303

Cai TT, Olsen TW, Yost RS, Silva JA (1997) Performance indices for tests of soil nutrient status: extractable P. Commun Soil Sci Plan 28:329–339

Campbell CA, Lafond GP, Biederbeck VO, Winkleman GE (1993) Influence of legumes and fertilization on deep distribution of available phosphorus (Olsen-P) in a thin black chernozemic soil. Can J Soil Sci 73:555–565

Cavigelli MA, Thien SJ (2003) P bioavailability following incorporation of green manure crops. Soil Sci Soc Am J 67:1186–1194
Chapin FS (1980) The mineral-nutrition of wild plants. Annu Rev Ecol Syst 11:233–260
Claassen N, Steingrobe B, Syring KM (2002) A mechanistic model to describe the effect of complexing root exudates on transport and uptake of soil nutrients. In: Plant nutrition, developments in plant and soil sciences, symposium 8, vol 92, pp 600–601. doi:10.1007/0-306-47624-X_291
Cobo JG, Dercon G, Cadisch G (2010) Nutrient balances in African land use systems across different spatial scales: a review of approaches, challenges and progress. Agr Ecosyst Environ 136:1–15
Commission regulation (EC) No 889/2008 of 5 September 2008 laying down detailed rules for the implementation of Council Regulation (EC) No 834/2007 on organic production and labelling of organic products with regard to organic production, labelling and control
Cornish PS (2009) Phosphorus management on extensive organic and low-input farms. Crop Past Sci 60:105–115
Daroub SH, Ellis BG, Robertson GP (2001) Effect of cropping and low-chemical input systems on soil P fractions. Soil Sci 166:281–291
Dawson JC, Huggins DR, Jones SS (2008) Characterizing nitrogen use efficiency in natural and agricultural ecosystems to improve the performance of cereal crops in low-input and organic agricultural systems. Field Crops Res 107:89–101
Demaria P, Flisch R, Frossard E, Sinaj S (2005) Exchangeability of phosphate extracted by four chemical methods. J Plant Nutr Soil Sci 168:89–93
Devliegher W, Verstraete W (1997) The effect of lumbricus terrestris on soil in relation to plant growth: effects of nutrient-enrichment processes (NEP) and gut-associated processes (GAP). Soil Biol Biochem 29:341–346
Dreesmann S (1993) Pflanzenbauliche Untersuchungen zu Rotklee- und Luzernegras-Grünbrachen in der modifizierten Fruchtfolge Zuckerrüben - Winterweizen -. Wintergerste. Dissertation Thesis. Faculty of Agriculture, University of Bonn, Germany, pp 116
Eichler-Löbermann B, Kohne S, Kowalski B, Schnug E (2008) Effect of catch cropping on phosphorus bioavailability in comparison to organic and inorganic fertilization. J Plant Nutr 31:659–676
El Dessougi H, zu Dreele A, Claassen N (2003) Growth and P uptake of maize cultivated alone, in mixed culture with other crops or after incorporation of their residues. J Plant Nutr Soil Sci 166:254–261
Elfstrand S, Bath B, Martensson A (2007) Influence of various forms of green manure amendment on soil microbial community composition, enzyme activity and nutrient levels in leek. Appl Soil Ecol 36:70–82
Evans J, Price A (2009) Influence of rates of reactive phosphate rock and sulphur on potentially available phosphorous in organically managed soils in the south-eastern near-mediterranean cropping region of Australia. Nutr Cycl Agroecosys 84:105–118
Fleige H, Strebel O, Renger M, Grimme H (1981) Die potentielle P-Anlieferung durch Diffusion als Funktion von Tiefe, Zeit und Durchwurzelung bei einer Parabraunerde aus Löß. Mitteilungen der Deutschen Bodenkundlichen Gesellschaft 32:305–310
Flisch R, Sinaj S, Charles R, Richner W (2009) Grundlagen für die Düngung im Acker- und Futterbau (GRUDAF) 2009. Agrarforschung 16:1–97
Franchini JC, Pavan MA, Miyazawa M (2004) Redistribution of P roots in soil through cover crop roots. Braz Arch Biol Technol 47:381–386
Frossard E, Achat DL, Bernasconi SM, Bünemann EK, Fardeau JC, Jansa J, Morel C, Rabeharisoa L, Randriamanantsoa L, Sinaj S, Tamburini F, Oberson A (2011) The use of tracers to investigate phosphate cycling in soil-plant systems. In: Bünemann E, Oberson A, Frossard, E (eds) Phosphorus in action—biological processes in soil phosphorus cycling. Springer Soil Biology Series, pp 59–61
Frossard E, Skrabal P, Sinaj S, Bangerter F, Traoré O (2002) Form and exchangeability of inorganic phosphate in composted solid organic wastes. Nutr Cycl Agroecosys 62:103–113

Gahoonia TS, Care D, Nielsen NE (1997) Root hairs and acquisition of P by wheat and barley cultivars. Plant Soil 191:181–188
Gahoonia TS, Claassen N, Jungk A (1992) Mobilization of phosphate in different soils by ryegrass supplied with ammonium or nitrate. Plant Soil 140:241–248
Gahoonia TS, Nielsen NE (2004) Root traits as tools for creating phosphorus efficient crop varieties. Plant Soil 260:47–57
Gallet A, Flisch R, Ryser JP, Frossard E, Sinaj S (2003) Effect of phosphate fertilization on crop yield and soil phosphorus status. J Plant Nutr Soil Sci 166:568–578
Garz J, Schliephake W, Merbach W (2000) Changes in the subsoil of long-term trials in Halle (Saale), Germany, caused by mineral fertilization. J Plant Nutr Soil Sci 163:663–668
Gilbert GA, Knight JD, Vance CP, Allan DL (1999) Acid phosphatase activity in phosphorus-deficient white lupin roots. Plant Cell Environ 22:801–810
Godlinski F, Leinweber P, Meissner R, Seeger J (2004) Phosphorus status of soil and leaching losses: results from operating and dismantled lysimeters after 15 experimental years. Nutr Cycl Agroecosys 68:47–57
Gosling P, Hodge A, Goodlass G, Bending GD (2006) Arbuscular mycorrhizal fungi and organic farming. Agric Ecosyst Environ 113:17–35
Gransee A, Merbach W (2000) P dynamics in a long-term P fertilization trial on Luvic Phaezom at Halle. J Plant Nutr Soil Sci 163:353–357
Guppy CN, McLaughlin MJ (2009) Options for increasing the biological cycling of P in low-input and organic agricultural systems. Crop Pasture Sci 60:116–123
Hamid N, Ali S, Malik KA, Hafeez FY (2007) Diagnosis of nutritional constraints of Azolla spp. to enhance their growth under flooded conditions of salt affected soils. Pak J Bot 39:161–167
Haneklaus S, Schnug E, Paulsen HM, Hagel I (2005) Soil analysis for organic farming. Commun Soil Sci Plan 36:65–79
Helal HM, Sauerbeck D. (1986) Effect of plant roots on carbon metabolism of soil microbial biomass. Z Pflanz Bodenk 149:181–188
Hinsinger P (2001) Bioavailability of soil inorganic P in the rhizosphere as affected by root-induced chemical changes: a review. Plant Soil 237:173–195
Hogh-Jensen H, Schjoerring JK, Soussana JF (2002) The influence of P deficiency on growth and nitrogen fixation of white clover plants. Ann Bot 90:745–753
Horst WJ, Waschkies C (1987) Phosphatversorgung von Sommerweizen (Triticum aestivum L.) in Mischkultur mit Weißer Lupine (Lupinus albus L.) (1987) Z Pflanz Bodenk 150:1–8
Jungk A, Claassen N (1989) Availability in soil and acquisition by plants as the basis for phosphorus and potassium supply to plants. Z Pflanz Bodenk 152:151–157
Jungk A, Claassen N, Schulz V, Wendt J (1993) Pflanzenverfügbarkeit der Phosphatvorräte ackerbaulich genutzter Böden. - Langfristige Feldversuche zur Nutzbarkeit des Bodenphosphors und zur Bewertung der Bodenuntersuchung. Z Pflanz Bodenk 156:397–406
Kautz T, Amelung W, Ewert F, Gaiser T, Horn R, Jahn R, Javaux M, Kemna A, Kuzyakov Y, Munch JC, Patzold S, Peth S, Scherer HW, Schloter M, Schneider H, Vanderborght J, Vetterlein D, Walter A, Wiesenberg GLB, Kopke U (2013) Nutrient acquisition from arable subsoils in temperate climates: a review. Soil Biol Biochem 57:1003–1022
Keller M, Oberson A, Annaheim KE, Tamburini F, Maeder P, Mayer J, Frossard E, Buenemann EK (2012) Phosphorus forms and enzymatic hydrolyzability of organic phosphorus in soils after 30 years of organic and conventional farming. J Plant Nutr Soil Sci 175:385–393
Kolbe H, Schuster M (2011) Bodenfruchtbarbeit im Öko-Betrieb. Untersuchungsmethoden. Sächsisches Landesamt für Umwelt, Landwirtschaft und Geologie, Dresden
Köster W, Nieder R (2007) Wann ist eine Grunddüngung mit Phosphor, Kalium und Magnesium wirtschaftlich vertretbar?, Beratung-mal-anders.de, Bodenfruchtbarkeit und Pflanzenernährung, Heft 1, 2. Auflage
Kratz S, Schnug E (2005) Schwermetalle in P-Düngern. Landbauforsch Volk 286:37–45 (special issue)

Kratz S, Schnug E (2006) Rock phosphates and P fertilizers as sources of U contamination in agricultural soils. In: Merkel BJ, Hasche-Berger A (eds) Uranium in the environment. Springer, Berlin, pp 57–67
Kraus M, Fusseder A, Beck E (1987) Development and replenishment of the P-depletion zone around the primary root of maize during the vegetation period. Plant Soil 101:247–255
Kuchenbuch RO, Buczko U (2011) Re-visiting potassium- and phosphate-fertilizer responses in field experiments and soil-test interpretations by means of data mining. J Plant Nutr Soil Sci 174:171–185
Li L, Zhang F, Li X, Christie P, Sun J, Yang S, Tang C (2003) Interspecific facilitation of nutrient uptake by intercropped maize and faba bean. NutrCycl Agroecosys 65:61–71
Liu J, Aronsson H, Blomback K, Persson K, Bergstrom L (2012) Long-term measurements and model simulations of phosphorus leaching from a manured sandy soil. J Soil Water Conserv 67:101–110
Loes AK, Ogaard AF (2001) Long-term changes in extractable soil P (P) in organic dairy farming systems. Plant Soil 237:321–332
Mahimairaja S, Bolan NS, Hedley MJ (1995) Dissolution of phosphate rock during the composting of poultry manure—an incubation experiment. Fert Res 40:93–104
Marschner P (2012) Marschner's mineral nutrition of higher plants, 3rd edn. Academic Press/Elsevier, Amsterdam 668p
Mat Hassan H, Marschner P, McNeill A (2010) Growth, P uptake in grain legumes and changes in soil P pools in the rhizosphere. In: 19th world congress of soil science, soil solutions for a changing world, Brisbane, Australia. Published on DVD, 1–6 Aug, pp 179–181
McDonald GK, Bennet EM, Potter PA, Ramancutty N (2011) From the cover: agronomic P imbalances across the world's. Proc Natl Acad Sci 108(7):3086–3091. doi:10.1073/pnas.1010808108
McLaughlin MJ, McBeath TM, Smernik R, Stacey SP, Ajiboye B, Guppy C (2011) The chemical nature of P accumulation in agricultural soils-implications for fertilizer management and design: an Australian perspective. Plant Soil 349:69–87
Modin-Edman AK, Oborn I, Sverdrup H (2007) FARMFLOW—a dynamic model for phosphorus mass flow, simulating conventional and organic management of a Swedish dairy farm. Agr Syst 94:431–444
Möller K (2009) Inner farm nutrient flows between arable land and permanent grassland via the stable in organic cropping systems. Eur J Agron 31:204–212
Murdock JT, Engelbert LE (1958) The importance of subsoil phosphorus to corn. Soil Sci Soc Am J 22:53–57
Nelson NO, Janke RR (2007) P sources and management in organic production systems. HortTechnology 17:442–454
Newman EI (1997) P balance of contrasting farming systems, past and present. Can food production be sustainable? J Appl Ecol 34:1334–1347
Nowak B, Pessl A, Aschenbrenner P, Szentannai P, Mattenberger H, Rechberger H, Hermann L, Winter F (2010) Heavy metal removal from municipal solid waste fly ash by chlorination and thermal treatment. J Hazard Matter 179:323–331
Oberson A, Frossard E (2005) Phosphorus management for organic agriculture. In: Sims JT, Sharpley AN (eds) Phosphorus: agriculture and the environment. ASA, CSSA and SSSA, Madison, pp 761–779
Oberson A, Frossard E, Bühlmann C, Mayer J, Mäder P, Lüscher A (2013) Nitrogen fixation and transfer in grass-clover leys under organic and conventional cropping systems. Plant Soil 371:237–255
Oberson A, Pypers P, Bünemann E, Frossard E (2011) Management impacts on biological phosphorus cycling in cropped soils. In: Bünemann E, Oberson A, Frossard, E (eds) Phosphorus in action—biological processes in soil phosphorus cycling. Springer Soil Biology Series, pp 431–458
Oberson A, Tagmann H, Langmeier M, Dubois D, Mäder P, Frossard E (2010) Fresh and residual P uptake by ryegrass from soils with different fertilization histories. Plant Soil 334:391–407

Oehl F, Frossard E, Fliessbach A, Dubois D, Oberson A (2004) Basal organic phosphorus mineralization in soils under different farming systems. Soil Biol Biochem 36:667–675
Oehl F, Oberson A, Probst M, Fliessbach A, Roth HR, Frossard E (2001) Kinetics of microbial P uptake in cultivated soils. Biol Fert Soils 34:31–41
Oehl F, Oberson A, Tagmann HU, Besson JM, Dubois D, Mader P, Roth HR, Frossard E (2002) P budget and P availability in soils under organic and conventional farming. Nutr Cyclin Agroecosys 62:25–35
Ohm M, Schüler M, Fystro G, Paulsen HM (2015) Redistribution of soil phosphorus from grassland to cropland in an organic dairy farm. Landbauforsch Appl Agric Forestry Res 65 (3/4):193–204
Pankhurst CE, Pierret A, Hawke BG, Kirby JM (2002) Microbiological and chemical properties of soil associated with macropores at different depths in a red-duplex soil in NSW. Aust. Plant Soil 238:11–20
Pierret A, Moran CJ, Pankhurst CE (1999) Differentiation of soil properties related to the spatial association of wheat roots and soil macropores. Plant Soil 211:51–58
Puustinen M, Tattari S, Koskiaho J, Linjama J (2007) Influence of seasonal and annual hydrological variations on erosion and phosphorus transport from arable areas in Finland. Soil Tillage Res 93:44–55
Rahmann G, Oppermann R, Paulsen HM, Weissmann F (2009) Good, but not good enough? Research and development needs in organic farming. Landbauforsch Volk 59:29–40
Richardson AE, Barea JM, McNeill AM, Prigent-Combaret C (2009) Acquisition of phosphorus and nitrogen in the rhizosphere and plant growth promotion by microorganisms. Plant Soil 321:305–339
Richardson AE, Simpson RJ (2011) Soil microorganisms mediating P availability. Plant Physiol. doi:10.1104/pp.111.175448
Riley H, Loes AK, Hansen S, Dragland S (2003) Yield responses and nutrient utilization with the use of chopped grass and clover material as surface mulches in an organic vegetable growing system. Biol Agric Hortic 21:63–90
Römer W (2006) Plant availability of P from recycling products and phosphate fertilizers in a growth-chamber trial with rye seedlings. J Plant Nutr Soil Sci 169:826–832
Römer W (2013) Phosphordüngewirkung neuer Phosphatrecyclingprodukte. Berichte über Landwirtschaft 91(1). http://dx.doi.org/10.12767/buel.v91i1.15.g49
Römer W, Lehne P (2004) Neglected P and K fertilization in organic farming reduces N-2 fixation and grain yield in a red clover-oat rotation. J Plant Nutr Soil Sci 167:106–113
Römer W, Schilling G (1986) P requirements of the wheat plant in various stages of its life cycle. Plant Soil 91:221–229
Sattari SZ, Bouwman AF, Giller KE, van Ittersum MK (2012) Residual soil phosphorus as the missing piece in the global phosphorus crisis puzzle. P Natl Acad Sci USA 109:6348–6353
Schachtschabel P, Koster W (1985) Relationship between the phosphate content in soil and the optimum phosphate fertilization in long-term field trials. Z Pflanz Bodenk 148:459–464
Schiemenz K, Eichler-Löbermann B (2011) Biomass ashes and their phosphorus fertilizing effect on different crops. Nutr Cycl Agroecos 87:471–482
Schmutz U, Rayns F, Firth C (2007) Balancing fertility management and economics in organic field vegetable rotations. J Sci Food Agric 87:2791–2793
Schnug E, Haneklaus S (2006) What will organic farming deliver? COR 2006 79:277–281 (Warwik, UK)
Schnug E, Rogasik J, Haneklaus S (2003) The utilisation of fertilizer P with special view to organic farming. Landbauforsch Volk 53:1–11
Schröder JJ, Smit AL, Cordell D, Rosemarin A (2011) Improved phosphorus use efficiency in agriculture: a key requirement for its sustainable use. Chemosphere 84:822–831
Schwertmann U, Huith M (1975) Erosionsbedingte Stoffverteilung in zwei hopfengenutzten Kleinlandschaften der Hallertau (Bayern). Z Pflanz Bodenk 138:397–405

Shrivastava M, Kale SP, D'Souza SF (2011) Rock phosphate enriched post-methanation bio-sludge from kitchen waste based biogas plant as P source for mungbean and its effect on rhizosphere phosphatase activity. Eur J Soil Biol 47:205–212

Simpson RJ, Oberson A, Culvenor RA, Ryan MH, Veneklaas EJ, Lambers H, Lynch JP, Ryan PR, Delhaize E, Smith FA, Smith SE, Harvey PR, Richardson AE (2011) Strategies and agronomic interventions to improve the phosphorus-use efficiency of farming systems. Plant Soil 349:89–120

Sorensen P, Rubaek GH (2012) Leaching of nitrate and phosphorus after autumn and spring application of separated solid animal manures to winter wheat. Soil Use Manage 28:1–11

Sorensen JN, Thorup-Kristensen K (2011) Plant-based fertilizers for organic vegetable production. J Plant Nutr Soil Sci 174:321–332

Steffens D (1987) Influence of Long-Term Application of Different P-Fertilizers on Phosphate Availability in the Rhizosphere of Rape. Z Pflanz Bodenk 150:75–80

Steffens D, Stamm R, Leithold G, Schubert S (2003) Phosphat-Mobilisierung durch Haupt- und Zwischenfrüchte nach Düngung von weicherdigem Rohphosphat im ökologischen Landbau. Justus-Liebig-Universität Gießen, Institut für Pflanzenernährung, Projektbericht Bundesprogramm ökologischer Landbau

Stinner W, Möller K, Leithold G (2008) Effects of biogas digestion of clover/grass-leys, cover crops and crop residues on nitrogen cycle and crop yield in organic stockless farming systems. Eur J Agron 29:125–134

Stumpe H, Garz J and Scharf H (1994) Wirkung der Phosphatdüngung in einem 40jährigen Dauerversuch auf einer Sandlöß-Braunschwarzerde in Halle. J Plant Nutr Soil Sc 157:105–110

Tarafdar JC, Claassen N (2003) Organic phosphorus utilization by wheat plants under sterile conditions. Biol Fert Soils 39:25–29

Tiunov AV, Scheu S (1999) Microbial respiration, biomass, biovolume and nutrient status in burrow walls of Lumbricus terrestris L. (Lumbricidae). Soil Biol Biochem 31:2039–2048

Vance CP (2001) Symbiontic nitrogen fixation and P acquisition. Plant nutrition in a world of declining renewable resources. Plant Physiol 127:390–397

Vance CP, Uhde-Stone C, Allan DL (2002) P acquisition and use: critical adaptations by plants for securing a nonrenewable resource. New Phytol 157:423–447

Vitousek PM, Naylor R, Crews T, David MB, Drinkwater LE, Holland E, Johnes PJ, Katzenberger J, Martinelli LA, Matson PA, Nziguheba G, Ojima D, Palm CA, Robertson GP, Sanchez PA, Townsend AR, Zhang FS (2009) Nutrient imbalances in agricultural development. Science 324:1519–1520

Watson CA, Bengtsson H, Ebbesvik M, Loes AK, Myrbeck A, Salomon E, Schroder J, Stockdale EA (2002) A review of farm-scale nutrient budgets for organic farms as a tool for management of soil fertility. Soil Use Manage 18:264–273

Watt M, Silk WK, Passioura JB (2006) Rates of root and organism growth, soil conditions, and temporal and spatial development of the rhizosphere. Ann Bot 97:839–855

Wechsung G, Pagel, H (1993) Akkumulation und Mobilisation von Phosphaten in einer Schwarzerde im Statischen Dauerversuch Lauchstädt - Betrachtung der P-Bilanz nach 84 Versuchsjahren. J Plant Nutr Soil Sc 156:301–306

Withers PJA, Hodgkinson RA, Bates A, Withers CL (2007) Soil cultivation effects on sediment and phosphorus mobilization in surface runoff from three contrasting soil types in England. Soil Till Res 93:438–451

Utilization of Phosphorus at Farm Level in Denmark

Leif Knudsen and Ewald Schnug

Abstract The Baltic Sea is one of the marine water bodies suffering most from eutrophication with phosphorus deriving from agricultural operations. The countries sharing the marine environment of the Baltic Sea are gathered under the so-called HELSINKI convention, which among others aims to reduce the phosphorus input. Although since more than two decades in function the convention shows no significant effect on the P discharge from agriculture. This seems not to be the result of lack of knowledge involved in P dynamics in soils and handling options for P fertilizers but lack of handling of otherwise efficient codes of good agricultural practice. This chapter shows how Denmark as the first HELCOM state succeeded to reduce P discharges from agriculture by implementing a mandatory advisory/recommendation system.

Keywords P-demand · P-balances · P-analyses

Legal Framework for Phosphorus Management in Denmark

In the recent discussion of issues of phosphorus (P) fertilization the problem of limited P resources and the risk of P losses to the aquatic environment are predominant. However, farmers, who are the ones handling P supply and balance in practice are mostly not aware of this controversy. Basically, farmers aim to optimize the economy on the whole farm including plant and animal production by adjusting fertilizer rates to the minimum requirement for maximum economic return in order to save costs. Fertilization takes a high utilization rate of nutrients into

L. Knudsen (✉)
SEGES P/S, Agro Food Park 15, Dk-8500 Aarhus N, Denmark
e-mail: lek@seges.dk

E. Schnug
Federal Research Centre for Cultivated Plants, Institute for Crop and Soil Science, Julius Kuehn-Institut, Bundesallee 50, D-38116 Braunschweig, Germany

E. Schnug and L.J. De Kok (eds.), *Phosphorus in Agriculture: 100 % Zero*,
DOI 10.1007/978-94-017-7612-7_11

account though such aim can be impeded by additional costs. These can be so high that they exceed the monetary value of the nutrients. In case of nitrogen (N) the costs for avoiding leaching by cultivating catch crops or specific crop rotations often exceed the costs when replacing agronomic measures by use of mineral N fertilizers. In case of P there also are situations where optimization of the income conflicts with a high utilization of P. The mean P losses by leaching or erosion are below 0.5 kg P per ha in Denmark compared to an N loss of up to 100 kg/ha with leaching, ammonia emission and denitrification. Consequently the monetary loss for farmers is only minor in case of P while it is substantial in case of N.

In Denmark P losses are generally low and policy measures ensure that the losses will be low on a long-term basis. This is necessary to prevent a P surplus in soils which is closely linked to an accumulation of P, which increases the potential risk of P losses on a long-term basis. The reduction of short-term P losses is required in order to obtain a good ecological status of some lakes. According to the EU Water Frame Directive riparian zones along watercourses and lakes are mandatory and are to be included in the legislation as is the off-take of P from wetlands from 2015 to 2021 (Naturstyrelsen 2015).

The fundamental problem of optimizing farm income with respect to adjust P rates is that on livestock farms the production of animal manure and the P demand of soils belonging to the farm or in the neighborhood show big discrepancies. In practice, economic aspects of extending animal production are more or less independent of the utilization rate of P provided by manure. An equivalent of about 4–7 % of the prices for milk and meat can be found in manure if the price level of single nutrients for mineral fertilizers is assumed. In case of P, the corresponding value is below 1 %. Thus the utilization of P is of minor interest for farmers when planning an extension of livestock production. A major problem of P in animal manure and a number of waste products is that the concentration of P and other nutrients is very low which means that transportation costs are high when compared to the value of nutrients.

Slurry from dairy cows or from pigs can be transported up to 7 and 12 km if the fertilizer value of P alone must pay for transportation costs. If the value of all nutrients is included the possible transportation distances for slurry are 48–55 km, respectively. Chicken manure is far more concentrated in nutrients compared to slurry from pigs or dairy cows. Here the possible transportation distance is 266 km. In practice the incitement for the farmer to transport animal manure over long distances to other farms is significantly lower. The prerequisite is that the buyer of the manure is willing to pay for the nutrients in animal manure. Prices depend on supply and demand for manure in the region. Usually in regions with a high animal density the buyer of manure only pays for costs for spreading. The costs in Table 1 show that under Danish conditions the distribution of manure in the area of a single farm needs to be optimized because the transportation costs are much lower than the fertilizer value of the manure though the farm size is big nowadays.

Most EU countries have an upper limit for the permitted number of livestock units per hectare in order to avoid too big differences between P supply by animal manure and P off-take with crop products. In Denmark the maximum allowed

Table 1 Concentration of P in animal manure and sludge, monetary values of nutrients and maximum transport distance before transportation costs exceed the value of nutrients (Petersen 2007; Poulsen 2013)

Source	Manure type	kg P/t	Monetary nutrient value (€/t[a])			Maximum distance of transport (km)	
			P	N, K	Total	P	N, P, K
Dairy cows	Slurry	0.7	0.92	5.49	6.40	7	48
Slaughtering pigs	Slurry	1.2	1.62	5.75	7.37	12	55
Chicken	Solid manure	7.0	9.38	26.5	35.9	70	266
Sludge	Solid	5	6.74	6.20	12.9	50	96
Triple-superphosphate	Mineral fertilizer	200	270	0.00	269		

[a]Calculated from prices of mineral fertilizer in October 2014; 70 % utilization of N and 100 % utilization of P and K

livestock density was limited to 2.3 animal units per ha for cattle and 1.7 for pigs in 1985. This figure was reduced several times in the past and since 1999 the limit for P in practice is an indirect consequence of the EU Nitrate directive, which states that not more than 170 kg of N may be applied by animal manure per hectare. In Denmark this limit has been only 140 kg of N per ha for pig, poultry and fur to avoid a too high surplus of P.

Recommendations for P Fertilization in Denmark

Farmers and advisers in Denmark do not pay special attention to P in their fertilizer planning. One reason for this may be that only very few advisers have ever seen severe P deficiency symptoms on production fields (Fig. 1). The reason is that the P

Fig. 1 A rare picture: severe P deficiency in a barley crop on the Jutland Peninsula (*Photo* Ewald Schnug)

level in soils is typically high enough to avoid an undersupply even if no P is fertilized or only low rates.

In comparison, in the 1970s and 1980s P was fertilized excessively because of serious P deficiency back in the 1940s and 1950s. However, since 2010 an increasing number of fields are reported to suffer from acute P deficiency. This is a consequence of a continuous undersupply with P, especially on arable farms in order to save costs for mineral P fertilizers. For several years this practice had no visible or measurable impact on crop yield through the hysteresis of crop response during depletion of soil P. Meanwhile the soil P status is critically low in an increasing number of fields in Denmark (Oversigt over Landsforsøgene 1987–2014).

The principle for estimating the P demand at field level in Denmark seems to be very much alike the principles in other countries. It includes different soil tests for P and estimates for the export of P with harvest products. Astonishingly basic values for the P demand vary by factor 3 between countries (Jordan-Meille et al. 2013). The principle of calculating the P demand in Denmark involves 4 steps which are described in detail below:

1. Determination of the correction factor for plant-available P in soil.
2. Calculation the P uptake Phosphorus uptake of the crop by aboveground plant parts.
3. Correction factors for different crop species.
4. Correction factors for the P balance in the entire crop rotation.

Determination of the Correction Factor for Plant-Available P in Soil

In Denmark the plant-available P content in soils is determined by extraction with sodium-ammonium-bicarbonate according to Olsen (Shwiekh et al. 2015). This method is the official standard method since 1987. Soil testing for P is not mandatory for farmers. The official recommendation is soil testing of pH, P, K and Mg every 5–7 years. Core soil samples from 3 to 5 ha are either mixed or geo-referenced samples are taken in regular grids of 1 ha (Rubæk and Sørensen 2011). The correction factors for the measured soil P are listed in Table 2.

Table 2 Recommended correction factors for the P demand based on soil analytical data (Olsen)

Olsen P (mg/100 g)	CAL P (mg/100 g)	Recommended P application
<2	<3	20–40 % more than off-take by harvest products[a]
2–4	3–4	P off-take by crops; slightly higher rates in sensitive crops
4–6	4–8	25–50 % of P off-take[a]
>6	>8	No P application[a]

[a]Verification of soil analytical data every 5–7 years

The interpretation results from soil P tests using Olsen extraction differs between countries. In Denmark the sufficiency range is between 2 and 4 mg/100 g P, in England and Wales it is 1.6–2.5 mg/100 g P and in France it is 3.1–3.8 mg/100 g P (Jordan-Meille et al. 2013). In each country the limits may have been deduced from trials, but interpretation of data gives leeway to different ranges.

In Denmark there are no written official guidelines for correction for other soil factors when interpreting the value of Olsen-P. But in practice it is taken in consideration that the limit is lower on well structured soils with a good root development and higher in soils with poor root development.

Calculation of the Uptake of P by Crops (Above Soil Parts Only)

This calculation is a standard calculation done from the expected yields and the concentrations of P in different crops. The expected yield is given from the farmer or average yields from the soil type are used. The concentrations of P in the main crops are measured regularly. In cereals the P concentrations are monitored each year for setting annually corrected default values for P concentration in grains for optimizing the composition and need of adding P in mineral concentrate to the feed. In grass and maize silage the concentrations are determined in the common analyses for feed composition conducted for the farmers. An interesting aspect is that there seems to be a significant decline in the P content of grain in Denmark. A statistical analysis of the development of P is shown in Table 3. In Table 4 off-take of P by

Table 3 Change in concentrations of P and protein from 2003 to 2014 in grain in Denmark. All *p*-values <0.0001 (Sloth 2014)

Crop	No. of samples	Annual change (2003–2014)	
		P (mg/year)	Protein (%/year)
Winter barley	357	−0.035	−0.11
Spring barley	382	−0.045	−0.09
Winter wheat	384	−0.030	−0.09

Table 4 Calculated P off-take (kg/ha) with average yields (t/ha DM) in Denmark on two soil types and standards for concentrations of P in the crops

Crop	P concentration (% d.m.)	Coarse sand[b]		Clay	
		Yield	P off-take	Yield	P off-take
Winter wheat[a]	0.31	4.9	13	8.6	23
Spring barley[a]	0.35	4.0	12	6.3	19
Winter oilseed rape[a]	0.76	2.7	19	4.2	29
Maize for silage	0.21	11.3	24	13.0	27
Grass for cut	0.33	7.8	26	8.4	28

[a]Not including eventually off-take of 2–3 kg P per ha by straw
[b]Without irrigation

Table 5 Equilibrium P content in soil (Pt) securing a zeroed P balance (calculated in Mark-Online; Bligaard 2014)

Crop	Yield[a] (dt/ha) in brackets: P off-take (kg/ha P)	Yield, straw/leaves (dt/ha) in brackets: P off-take (kg/ha P)	P recommendation (kg/ha P) if Olsen P 2–4 mg/100 g P	Model calculated equilibrium for Olsen-P (mg/100 g P)
Spring barley	46 (17)	25 t/ha (2)	22	3.2
Winter wheat	63 (20)	30 (2)	20	2.5
Winter oil seed rape	31 (27)	25 (2)	27	2.5
Potatoes (starch)	450 (26)	–	41	3.5
Sugar beets	559 (23)	–	38	3.5
Maize (whole crop)[a]	9300 FU/ha (29)	–	44	3.5
Grass, cut[a]	7200 FU/ha (33)		33	2.5

[a]Yield: cereal: 85 % d.m.; oilseed rape seeds: 91 % d.m., potatoes and sugar beet fresh weight with 24 % d.m.; maize and grass in "Feeding Units (FU)" (maize 1.15 kg DM/FU; grass 1.3 kg DM/FU)

harvest products of characteristic Danish crops is shown for 3 different soil types. The yields used in the calculations are based on the average yield in Denmark for a 5-year period. Data in Table 5 show a huge difference in P off-take for the two main soil types in Denmark representing the lowest and highest average yields.

Correction for Crop Type

The Danish recommendation system does have a simple correction for crop type. This correction is based on a mix between results from experiments and a logical correction based on the knowledge about P behavior in soils: no correction for permanent crops, grass, autumn sown crops like winter cereals, +5 kg/ha P for spring sown crops (except row crops) like spring barley, +15 kg/ha P for spring sown row crops like maize, sugar beet and potatoes. The assumption is that spring sown crops have a higher need for available P than autumn sown or permanent crops because of low temperatures in the germinating period in spring. A row sown crop utilizes less P from the soil in the beginning of the growing season because of a lower root concentration in the soil volume.

Correction for P Balance in the Crop Rotation

All results from experiments with applying P fertilizer show that the response for P in one year is usually not profitable for the farmer. The P demand is based on the premise that a certain P level needs to be maintained in soils on a long-term basis. Supposed a need for P on a field has been calculated from step 1 to 3 then the P supply for the whole crop rotation still needs to be taken in consideration. The correction addresses especially surplus P from previous years which does not show up in the measured P content of the soil. If the P level in the soil is sufficiently high the total P demand over a crop rotation period equals the total P export with the crops. However, this requires keeping track on yields and P supplies in every year.

Calculation P Demand with the Fertilizer Planning Software Mark-Online

In Denmark fertilizer planning is mandatory for farmers. The software from the Knowledge Centre For Agriculture Mark-Online is used for planning of more than 80 % of the agricultural land (Bligaard 2014). The program contains a model for P demand calculations on field level. In general, the model is based on the same principles as described above. The model keeps track of the site-specific P balance from the annually supply in mineral and organic fertilizers and the output by crop off-take. When using the above described principle the calculated P demand will equal the output. If the P status in the soil (Pt) is lower than the target value, the demand will be higher than the export and lower, if Pt is higher than the target value Pt. By simulation of crop rotations for a long period of time the equilibrium Pt in the soil, where the demand equal the export can be calculated (Table 5).

Effect of the P Form in Fertilizers

In practice not much attention is paid to the efficiency of different types of mineral P fertilizers. Most straight P mineral fertilizer products are water-soluble including that in compound NPK fertilizers and diammonium-phosphate. For animal manure P availability in the first year is assumed to be 90 %. For waste products like sludge, ashes of straw it is well known that the P availability in the first year is highly variable. But if the level of available P in the soil is sufficiently high the first year P effect is of minor relevance. The P supply of these sources is assumed to be sufficient to maintain the available pool on a long-term basis. Otherwise soil tests should reveal the change in available P in soil if carried out every 5–7 years.

Use of Waste Products for P Fertilization

The statistics for the use of P in Danish agriculture shows a supply of 3000 t per year as sludge from waste plants at the municipalities and 2500 t from other waste products. Compared to a total P supply by animal manure of more than 40,000 t P the amount of P supplied with waste products is small. But locally it may be a relevant factor. According to the Danish legislation the use of sewage sludge is legal, when the sludge is documented to comply with different quality parameters as heavy metals, organic pollutants and hygienic conditions. Most of the sludge is applied in autumn and ploughed into the soil before sowing of winter cereals. More than 50 % of the total production of sewage sludge in Denmark is spread on agricultural land. Usually the upper dose of sludge per ha results from the upper limit of 30 kg/ha P per year in the legislation. However, it is permitted to add up this quantity of 30 kg P for a 3 year period, so that the limit is 90 kg/ha P. The availability of P in sewage sludge is not much discussed. In general, the Danish soils have a sufficient P level, and P fertilization with sewage sludge is, without further knowledge regarded as a method to keep a sufficient level. This might be justified in terms of providing the crop with sufficient amounts of P for growing but definitively not from the viewpoint of a sustainable use of P as a limited resource (Schnug and Haneklaus 2016).

Different industrial waste products are supplied on agricultural land, which contain a high amount of P. The most important products are waste products from the production of enzymes (Novozymes 2015) and waste products from processing starch potatoes. In biogas plants animal manure is degassed and the digestates contain varying amounts of P. Data from 2002 showed that about 400,000 t of different waste products were treated in biogas plants. The level of imported P in waste products from biogas plants is probably about 300–600 t. In relation to the total amount of P in animal manure this quantity is relatively small.

P Status in Danish Soils

Most Danish soils are tested regularly for pH, P, K and Mg. As described above the standard test for P in Denmark is Olsen P (Shwiekh et al. 2015). This method has been the official method since 1987. Most soil sampling is handled in the Danish Agricultural Advisory Service which is owned and organized by the Danish farmers. Each year SEGES collects the results for establishing regional statistics. The data are published annually in "Oversigt over Landsforsøgene". From this statistic the main trends over time in the development of the Olsen P content of soils can be calculated. In Fig. 2 the mean Olsen P content is shown for the entire country and for the eastern islands (Sealand, Lolland-Falster). On the national scale the statistic is based on about 50,000–110,000 samples; for the eastern islands 7000–18,000 samples accrue each year. The trend shows a decrease in Olsen P

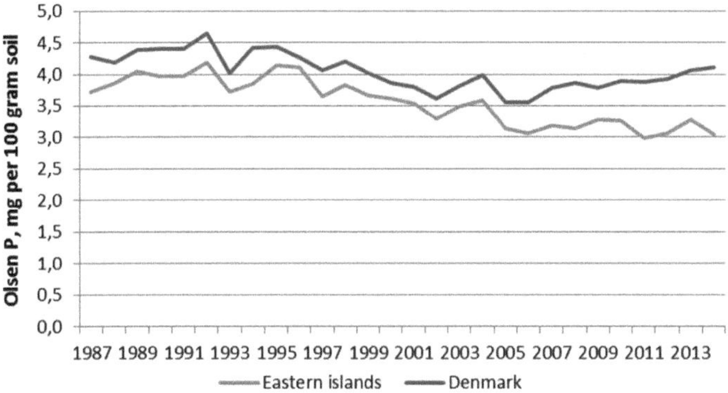

Fig. 2 Trends in Olsen P (mg/100 g P) in average of Denmark and for the eastern islands alone [data from Oversigt over Landsforsøgene (1987–2014)]

from 1990 to 2005 and from then onwards an up going trend until 2014. The decrease may be difficult to explain because the national surplus of P in this period was 10–15 kg/ha P (Fig. 2). Rubæk et al. (2005) interpret the missing correlation between the trend in Olsen P and the P surplus by a change in the representativeness of the samples, the systematic change in the time for soil sampling and fertilization with P, the change of the laboratory, the redistribution between soluble and slow soluble P pools in soils, the different effect of P in animal manure compared to mineral P, the redistribution of P from the upper layer (0–25 cm), where soil samples are taken and deeper zones and the redistribution of P by soil erosion. In Fig. 2 the trend for the eastern islands (Sealand and Lolland-Falster) is shown separately. This area in Denmark is characterized by a low animal density. In this region the P balance has been negative for a number of years and a decrease in Olsen P is observed. In this area more than 15 % of the soil samples have an Olsen P content lower than recommended. On average for Denmark the Olsen P content is in the upper range of the recommendation P (Table 2). This means that the P balance in Denmark might turn negative for a number of years without risk of significant yield losses.

P Balance on National, Regional and Farm Level in Denmark

National Level

Generally the available soil P status in Denmark is high as a consequence of a high surplus of P in the last part of the 20th century (Fig. 3). The P surplus was calculated from the total input and output from Danish agriculture. This procedure provides more reliable data than the calculation of the field balance alone. The

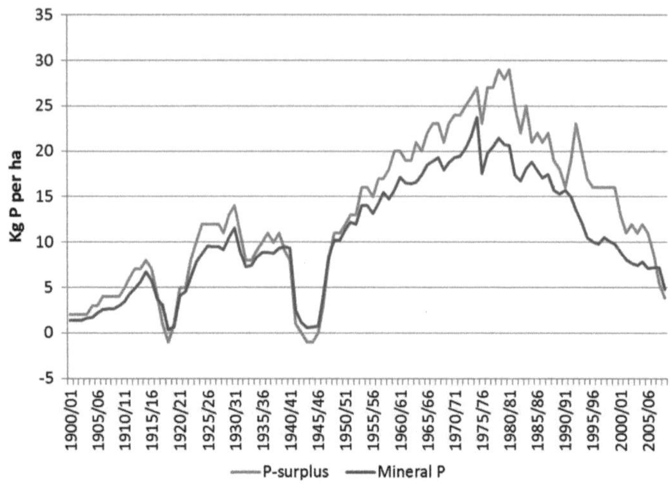

Fig. 3 Surplus of P and supply of P with mineral fertilizers from 1900 to 2010 [data from Kyllingsbæk (2005) and Vinther (2012)]

surplus increased up to a surplus of 30 kg/ha P in 1980. Since 1980 the surplus of P decreased continuously and significantly. In 2010 the balance was only a surplus of 5 kg/ha P. The surplus depicts solemnly the use of mineral fertilizer P. If the use of mineral fertilizer would have been zero, there would have been virtually no surplus at all. This reveals that in general the use of P from animal manure and waste products would have been able to cover the amount of P exported from the fields. From the data presented in Fig. 3 it can be calculated that the total P accumulation in Danish soils from 1900 to 2010 amounts to 1400 kg/ha P. The total content of P in Danish agricultural soils up to a depth of 0.75 m depth is estimated to equal 4600 kg/ha P on agricultural and 2700 kg/ha P on forest soils, respectively (Rubæk et al. 2000). The difference of 1900 kg/ha P reflects very well the calculated values from the national P balance (Miljøministeriet 2003).

The decrease of the surplus in the Danish national P balance was mainly caused by a reduction in the use of mineral fertilizers. The supply with P from animal manure was almost constant until the beginning of the 2000s but decreased since then significantly. Reason for this development is a reduction of mineral P in feed additives. Before 2000 about one third of the excretion of P from animals derived from mineral P feed additives. The amount of mineral P additives was reduced from 18,000 t in 2001 to 10,000 t in 2008 and is expected to be further shortened to 6000 t by 2015. This decline corresponds to a reduction of the surplus in the national balance of approximately 4 kg/ha P (Fig. 4). The reduction of mineral P in feed additives, especially those in pig feed, was possible by the addition of synthetic phytase which improves the utilization of the natural P content in cereals. The natural phytase is deactivated by heat treatment of the feed in order to combat salmonella infections.

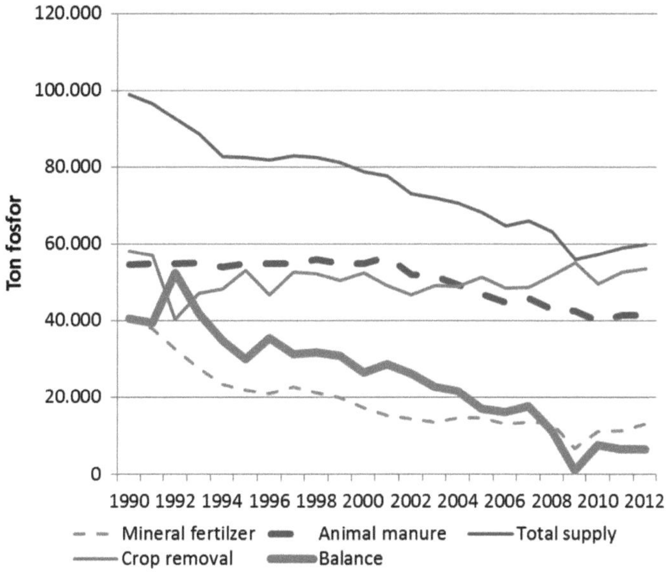

Fig. 4 P balance on field level for Denmark from 1990 to 2012. Data from Blicher-Mathiesen et al. (2013). The supply with 5000 t P by waste products is not shown, but included in the total P supply

P Balance on a Regional Level in Denmark

The P surplus is not equally distributed throughout the country because of regional differences in livestock density. Animal production is generally higher on sandy soils with low productivity compared to more high yielding clay soils. This is the case for milk and pig production. The reason is that on clays soils farmers have been able to make a living without animal production. SEGES calculated the difference between P off-take by crops and the P supply with animal manure for each of the 98 municipalities in Denmark (Fig. 5). The P off-take by harvest products is based on the average yield of crops per soil type and the actual crop distribution on crops on soil types for each municipality. The P supply per ha has been calculated from the recorded fertilizer accounts.

The municipalities in the western part of the country have only a small surplus (0–5 kg/ha P) if P supplied by animal manure is compared to off-take by crops. From this it can be stated that there seems to be no need for long distance transportation of animal manure to obtain a fair P balance in Denmark. It is feasible to redistribute all P within a municipality and thus within in relatively short distances. In the eastern part of the country P off-take is significantly higher than the supply with animal manure. In this part of the country there is a need for using P from waste products or mineral fertilizers.

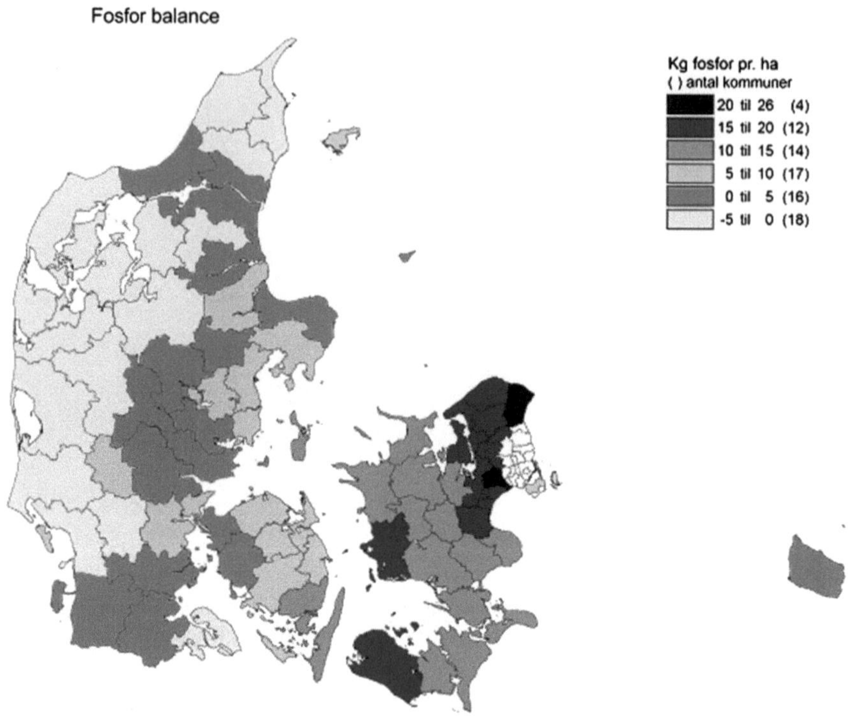

Fig. 5 Difference between P off-take and P supply with animal manure for the municipalities in Denmark in 2011

Farm Balances of P

Farm balances for nutrients have been calculated for numerous farms by SEGES. These balances show a correlation between the number of livestock units per ha and the P surplus. In Fig. 6 the surplus of P for farms with pig production are shown for coarse sandy soils (83 farms) and for loam soils (333 farms) in the period from 2003 to 2008 (Hvid 2010). The P surplus is higher on sandy soils as yields are lower and consequently the P off-take by crops, too (Fig. 6). In general, the Danish legislation permits manure applications which equal 1.4 livestock units per ha, and with values higher than 1.7 for cattle and 2.3 for cattle the derogation rules in the EU nitrate directive become legal. One live stock unit is in Demark defined as number of animals, which gives a production of 100 kg N at store. The actual maximum permitted P surplus can be calculated on the basis of current legislation taking into account the permitted number of animals per ha, standards for P excretions per animal and the expected P off-take by crops. For 2015 a calculation example is shown in Table 6. Particularly on farms with production of chicken and fur (mink) there is a potential high surplus of P (Table 6). In practice the P surplus is lower as the number of livestock units per ha often is lower than legally permitted.

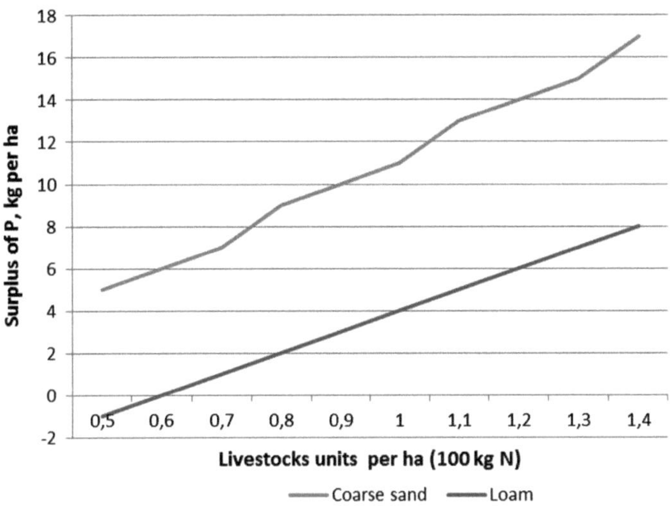

Fig. 6 Correlation between the number of livestock units per ha and the P surplus on farm level in relation to soil type [2003–2008; data from Hvid (2010)]

Table 6 Calculated maximum permitted P surplus (kg/ha) following the legislation in Denmark in 2015

	All farms[a]	Dairy farms	Sows + piglets farms	Pig, finishers (30–105 kg)	Chicken	Mink	Dairy cows (1.7 LU/ha)	Dairy cows (2.3 LU/ha)
Kilogram P per livestock unit (LU)			25	22	34	30	14	
Permitted maximum P (kg/ha P)			35	31	47	42	24	33
	P off-take by crops (kg/ha)		Calculated maximum P-surplus (kg/ha)					
Coarse sand, not irrigated	15	22	20	16	32	27	2	11
Coarse sand, irrigated	20	27	15	11	27	22	−3	6
Loamy soils	24	25	11	7	23	18	−1	8

[a]Except dairy farms

Conclusions for Improving the P Balance of Farms

Usually a P surplus is related to the quantity of animal manure. The simplest way to reduce the P surplus is to spread the animal manure on a bigger area. If this is not feasible on the land owned by the farmer, manure can be exported to neighbors. It will typically be a win-win situation because both the farmer with the animal production and the farmer who is going to receive the manure can have an economic profit out of such agreement.

In areas with a very high production of animal manure it can be necessary to export manure over long distances. In such a situation a separation of the manure can make sense. The P in animal manure is correlated to the dry matter fraction. It is well documented that centrifugation with decanter centrifuges is very effective to concentrate P in the solid fraction and create a liquid fraction very low in P concentrations.

References

Blicher-Mathiesen G, Rasmussen A, Grant R, Jensen PG, Hansen B, Thorling L (2013) Landovervågningsoplande 2012. NOVANA. Aarhus Universitet, DCE, Nationalt Center for Miljø og Energi, Videnskabelig rapport fra DCE, Nationalt Center for Miljø og Energi no. 74, p 154 (http://dce2.au.dk/pub/SR74.pdf)

Bligaard J (2014) Mark Online, a full scale GIS-based Danish Farm Management Information System. https://ideas.repec.org/a/ags/ijofsd/198969.html

Hvid SK (2010) Sammenligningstal for næringsstofoverskud på bedrifter med grønt regnskab 2003–2008. Landbrugsinfo. https://www.landbrugsinfo.dk/Planteavl/Goedskning/Naeringsstoffer/Sider/pl_10_067.aspx#

Jordan-Meille J, Rubæk GH, Ehlert PAI, Genot V, Hofman G, Goulding KWT, Recknagel G, Provolo G, Barraclough PB (2013) Review of phosphorus fertiliser advice in Europe. Soil testing, calibration and recommendations. Proceedings International Fertilizer Society 735

Kyllingsbæk A (2005) Nutrient balances and nutrient surpluses in Danish agriculture 1979–2002: nitrogen, phosphorus, potassium DJF Report Nr. 116, Research Center Foulum, Danish Institute of Agricultural Sciences, p 100

Miljøministeriet (2003) Fosfor i dansk landbrug Omsætning, tab og virkemidler mod tab. Rapport fra Fosforgruppen (P-U-1) Forberedelse af Vandmiljøplan III (http://www.vmp3.dk/Files/Filer/Rap_fra_t_grupper/fosfor-i-dansk-landbrug-okober-2003.pdf)

Naturstyrelsen (2015) Vandområdeplaner 2009–2015. http://naturstyrelsen.dk/vandmiljoe/vandplaner/vandplaner-2009-2015/

Novozymes (2015) http://www.novozymes.com/en/news/image/Pages/production.aspx

Oversigt over Landsforsøgene (1987–2014) Videncentret for Landbrug. Planter & Miljø, Agro Food Park 15, Dk-8500 Aarhus N, Denmark

Petersen J (2007) FarmTest - Transport af gylle, Maskiner og planteavl no. 61. https://www.landbrugsinfo.dk/Tvaerfaglige-emner/FarmTest/Maskiner-og-planteavl/Sider/FarmTest__Transport_af_gylle.aspx

Poulsen HD (ed) (2013) Normtal for husdyrgødning –2013, 33 sider. (Versionen er opdateret 1. oktober 2013). http://anis.au.dk/normtal/

Rubæk GH, Sørensen P (2011) Jordanalyser – kvalitet og anvendelse. DCA-rapport nr. 002, Nationalt Center for Fødevarer og Jordbrug

Rubæk GH, Djurhuus J, Heckrath G, Olesen SE, Østergaard HS (2000) Er danske jorde mættede med fosfor? In: Jacobsen OH, Kronvang B (eds) Tab af fosfor fra landbrugsjord til vandmiljøet. DJF rapport no. 34, pp 17–30

Rubæk G, Heckrath G, Knudsen L (2005) Fosfor i dansk landbrugsjord. Grøn Viden, Markbrug, Danmarks JordbrugsForskning no. 305

Schnug E, Haneklaus S (2016) The Enigma of fertilizer phosphorus utilization. Springer, Dordrecht

Shwiekh R, Schick J, Kratz S, Rückamp D, Schnug E (2015) Comparative investigation and inter-calibration of different soil P tests. J für Kulturpflanzen 67:61–72

Sloth NM (2014) Personal communication, data compiled from Oversigt over Landsforsøgene (annually) Videncentret for Landbrug. Planter & Miljø, Agro Food Park 15, Dk-8500 Aarhus N, Denmark

Vinther (2012) Oversigt over Landsforsøgene (annually) Videncentret for Landbrug. Planter & Miljø, Agro Food Park 15, Dk-8500 Aarhus N, Denmark

Trace Element Contaminants and Radioactivity from Phosphate Fertiliser

Matthew Taylor, Nick Kim, Geerd Smidt, Chris Busby, Sam McNally, Brett Robinson, Sylvia Kratz and Ewald Schnug

Abstract A risk ranking model was developed to provide a systematic evaluation of the range and quantity of 28 elemental contaminants applied to land in New Zealand and applied to mineral P fertilisers. The methodology is transparent, flexible and robust and allows contamination issues to be ranked according to their real or potential impact. The quantitative ranking model is based on the relative importance of each element in relation to accumulation in soil, transfer to water or uptake by plants, toxicity to soil organisms, plants and people, and the contribution of any radioactive isotopes. The highest risk score for potential environmental significance of P fertiliser borne trace element contaminants was found for uranium, followed (in decreasing order) by cadmium, mercury, boron, fluoride, selenium, arsenic, silver and rare earth elements. The lowest score (rank 28) was attributed to strontium.

Keywords Priority analysis · Trace element fate · Environmental management · Cross-sphere transfer

M. Taylor (✉)
Waikato Regional Council, PO Box 4010 Hamilton East, Hamilton 3247, New Zealand
e-mail: matthew.taylor@waikatoregion.govt.nz

N. Kim
Institute of Food, Nutrition and Human Health, Massey University, Wellington, New Zealand

G. Smidt
School of Engineering and Science, Jacobs University, Bremen, Germany

C. Busby
Green Audit, Aberystwyth, United Kingdom

S. McNally
Faculty of Science and Engineering, Waikato University, Hamilton, New Zealand

B. Robinson
Faculty of Agriculture and Life Sciences, Lincoln University, Canterbury, New Zealand

S. Kratz · E. Schnug
Institute for Crop and Soil Science, Julius Kühn-Institut (JKI), Federal Research Centre for Cultivated Crops, Brunswick, Germany

E. Schnug and L.J. De Kok (eds.), *Phosphorus in Agriculture: 100 % Zero*,
DOI 10.1007/978-94-017-7612-7_12

Introduction

There is general concern about the input of contaminant trace elements to soil as these can have negative effects on soil organisms, plants and higher animals (de Vries et al. 2007). Contaminants can accumulate in soil, be taken up into the food chain, be transported to water, or be volatilised. This raises the potential for higher trace element concentrations in some food products. Excessive levels of trace elements in food can have detrimental effects on human health, market access and trade, and the ability to change from one land use to another (CWG 2008). Once contaminants are mobilised they can have impacts on life at considerable distances from their origin. Sometimes the mobilisation process takes years or decades before the full impact is seen. Therefore, it is important to employ the precautionary principle when managing the impact of trace metal contaminants in the environment.

This chapter will address identification of elements of key concern, identification of knowledge gaps, and determine the most effective means of managing the environmental and health risks.

Types of Phosphate Fertiliser

In November 1802, Alexander von Humboldt studied guano and its fertilising properties at Callao in Peru, and his subsequent writings on this topic made the subject known in Europe. Bird guano has a fertiliser analysis of 11–16 % nitrogen, 8–12 % equivalent phosphoric acid, and 2–3 % equivalent potash. Several different types of phosphate fertiliser have been developed since the formation of the modern understanding of plant nutrition, which dates to the 19th century and the work of Justus von Liebig. Bones, which contain the element phosphorus, were used as an agricultural fertiliser. In 1834, Sir John Bennett Lawes sold the first practical superphosphate fertiliser which he made by dissolving phosphate rock in sulphuric acid. In 1842, he opened the Deptford factory in England, making fertiliser from bones and sulphuric acid. However, each type of fertiliser still contains greater or lesser amounts of trace element impurities (Table 1). In the past, ground rock phosphate has been used as a source of P for acid soils but, due to the low availability of P in this native material, high transportation costs, and small crop responses, other phosphate fertilisers were developed. Nowadays, rock phosphate is the raw material used in the manufacture of most commercial phosphate fertilisers.

Phosphate Rock

There are two major types of phosphate rock

Igneous (e.g. from the Kola peninsula in Russia)
Sedimentary (e.g. Morocco).

Table 1 Average total concentrations (mg/kg) of selected elements in common phosphate fertilisers and phosphate rock

Element (mg/kg)	DAP	MAP	NPK	PK	SP	TSP	PR
Ag	1.4	0.1	0.1	0.1	1.7	0.4	0.2
As	14	15	13	16	141	26	9
Au	–	–	0.002	–	–	–	–
B	257	440	374	799	432	423	537
Ba	165	23	226	169	274	74	844
Be	0.9	0.5	0.3	1.1	0.4	3.0	1.7
Bi	0.03		0.03	0.04	0.08		0.09
Cd	22	47	6	6	12	14	28
Ce	132	223	624	39	634	365	301
Co	87	5	3	26	4	3	6
Cr	159	35	43	105	58	146	126
Cs	0.66	1.00	0.65	0.67	0.32	0.36	0.40
Cu	33	27	92	25	57	42	22
Dy	4.4	11.1	15.4	4.8	14.5	5.1	–
Er	3.2	6.9	6.3	3.4	7.5	4.5	–
Eu	2.5	5.7	9.9	1.1	9.4	4.0	10.7
F	21,000	19,000	23,000	–	14,150	21,000	29,856
Fe	5130	12,316	2174	8082	5096	5167	7989
Ga	–	–	0.37	–	–	–	–
Gd	5	14	20	5	19	4	101
Ge	–	–	0.09	–	–	–	–
Hf	–	–	0.2	–	–	–	1.0
Hg	0.48	0.05	0.14	0.05	0.69	0.83	3.41
La	62	96	244	26	260	179	606
Li	3.0	–	1.6	1.5	1.6	2.7	15.8
Lu	0.5	1.2	0.7	0.5	0.9	1.1	2.3
Mg	12,534	7228	8487	14,910	3378	5665	6390
Mn	137	631	339	855	825	172	502
Mo	5	6	23	37	6	7	65
Nb	3	8	66	1	159	1	–
Nd	54	112	203	22	118	156	156
Ni	42	27	18	18	27	31	27
Pb	5	5	5	9	20	17	12
Pd	–	–	0.004	–	–	–	–
Pr	5	17	53	5	30	3	–
Pt	–	–	0.005	–	–	–	–
Rb	14	12	20	22	10	6	81
Sb	2.6	0.3	1.3	1.2	2.2	1.7	3.4
Sc	3.3	8.8	7.62	3.12	9.66	9.75	11.70

(continued)

Table 1 (continued)

Element (mg/kg)	DAP	MAP	NPK	PK	SP	TSP	PR
Se	1.9	2.0	1.0	2.6	3.6	4.5	6.1
Sm	12	23	33	2	38	25	41
Sn	10.9	3.4	1.3	0.3	3.8	2.7	2.3
Sr	432	207	1115	749	2774	933	2101
Ta	–	–	0.1	–	–	–	0.7
Tb	1	2	3	1	3	2	7
Te	–	–	0.01	0.10	–	–	–
Th	3	13	19	3	15	5	7
Ti	176	750	197	421	219	289	218
Tl	0.6	0.8	0.3	0.5	0.9	0.4	0.7
Tm	0.4	1.0	0.7	0.4	0.9	0.7	18.1
U	117	72	36	55	52	118	92
V	209	93	71	79	68	187	69
W	–	–	0.04	–	–	–	–
Y	53	101	82	59	103	74	150
Yb	3	7	4	3	6	6	18
Zn	245	424	377	147	207	365	245

DAP diammonium phosphate, *MAP* monoamonium phosphate, *NPK* blended nitrogen phosphorous potassium fertilisers, *PK* blended potassic phosphate, *SP* superphosphate, *TSP* triple superphosphate, *PR* phosphate rock

Sources Authors (Taylor, Kim, Smidt, Kratz) unpublished data, Al-Shawi and Dahl (1999a, b), Azouazi et al. (2001), Barisic et al. (1992), Bech et al. (2010), Bramley (1990), Camelo et al. (1997), Cevik et al. (2010), Chandrajith et al. (2010), Charter et al. (1995), Cronin et al. (2000, 2003), EFMA (2000), Erikssen (2001), Gargouri et al. (2011), Germann et al. (1984), Hakam and Choukri (2008), Kpmblekou and Tabatabai (1994), Lambert et al. (2007), Lauria et al. (2006), Loganathan et al. (2003), Lottermoser (2009), Manohran et al. (1996), Mazzilli et al. (2011), McBride and Spiers (2001), McLaughlin et al. (1996), Menzel (1968), Mermut et al. (1996), Mirlean and Roisenberg (2007), Molina et al. (2009), Notholt et al. (1989), Ogunleye et al. (2002), Pantelica et al. (1997), Raven et al. (1997), Righti et al. (2005), Roemer et al. (2010), Roselli et al. (2009), Saueia et al. (2005), Senesi et al. (1999), Syers et al. (1986), Takeda et al. (2006), Tankard (1974), Tayibi et al. 2009, Uprety et al. (2009), Waheed et al. (2011), Wassef et al. (1977), Williams (1974), Yamazaki and Geraldo (2003), Zhao and Wang (2010), Zielinski et al. (2000, 2006)

The mineral fluorapatite ($Ca_{10}(PO_4)_6(F,OH)_2$) predominates in igneous phosphate rocks and the mineral francolite ($Ca_{10}(PO_4)_{6-x}(CO_3)_x(F,OH)_{2+x}$) predominates in sedimentary rocks.

Sedimentary deposits are generally associated with matter derived from living creatures and thus contain organic compounds. These phosphates are interposed with sedimentary and gangue minerals and thus sedimentary phosphate rocks have differing compositions within the same source. In addition, most phosphate ores have to be concentrated or beneficiated before they can be used or sold on the international phosphate market. Different techniques may be used at the beneficiation stage, to treat the same ore for removal of the gangue and associated impurities. This gives rise to further variations in the trace element composition of the finished ore concentrate product (EFMA 2000).

Phosphoric Acid Production

There are two methods of phosphoric acid production called the wet and dry (or thermal) methods.

In the older dry process phosphoric acid was first produced by treating phosphate rock in an electric furnace. In an attempt to make it less expensive, phosphate rock was treated with phosphoric acid, in a so called "wet" process, to produce orthophosphoric acid and gypsum.

However, the dry process has been abandoned by the fertiliser industry because of the amount of energy needed and the resulting cost (EFMA 2000) although it could produce very pure phosphoric acid.

In the wet process, phosphate rock is fed to a series of reactors along with recycled phosphoric acid already made from this wet process. The phosphoric acid-rock slurry is mixed with a leaching agent, sulphuric acid. After completing the reaction series, the process stream is washed with evaporator condensate while being forced through a filter. The filtercake, composed primarily of phosphogypsum is a waste by-product. The phosphoric acid product is concentrated to 75–100 % by a series of evaporators and then cooled before storage.

The impurities in the original phosphate rock are proportioned between the phosphoric acid and the phosphogypsum. The proportion of these elements transferred depends on the type of process and the rock used, e.g. uranium mainly goes into the phosphoric acid but any radium, which is a component of the uranium decay series, is co-precipitated with the phosphogypsum (EFMA 2000).

Sulphuric acid used to acidify the raw phosphate rock may contain mercury as a contaminant (Mirlean et al. 2008). While the sulphuric acid contributes the nutrient sulphur to the resulting fertiliser it also adds mercury. In addition, treatment with sulphuric acid releases about half the fluoride from the fluorapatite rock in gaseous form, mostly as hydrogen fluoride. The reaction of hydrofluoric acid with silica present in the rock also produces silicon tetrafluoride gas (Czarnowski and Krechniak 1990). This ends up in the cooling water or cooling air, depending on which system is used. Where cooling air is used in modern factories, a gas washing systems or scrubber is used to remove gaseous fluoride emissions.

Superphosphate

Superphosphate is unique in that it does not use phosphoric acid in the manufacturing process. In 1834 Sir John Bennett Lawes sold the first practical superphosphate fertiliser, which he made by dissolving phosphate rock in sulphuric acid. Reacting concentrated sulphuric acid with crushed phosphate rock produces orthophosphoric acid, calcium sulphate and hydrofluoric acid. It has a phosphorus content of 7–9.5 % (after USEPA 1979).

There exist the same issues of mercury contamination from sulphuric acid used to acidify the raw phosphate rock (Mirlean et al. 2008) and reaction of hydrofluoric acid with silica present in the rock to produce silicon tetrafluoride gas (Czarnowski and Krechniak 1990) as described under the phosphoric acid production section.

The most common phosphate fertiliser used in New Zealand is superphosphate. The development of aerial topdressing in New Zealand allowed superphosphate to be spread economically over large areas that could not economically be reached by other means.

Triple Superphosphate

In the 1870s triple superphosphate was made in Germany using phosphoric acid. In 1897 Georg Schueler, from Stettin filed the patent: "Verfahren zur Darstellung von hochprocentigem Doppelsuperphosphat", German Patent DE 104,364; filed Nov. 12, 1897; pat. July 10, 1899, on the use of phosphoric acid, instead of sulphuric acid, to increase the content in soluble phosphate.

Triple superphosphate is produced by reacting concentrated phosphoric acid with crushed phosphate rock. Triple superphosphate differs from superphosphate in that calcium sulphate is not formed in the process. So the phosphorus content of triple superphosphate (about 17–23 % P) is therefore greater than that of superphosphate (about 7–9.5 % P).

Phosphate enriched fertilisers, such as triple superphosphate, have higher levels of uranium compared to single superphosphate (196 and 90 mg/kg U respectively) made from the same source of phosphate rock (Guzman et al. 1995). The uranium content of triple superphosphate comes from the sum of uranium in phosphate rock and the phosphoric acid which is reacted with the phosphate rock (Erdem et al. 1996).

Ammonium Phosphate/Diammonium Phosphate

Ammonium phosphates are used as an ingredient in some fertilisers as a source for nitrogen. There are several different variations on the original process but ammonium phosphates are formed by mixing concentrated solutions of ammonia and phosphoric acid or by adding excess ammonia to phosphoric acid and evaporating to produce a series of water-soluble ammonium phosphate salts, e.g. $NH_4H_2PO_4$, $(NH_4)_2(HPO_4)$, $(NH_4)_2HPO_4$. Pure ammonium phosphate tends to cake badly in storage but the reintroduction of impurities precipitated and filtered off during production, such as gelatinous aluminium compounds, improve granulation by coating the diammonium phosphate particles (Houston et al. 1955). However, precipitation of these Al and Fe impurities may also coprecipitate trace elements, which are reintroduced to the diammonium phosphate.

Ammonium Polyphosphate

The phosphoric acid produced by either the wet or the dry process is frequently heated, driving off water and producing a superphosphoric acid. The phosphate concentration in superphosphoric acid is usually about 75 %. The P in this acid is present as both orthophosphate and polyphosphate. Polyphosphates consist of a series of orthophosphates that have been chemically joined together by condensation. Upon contact with soils, polyphosphates revert back to orthophosphates over time (Torres-Dorrante et al. 2005). Ammonia can be added to the superphosphoric acid to create liquid or dry materials containing both nitrogen (N) and P with the formula $[NH_4\ PO_3]_n$ (Gowariker et al. 2009).

Blended Fertilisers

Other fertilisers can be made by blending the different types of phosphate fertiliser with nitrogenous and potassic fertilisers (NPK fertilisers) and deliberate enrichment with essential trace elements. These blended fertilisers retain the contaminants found in their constituents. These are the main types of fertiliser used today in Germany.

Phosphogypsum

Even though it is possible to produce relatively contaminant free phosphate fertiliser from phosphoric acid, there remains the issue of the phosphogypsum by-product and it would be thoughtless to overlook this important waste product. Phosphogypsum (PG) is a by-product of the wet production of phosphoric acid and contains many trace elements of environmental concern.

For each ton of phosphoric acid, approximately 4–5 tons of phosphogypsum are produced (Rutherford et al. 1994; EFMA 2000). In 2008, worldwide production of PG was on the order of 180 million metric tons (S. Jasinski, US Geological Survey, written communication, 2008 in Zielinski et al. 2011). A proportion of the trace elements found in phosphate rock are retained in the gypsum and the rest are transferred to the phosphoric acid. The phosphogypsum is "disposed" either into water or land.

Water disposal consists of the phosphogypsum being pumped through an outfall into rivers or the sea. Disposal into rivers and small bodies of water, such as the Mediterranean Sea, are no longer politically acceptable in Europe (EFMA 2000). Relatively few phosphoric acid manufacturers discharge phosphogypsum directly to waterways, although some plants in Morocco discharge into the Atlantic Ocean.

Land "disposal" consists of building a gypsum stack. The phosphogypsum is transported from fertiliser factory to an open air storage area and stacked. These stacks can be very large, up to 360 ha and up to 90 m high (Perpich et al. 2005). In 1989, the USEPA identified 63 stacks in the USA covering areas from 5 to 300 ha, with height of between 3 and 60 m (USEPA About phosphogypsum, http://www.epa.gov/radiation/neshaps/subpartr/about.html, last accessed Aug 4, 2015). By 2005 there were 24 phosphogypsum stacks spread over 3600 ha in Florida alone. Perpich et al. (2005) estimated these stacks held 1.1 Gt of phosphogypsum and 189 GL of process water.

The main environmental concerns associated with phosphogypsum land storage are (EFMA 2000; Korcak 1998; Eisenbud and Gesell 1997; Rutherford et al. 1994):

Movement of fluoride, sulphate, total dissolved solids, certain trace elements, and radionuclides from the U-238 decay series below phosphogypsum stacks into groundwater
Radon-222 exhalation, which may pose a health risk to workers on the site or people living close to stacks
Radon-222 exhalation from soil into residential homes when agricultural land previously treated with phosphogypsum is converted to residential usage
Acts as a potential source of undesirable trace elements to the wider environment
Acidification; the pH of phosphogypsum is 4.5–5.

The radionuclides and other trace contaminants found in phosphogypsum inhibit its use in saleable products. So the stack is likely to remain indefinitely and require indefinite management, including a lining system, collection ditches, and a recycling system for water used for transporting the phosphogypsum and for stormwater (EFMA 2000). Otherwise, as a phosphogypsum stack is not totally watertight, effluents coming from leaching may pollute the environment (Perez-Lopez et al. 2010). In addition to the radionuclides and other undesirable contaminants found in phosphogypsum, it also has high compressibility and is relatively highly soluble, so is physically unstable. In 1994, a sinkhole extending from the surface to beyond the phosphogypsum base, formed in a 66 m high, 160 ha phosphogypsum stack, in the IMC-Agrico plant in Polk County, Florida, USA, as underlying limestone was dissolved by acidic water leaching through the stack and into the groundwater. Between 50,000 and 170,000 m^3 of contaminated water was able to flow into the Florida drinking water aquifer and cost millions of dollars to clean up. Another expensive cleanup operation in Florida was at Pine Point, where advanced domestic wastewater treatment plants incorporating denitrification were constructed, at times fully treated wastewater was dispersed 160 km offshore in the Gulf of Mexico using two 40-million-litre tankers, and new reverse osmosis technology developed to treat 7800 l/s of effluent to discharge standard (Perpich et al. 2005).

There have been some attempts to utilise phosphogypsum. A little phosphogypsum is mixed with mine tailings and backfilled into old, exhausted mine pits. This works where the mine tailings are highly calcareous clays, which neutralises the acid remaining in the phosphogypsum. Unfortunately, this situation is very rare. Another possibility is refining the phosphogypsum to retrieve the sulphur for use in sulphur-fertilisers. A few companies use the Merseburg process to convert the sulphur in gypsum to ammonium sulphate and a calcium carbonate by-product. When applied to phosphogypsum, radionuclide's are reported to accumulate in the calcium carbonate leaving the ammonium sulphate almost radionuclide free (Maxwell 2005; Burnett et al. 1996).

Phosphogypsum still contains about 0.4 % phosphorous and is applied as a soil amendment to improve the physical properties of soils and to provide sulphur, calcium and phosphorous despite its radioactivity (Rutherford et al. 1994). E.g. more than 1/3 of the total annual phosphogypsum production (4.5 Mt) in Brazil is used as soil amendment even though the measured activity often exceeds the USEPA guideline of 370 Bq/kg (Santos et al. 2006; Saueia and Mazzilli 2006; USEPA 2005).

General Comment

No one type of phosphate fertiliser currently on the market is completely free of contaminants. Regardless of the process used to make the phosphate fertiliser, the use of low contaminant phosphate rock will result in fertiliser with low levels of contaminants and the use of higher contaminant phosphate rock will result in fertiliser and/or waste products with higher levels of contaminants. The challenge to fertiliser manufacturers is to devise ways to remove contaminants during the manufacturing process. The contaminant loads and impact on the environment of different types of fertiliser currently in use are assessed in Section "Risk Analysis of Trace Elements in Phosphate Fertilisers and Phosphate Rocks".

Risk Analysis of Trace Elements in Phosphate Fertilisers and Phosphate Rocks

A Quantitative Ranking Model

A contaminant may accumulate in the soil or be taken up into the food chain or transferred to water, eventually finding its way to the sea. Any of these scenarios could result in environmental damage and it is useful to rank contaminants found in phosphate fertilisers in order of risk to set response priorities.

The quantitative ranking model used in this section is based on the relative importance of each element in relation to each of seven factors. These factors are the questions about the potential impact on human health, animal welfare and trade and the wider environment. This choice of factors was designed to maintain the life-supporting capacity of air, soil and water while providing to the needs of existing and future generations.

The factors considered were:

The quantity of contaminant, as described by the trace elements in fertiliser to background soil ratio (F:BG ratio),
The mobility of the contaminant (two factors) from the soil-plant transfer factor and the sorption distribution coefficient (K_d),
The toxicity of a contaminant (three factors) to microorganisms, plants and animals as described by internationally accepted guideline values and
The Radioactivity of the Contaminant.

Scoring was based on measurements. For each diffuse contamination issue, each of the seven factors was assigned a score of 1–5 with 5 having the highest contaminant: background soil ratio, or being the most likely to transfer from soil to the wider environment, or having the highest toxicity. The ranking can always be revisited at any time in the future as new information comes to light.

Quantity of a Contaminant

This risk analysis was confined to those elements measured in identifiable background soils and with a suitable pool of published fertiliser measurements. An element in phosphate fertiliser could only "become of concern" if it is present in quantities several times that of background soils. The fertiliser to background soil ratio for 57 elements in common phosphate fertilisers was estimated by the following:

Concentrations of (total) trace elements in 760 common phosphate fertilisers and phosphate rock were compiled from 55 sources (Table 1). Outliers were excluded and average values calculated for each type of fertiliser.

The average values for background soils of the world were derived from 11 sources (Table 2).

The average value for each type of fertiliser was divided by the background soil value. Table 3 presents the fertiliser to background soil ratio for each trace element for common phosphate fertilisers and 24 trace elements. The greater the ratio the higher the potential risk of accumulation in soils or increased loss in leaching or uptake by plants. Conversely, fertiliser to background soil ratios of 1 or lower are not considered an environmental risk.

Cadmium, fluoride, and uranium have high fertiliser to background soil ratios in nearly every type of fertiliser. High ratios were also found for silver, arsenic, boron, mercury, molybdenum, selenium, strontium and the 17 rare earth elements (REE), such as lanthanum, in some types of fertiliser. These elements were subject to further assessment of factors relating to the element's potential to transfer to the food chain and potential toxicity to biota (Table 4).

Mobility of a Contaminant

Once a contaminant is in the soil it may be transferred, into the food chain via plants or water. Contaminants can also accumulate to higher concentrations (bio-magnification) as they move up the food chain.

The likelihood of transfer to water can be estimated using partition coefficients (K_d; also known as the sorption distribution coefficient). In soil, metal contaminants interact with ligands in soil solution and with surface sites on the soil. Reactions in which the trace element is bound to the solid matrix are referred to as sorption reactions and metal that is bound to the soil is said to be sorbed. The partition coefficient is the ratio of sorbed trace element concentration (expressed in mg/kg soil) to the dissolved metal concentration (expressed in mg/L water) at equilibrium (Allison and Allison 2005).

Plants inadvertently uptake nonessential elements along with essential elements. The likelihood of transfer of trace elements into the food chain can be estimated from soil/plant transfer coefficients. Transfer coefficients are defined for this ranking exercise as the ratio between the content of the editable portion of the plant and that of the soil. The median of coefficients from 20 studies were correlated and median values determined (Table 4).

K_d coefficients are influenced by multiple interrelated geochemical parameters and are site specific e.g. due to interactions with pH, clay content, soil type, and metal concentration. Thus any element has a wide range of values, typically about 2.5 orders of magnitude (Sheppard et al. 2009; USEPA and USDE 1999). However, uranium has a larger scatter of about 4 orders of magnitude from pH 4 to 9 due to its geochemistry. The hydrolysis of uranium under oxic conditions is dependent on the concentration of total dissolved uranium and dissolved hexavalent uranium can exist as polynuclear hydroxyl complexes. Under different geochemical conditions, uranium will absorb to soil at less than 5 % (very small K_d) or up to more than 95 % (very large K_d) of its original dissolved concentration (USEPA and USDE 1999).

However, despite the wide range of values, K_d coefficients are an indicator of relative mobility and can be ranked. Estimates of K_d coefficients from 15 studies were correlated and median values determined (Table 4). Uranium was treated as a special case with both extremely high and low K_d for alkaline and acid soils respectively. Studies of contaminated sites were not included as higher concentrations of a metal will decrease the apparent K_d coefficient (due to a lowered affinity of the solid phase for the metal as it moves toward saturation—Sauvé et al. 2000). There was limited data for fluoride (Baes et al. 1984).

Table 2 Average total concentrations (mg/kg) of trace elements in background soils

Element (mg/kg)	New Zealand	World	Sweden	Japan	Brazil	USA	Sankt-Petersburg	Australia	Resolution Island	Belgium	Average
Ag	0.13	–	0.11	0.10	0.05	–	–	–	–	–	0.10
As	4.8	4.4–9.3	3.8	–	–	7.2	2.6	–	–	7.0	6.1
Au	–	0.002	–	–	0.002	–	–	–	–	–	0.002
B	4	22–45	5	–	–	33	–	–	–	–	27
Ba	328	175–520	608	350	–	580	202	–	–	–	409
Be	–	1.9	1.36	1.40	–	0.92	–	–	–	–	1.40
Bi	0.90	0.70	0.16	0.33	–	–	–	–	–	–	0.52
Cd	0.11	0.37–1.10	0.17	0.33	0.18	–	0.20	0.01	0.08	0.18	0.36
Ce	36	49	60	52	89	75	–	–	–	–	60
Co	23	5–12	7	18	17	9	4	–	–	0.05	10
Cr	70	12–83	22	58	86	54	13	–	312	25	68
Cs	9	–	2	5.4	4.6	–	–	–	–	–	5.8
Cu	33	13–24	17	48	109	25	18	13	45	10	29
Dy	–	1	4.1	3.9	5.6	–	–	–	–	–	3.6
Er	–	2	2.2	2.2	3.1	–	–	–	–	–	2.3
Eu	–	1	1	1.2	1.5	–	–	–	–	–	1.17
F	210	130–550	–	269	430	–	–	–	–	–	313
Fe	42,800	–	–	–	–	–	13,900	–	–	–	28,350
Ga	18	1	9	20	31	17	–	–	–	–	16
Gd	–	2	3	4.2	5.5	–	–	–	–	–	3.8
Ge	1.41	1.2	19	–	1.9	1.2	–	–	–	–	4.94
Hf	7	3	8	2.5	12.7	–	–	–	–	–	6.6
Hg	0.11	0.05–0.26	0.04	–	0.053	0.09	–	–	0.18	0.17	0.11
Ho	–	–	0.87	0.73	1	–	–	–	–	–	0.93

(continued)

Table 2 (continued)

Element (mg/kg)	New Zealand	World	Sweden	Japan	Brazil	USA	Sankt-Petersburg	Australia	Resolution Island	Belgium	Average
I	14.50	1.7–4.0	–	13	1.2	–	–	–	–	–	5.0
In	0.40	–	–	0.09	0.11	–	–	–	–	–	0.20
La	22	26	33	23	34	37	–	–	–	–	29
Li	4	1–56	17	13	24	24	–	–	–	–	26
Lu	–	0.34	0.39	0.31	0.52	–	–	–	–	–	0.39
Mn	1315	270–525	411	–	535	550	118	–	1630	–	597
Mo	2.0	1.3–2.0	0.6	1.3	1.6	1.0	1.1	2.5	–	–	1.6
Nb	11.9	12	12	10	25.00	11	–	16.4	–	–	14
Nd	19.5	19	29	22	32	46	–	–	–	–	28
Ni	14	12–34	13	26	25	19	15	10	179	4	29
P	836	–	–	–	–	–	–	–	–	–	836
Pb	18	22–44	18.0	24.0	22.0	19.0	19.1	14.9	–	21.5	23
Pd	–	–	0.04	–	0.003	–	–	–	–	–	0.02
Pr	17	8	8	5	8	–	–	–	–	–	9.2
Pt	–	–	–	–	0.002	–	–	–	–	–	0.002
Rb	55	50	116	70	18	67	–	–	–	–	63
Sb	0.91	0.62	0.25	1	–	1	–	–	–	–	0.64
Sc	–	5–10	10	21	–	9	–	–	–	–	10.1
Se	1.49	0.25–0.70	0.23	–	0.47	0.39	–	–	–	–	0.50
Sm	–	3.1	4.50	4.4	6.70	–	–	–	–	–	4.68
Sn	3.0	1.8	1.80	2.4	–	1	1.40	–	–	–	1.99
Sr	94	87–210	163	190	240	–	162	18	–	–	146
Ta	3	1	1.10	1.70	2.30	–	–	0.83	–	–	1.75
Tb	–	0.4	0.5	0.74	0.9	–	–	–	–	–	0.63

(continued)

Table 2 (continued)

Element (mg/kg)	New Zealand	World	Sweden	Japan	Brazil	USA	Sankt-Petersburg	Australia	Resolution Island	Belgium	Average
Te	2	–	–	–	–	–	–	–	–	–	1.58
Th	11	8	8	9	11	9	–	9.5	–	–	9.5
Ti	5584	2300–26,000	3700	–	15,480	2900	1522	4210	–	–	6663
Tl	1.59	0.6	0.23	0.49	0.36	–	–	–	–	–	0.65
Tm	–	0.46	0.32	0.30	0.50	–	–	–	–	–	0.40
U	6	4	4	2	3	3	–	2	–	–	3.34
V	138	18–115	69	180	3200	80	16	–	–	–	341
W	–	1	1.30	1.30	1.40	–	1.00	–	–	–	1.2
Y	18	12	27	21	27	25	17	–	–	–	21
Yb	–	2.1	2.9	2.1	3.2	3.1	–	–	–	–	2.7
Zn	69	45–100	65	89	73	60	43	34	110	35	64
Zr	263	300	308	92	421	230	140	–	–	–	251

Sources Authors (Taylor) unpublished data, Lottermoser (2009), Kabata-Pendias and Mukherjee (2007), Doelsch et al. (2006), Licht (2005 quoted by Kabata-Pendias and Mukherjee 2007), Umfitseva and Terekhina (2005 quoted by Kabata-Pendias and Mukherjee 2007), Takeda et al. (2004), Burt et al. (2003), Eriksson (2001), Kabata-Pendias and Pendias (2001), Tack et al. (1997), Shaklette and Boerngen (1984 quoted by Kabata-Pendias and Mukherjee 2007)

Table 3 Ratio of average trace element concentrations in phosphate fertiliser to average background soil concentrations

Element	DAP	MAP	NPK	PK	SP	TSP	PR	Typical value for Table 4
Ag	14	1	1	1	18	4	3	6
As	2	2	2	3	23	4	1	6
Au	–	–	1	–	–	–	–	1
B	10	16	14	30	16	16	20	17
Ba	0.40	0.06	0.55	0.41	0.67	0.18	2.1	1
Be	0.6	0.4	0.2	0.8	0.3	2.2	1.2	1
Bi	0.06	–	0.06	0.08	0.15	–	0.17	0.1
Cd	61	130	18	16	33	40	77	54
Ce	2	4	10	1	11	6	5	6
Co	9	0.5	0.3	3	0.4	0.3	1	2
Cr	2	1	1	2	1	2	2	1
Cs	0.1	0.2	0.1	0.1	0.1	0.1	0.1	0.1
Cu	1	1	3	1	2	1	1	1
Dy	1	3	4	1	4	1	–	3
Er	1	3	3	2	3	2	–	2
Eu	2	5	8	1	8	3	9	5
F	67	61	73	–	45	67	95	68
Fe	0.2	0.4	0.1	0.3	0.2	0.2	0.3	0.2
Ga	–	–	0.02	–	–	–	–	0.02
Gd	1	4	5	1	5	1	26	6
Ge	–	–	0.02	–	–	–	–	0.02
Hf	–	–	0.03	–	–	–	0.2	0.09
Hg	4	0.5	1	0.4	6	8	31	7
La	2	3	8	1	9	6	21	7
Li	0.1	–	0.1	0.1	0.1	0.1	1	0.2
Lu	1	3	2	1	2	3	6	3
Mn	0.2	1.1	0.6	1.4	1.4	0.3	0.8	1
Mo	3	4	14	23	4	4	40	13
Nb	0.2	0.6	5	0.1	11	0.05	–	3
Nd	2	4	7	1	4	6	6	4
Ni	1	1	1	1	1	1	1	1
Pb	0.2	0.2	0.2	0.4	1	1	1	0.5
Pd	–	–	0	–	–	–	–	0.2
Pr	1	2	6	1	3	0.3	–	2
Pt	–	–	2	–	–	–	–	2
Rb	0.2	0.2	0.3	0.4	0.2	0.1	1	0.4
Sb	4	1	2	2	3	3	5	3
Sc	0.3	1	1	0.3	1	1	1	1

(continued)

Table 3 (continued)

Element	DAP	MAP	NPK	PK	SP	TSP	PR	Typical value for Table 4
Se	4	4	2	5	7	9	12	6
Sm	3	5	7	0.3	8	5	9	5
Sn	5	2	1	0.2	2	1	1	2
Sr	3	1	8	5	19	6	14	8
Ta	–	–	0.1	–	–	–	0	0.2
Tb	2	4	5	1	5	3	10	4
Te	–	–	0.01	0.1	–	–	–	0.0
Th	0.3	1	2	0.3	2	1	1	1
Ti	0.03	0.11	0.03	0.06	0.03	0.04	0.03	0.0
Tl	1	1	0.4	1	1	1	1	1
Tm	1	2	2	1	2	2	46	8
U	35	22	11	17	15	35	28	23
V	1	0.3	0.2	0.2	0.2	1	0.2	0.3
W	–	–	0.03	–	–	–	–	0.03
Y	3	5	4	3	5	4	7	4
Yb	1	3	2	1	2	2	7	3
Zn	4	7	6	2	3	6	4	4
Zr	0.1	0.1	0.1	0.1	0.1	0.1	3	1

As there were no soil: plant transfer factors or soil: water K_d values for some REE, and the similarity of chemical behaviour of all the REE, the mean of the combined cerium, europium, lanthanium and neodymium data was used.

Toxicity

The toxicity of a contaminant to the environment was assessed by comparing ecological screening values (ESV) for micro-organisms and plants. There were limited data for some elements. These were (data sources in parenthesis) lanthanum and REE (Barry and Mehann 2000), and uranium (Sheppard et al. 1983; Feldman et al. 1967).

The toxicity of a contaminant to human health was assessed by comparing maximum permitted values (MPV) or ceiling concentrations (NZMFE 2010; US Department of Health and Human Services 2003; NZWWA 2003; US National Research Council 1980; Friday 1998, Table 4). There was limited data for some elements. These were (data sources in parenthesis) boron (NZMFE 2010), fluoride (Choi et al. 2012; Meenakshi and Maheshwari 2006; World Health Organization 2011), lanthanum and REE (Kopittke et al. 2009), and uranium (Busby and Schnug 2008; Zielinski et al. 2000).

Table 4 Measurements of factors used in the risk analysis and ranking

Element	F:BG ratio (Table 3)	Median transfer factor[1]	Median K_d[2]	Plant ESV (mg/kg)[3]	Micro-organisms ESV (mg/kg)[4]	Human MPV (mg/kg)[5]	Naturally radioactive
Ag	6	0.01	290	2	50	500*	Stable
As	6	0.002	631	10	20	20	Stable
B	6	4.5	40	0.5	20	5200*	Stable
Cd	54	0.113	273	4	20	1	Stable
F	68	0.060	20,000*	200	30	500*	Stable
Hg	7	0.010	6305	0.3	30	1	Stable
La	7	0.004	6100	1500*	50	500*	Some isotopes
Other REE[7]	2–8	0.0067	3800	1500*[6]	506	500*[6]	Some isotopes
Mo	13	0.555	40	2	200	254	Stable
Se	6	0.022	34	1	100	0.81	Stable
Sr	8	0.32	394	1000	600	5000	Stable
U (low K_d)	23	0.008	0.4	5	10*	20*	α emitter
U (high K_d)	23	0.008	1,000,000	5	10*	20*	α emitter

[1]The Transfer factor is definided here as the ratio between the content of the edible portion of the plant and that of the soil. *Sources* Andrews and Cooke (1989), Baes et al. (1984), Boikat et al. (1985), EPA (1999), Fakayode and Onianwa (2002), Gerzabek et al. (1998), Handi et al. (2000), Hassoun (2011), IAEA (2010), Jha et al. (2008, 2009), Kabata-Pendias and Mukherjee (2007), Kabata-Pendias and Pendias (2001), Mollah and Begum (2001), Oyedele et al. (2006, Roca et al. (1997), Sauvé et al. (1996), Uchida and Tagami (2007), Warren et al. (2003), Zheng et al. (2008)

[2]*Sources* Allison and Allison (2005), Baes et al. (1984), Chen et al. (2007), USEPA and USDE (1999), USEPA (1999), IAEA (2010), Lemarchand et al. (2005), Nakamaru and Uchida 2008, Norden et al. (2010), Sauvé et al. (2000), Serne (2007), Sheppard et al. (2009), Sheppard (2011), Vandenhove et al. (2007), Whicker et al. (2007)

[3]Jha et al. (2009), Friday (1998), Efroymson et al. (1997a) except*

[4]Friday (1998), Efroymson et al. (1997b) except *

[5]NZ MFE (2010), US Department of Health and Human Services (2003), NZWWA (2003), US National Research Council (1980), Friday (1998), except *

[6]Toxicity data for rare earth elements (REE) is very limited. Because of the similarity of chemical behaviour of all the REE, toxicity factors for the REE were based on the analysis of lanthanum

[7]Because of the similarity of chemical behaviour of all the REE, the mean of the combined cerium, europium, lanthanium and neodymium data was used

*Low confidence due to limited data, see text for details

ESV Ecological Screening Value

MPV Maximum Permitted Value

Radioactivity

Health effects from radiation are conventionally modelled on the quantity Absorbed Dose. This is usually presented as a quantity termed Effective Dose Equivalent which has the unit Sievert (Sv). The Effective Dose Equivalent following exposure to uranium and its compounds is calculated by the risk model of the International Commission in Radiological Protection (ICRP 1996, Table 7). The method calculates Absorbed Dose in the conventional way as Energy per unit Mass and then converts this to Sieverts by multiplying by 20 see above). The algorithm which calculates the Committed Effective Dose Equivalent per unit intake dilutes the energy released from the radioactive decay of uranium-238 or the other isotopes uranium-234 or -235 into the mass of tissue which contains the uranium. This mass is generally more than a kilogram and so the effects of local enhancement of ionization (such as would be caused by uranium particulates or uranium ions bound to chromosomes) are not assessed. This averaging procedure has been criticised by a number of groups (CERRIE 2004a, b; IRSN 2005; ECRR 2003, 2010) as underestimating the effects of radiation on local tissue volumes. The Committed Effective Dose per unit intake is a function of the biological half life of the uranium, and the concentration in the organ of interest is a function of the route of exposure. These parameters have been obtained from various studies of humans and animals and are factored into the ICRP methodology.

The doses from inhalation are conceded to be much greater than the doses from ingestion. A 2 mg inhalation of slow dissolving uranium-238 contains 24.8 Bq and thus will deliver an effective dose of 0.2 mSv whereas an ingestion of 2 mg will give almost 1/200th of this. Note that for natural uranium, uranium-238 is only 49 % of activity. The small amount of uranium-234 is as active and is in equilibrium with uranium-238, which roughly doubles the overall alpha activity. In addition, there are the two beta-emitting intermediates thorium-234 and protoactinium-234 m (the "m" meaning metastable). Uranium-235 contributes about 2 % of the activity as it is such a small proportion of natural uranium.

Significant quantities of uranium in soil or fertiliser may contribute to an inhalation hazard as a result of dust resuspension (especially for farmers ploughing fields) and increased external radioactivity. The gamma radiation dose over a well mixed soil was 47 μSv/year for each mg/kg of uranium (NCRP 1987). Thus living on a soil containing the average of 3.3 mg/kg uranium (Table 2) will incur annual external dose of 155 μSv. However, for phosphate fertiliser containing 100 mg/kg U (Table 1), the dose rate will be about 30 times higher, about 4700 μSv (compared with the normal average background gamma radiation dose of about 600 μSv).

Doses to children have also been calculated by ICRP and are up to 5 times higher than the adult doses for the same exposure. The ECRR has published different dose conversion factors (see below) which are considerably higher (ECRR 2010).

Secondary Photoelectron Effect

Heavy elements may cause enhanced photoelectron ionization due to enhanced absorption of natural background radiation.

The quantity employed in radiation protection, absorbed dose, is defined as

$$D = \Delta\epsilon/\Delta m$$

i.e. energy is dispersed into a mass. Hitherto, the mass into which the energy has been diluted is that of living tissue; ICRU provide tables of absorption coefficients for different living tissue, adipose, bone, muscle etc., which can be employed for calculations involving doses, but generally all these denominator quantities have the absorption characteristics of water (ICRU35 1984). The absorption of electromagnetic (photon) radiation is due to a number of processes, the main three being pair-production, Compton scattering and photoelectron production. For elements of atomic number greater than about 30, and for photon energies of less than about 500 keV, the photoelectric effect predominates. Even for the low atomic number elements that make up living systems, below 200 keV there is fairly quantitative conversion of incident photon radiation (and induced photon radiation from second order and third order processes) into photoelectrons. These are fast electrons which are indistinguishable from beta radiation and have the energy of the incident photon less that of their binding energy (which is generally far less than the incident photon energy and can be ignored). The absorption of photon radiation by elements is proportional to the fourth or fifth power of the atomic number Z. Thus the predominant absorber in water is the Oxygen atom $Z = 8$ and it is reasonable to give the effective atomic number of water as 7.5. Of course, there are essential elements in tissue with higher atomic numbers than this, but interestingly, apart from iodine ($Z = 53$) few elements with $Z > 26$ (e.g. iron). The incorporation of high Z elements into living systems would generally be harmful since it could increase the radiation dose, and therefore such developments may have been lost through evolutionary selection. Iodine is an exception, but it should be noted that the main sites for radiation damage in terms of sensitivity are the main sites for iodine concentration, the thyroid gland and the blood. Busby and Schnug (2008) suggested that the metabolic and cell repair status controls exercised by the thyroid gland are the reasons why iodine has been incorporated into living systems and is employed as a kind of radiation-repair control mechanism.

A problem in radiation protection arises when high Z elements are incorporated into living tissue, since the enormously greater absorption of photon radiation by such material will result in enhanced doses to tissues adjacent to the high Z material. The problem was first addressed in 1947 in relation to X-rays of bone (Speirs 1949) and has been studied in the past in relation to prostheses. More recently, interest has shifted to the use of high Z material to enhance photon radiotherapies for tumour destruction where it has been shown to be effective e.g. gold nanoparticles have been successfully employed (and patented) for radiotherapy enhancement (Hainfeld et al. 2004).

Despite this knowledge, the enhancement of photon radiation by high Z contaminants has not been addressed in radiation protection. The situation may have arisen out of the fact that prosthetic materials are not intrinsically radioactive and contamination from high Z elements like lead (Z = 82) are considered under the heading of chemical toxicity.

There are two circumstances where the Secondary Photoelectron Effect (SPE) would have significant radiological implications. These are for elements that bind to DNA and for internal particulates. The uranyl ion has a very high affinity for DNA phosphate (Nielsen et al. 1992). This affinity has been known since the 1960s when it was first employed as an electron microscope stain for imaging chromosomes (Huxley and Zubay 1961). For internal particulates, the SPE effect will be much greater as the particle size is reduced, since for massive high Z contamination e.g. prostheses, most of the photoelectrons are lost inside the bulk material. The emergence of the photoelectrons into tissue is a function of the mean electron path in the material, and the absorbed dose in local tissue is a function of the electron range and thus its energy.

The SPE effect with uranium is likely to cause enhanced photoelectron ionization at the DNA due to enhanced absorption of natural background radiation.

Considering the potential damage of alpha emissions plus the secondary photon effect, uranium was given a score of 5 in the model.

Many of the rare earth elements have natural radioactive isotopes. However, only a small proportion of the natural abundance consists of radioactive isotopes and these isotopes have very long half lives (greater than 10,000,000,000 years) compared with uranium. Radioactive rare earths were given a score of 2.

It should be noted that group 2 metals, such as strontium-90 appear to have a high affinity for phosphate. The strontium-90 released to the environment in the 1960s may concentrate (as a result of rainfall scrubbing of weapons fallout) in phosphate rich soils. The affinity of radio-strontium (and uranium) for phosphate may also explain the anomalous enhancements of dose found by some epidemiological and animal studies (see below). Both strontium and uranium bind strongly to chromosomes since the cellular chromosomal DNA is maintained by a phosphate backbone. However, as this hypothesis is currently unproven, strontium was treated as naturally non-radioactive for this study.

Nonradioactive elements were given a score of 1.

Ranking

The ranking exercise (Table 5) showed uranium (rank 1 and 3) and cadmium (rank 2) to be the elements of most concern, consistent with studies on accumulation in soil (Taylor et al. 2007) and enhanced uranium in water in agricultural areas (Schnug et al. 2008; Zielinski et al. 2000). Mercury (4) and boron (5) are not usually considered a fertiliser derived issue but are toxic to some plants, microorganisms and animals at high concentrations. Boron is highly mobile in soil.

Surprisingly low is the ranking for fluoride (6) as fertiliser associated fluoride has been reported to accumulate in soil (Loganathan et al. 2001), and it can be highly toxic to some microorganisms. However, it is considered a moderate health risk to humans and plants. Selenium (7) is readily transferred to water and is toxic to plants and humans at high concentrations. Molybdenum (8) is also readily transferred to water and its antagonism with copper is well documented (e.g. Kabata-Pendias and Pendias 2001). Arsenic's (8) toxicity to biota is well known but it is generally not as readily taken up by plants or transferred to water as some of the other elements. Silver (9) is more readily taken up by plants than arsenic but is less toxic to microorganisms and humans. The 17 Rare Earth Elements (10 =), are usually considered a low health risk to plants and humans but can be toxic to microorganisms and aquatic organisms (e.g. Lantanum-Barry and Mehann 2000; Chu et al. 2003). Although strontium (28) is present in reasonable amounts in phosphate fertilisers, it is relatively benign to biota and unlikely to cause negative impacts on the environment. The same cannot be said for the radionuclide strontium-90 (see above) which may concentrate in phosphate rich soils and could constitute a vector for human exposure. This aspect deserves further research consideration. The ranking exercise suggests further study of boron, mercury and molybdenum and their fate after application to soil is advisable.

If these elements of concern are strongly retained by the soil, they will accumulate. The ratio of the concentrations of acid extractable (i.e. method-USEPA 200.4) elements in fertilised soils to the concentrations in background soils may be used to identify accumulation with the greater the ratio the greater the accumulation compared to background. Conversely, a ratio less than 1 indicated depletion of an element. Figure 1 shows data on trace elements in New Zealand soils from 383 fertilised and 87 unfertilised sites. The 6 elements with the highest ratios are all associated with phosphate fertiliser. Calcium and phosphorous are part of the fertiliser matrix. Cadmium has the greatest fertiliser to background ratio so more in proportion is going on to the soil. Although boron and selenium were elevated in phosphate fertilisers (Table 4), they do not appear to be accumulating significantly in the soil. It is likely that these contaminants are moving from the soil to ground and surface water.

Uranium, REE and fluoride are relatively immobile under New Zealand conditions but under alkaline conditions in Europe, uranium may be mobilised as carbonate complexes. The soil system is constantly changing. Elements that accumulate in the soil may be mobilised by changes in soil processes, such as acidification or changes in redox potential. An element strongly adsorbed to the soil may remain relatively biologically inactive at neutral to alkaline pH but become much more active and mobile at low pH, e.g. cadmium.

Table 5 Risk analysis scores and ranking of potentially environmentally significant trace element contaminants of phosphate fertilisers

Element	F: BG ratio	Transfer factor	Kd	Plant ESV	Microorganism ESV	Human MPV	Naturally radioactive	Score	Rank
Ag	1	2	4	4	3	3	1	288	9
As	1	1	4	3	4	4	1	192	8
B	1	5	5	5	4	1	1	500	5
Cd	5	3	4	4	4	5	1	4800	2
F	5	2	2	2	4	3	1	480	6
Hg	1	2	3	5	4	5	1	600	4
La	1	1	3	1	3	3	2	54	10
Other REE	1	1	3	1	3	3	2	54	10
Mo	2	3	5	4	1	3	1	360	8
Se	1	2	5	4	2	5	1	400	7
Sr	1	3	4	1	1	1	1	12	28
U (low K_d)	3	1	5	4	5	4	5	6000	1
U (high K_d)	3	1	1	4	5	4	5	1200	3

Potential Issues for Resource Managers

Potential Issues

Resource managers, environmental decision makers and the agricultural sector need to be aware that the potential for accumulation of trace elements from phosphate fertilisers is a problem for present and future soil, sediment and water quality. The potential problems that can arise from the accumulation of trace elements in soil are wide-ranging. Some come about as a direct result of concentrations passing toxic thresholds, whereas others result from more subtle secondary effects. Looking at the soil resource and agricultural sustainability we see the direct effects of toxicity to biota:

- Poisoning of soil organisms—from microbes to invertebrates.
- Toxicity to plants (grass, herbage or crops).
- Toxicity to grazing animals.

Poisoning of soil organisms, from microbes to invertebrates, may have a range of consequences. One of the most serious for agriculture is reduction in nitrogen fixation by soil microbes, which can result in increased fertiliser nitrogen inputs and therefore increased costs.

Toxicity to plants (grass, herbage or crops) and animals may restrict growth, or result in death in extreme cases. It also may result in secondary poisoning through

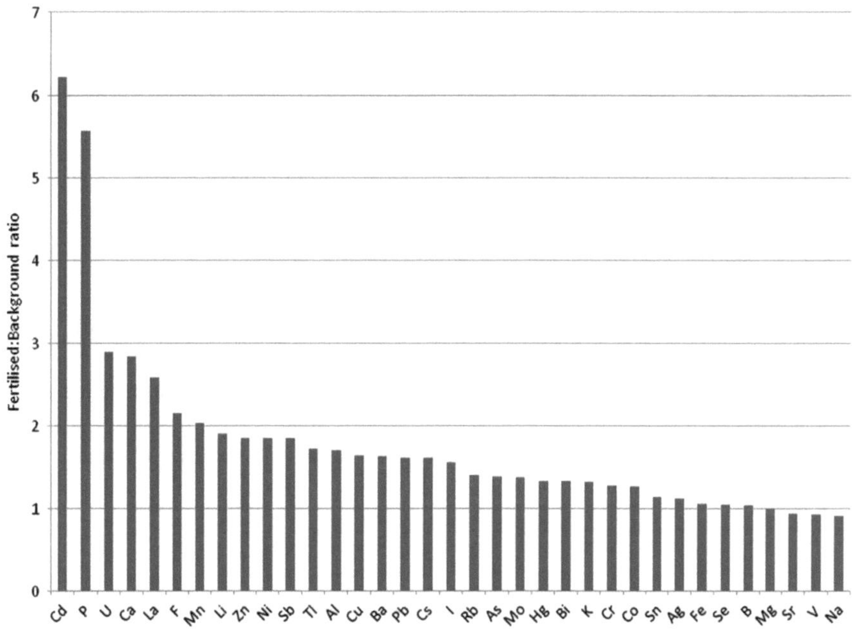

Fig. 1 Fertilised:Background ratio for mean concentrations of acid extractable elements (except for F which is total F) in New Zealand soils

excess accumulation in terrestrial or aquatic food chains that flow all the way up to humans.

An indirect effect is the competition for uptake by plants and organisms and the induction of deficiency in another (essential) element (see Section "Introduction"). One thing that is often overlooked is the competition for adsorption sites on the soil matrix between elements. The increase in the concentration of a contaminant may induce deficiency in another (essential) element, resulting in production losses.

The main potential human health and trade issues revolve around the breaking of food standards in crop and animal products, leading to increased dietary intakes in the human population. Food standards are constantly being revised as scientific knowledge increases. However, nearly all revisions have resulted in the lowering of the standards. This has ongoing impacts on soil guidelines and limitation of land use flexibility.

Protecting human health is clearly a key risk area. A higher level of contaminants in crops grown in or on the soil, and in animal products is likely to result in increased dietary intakes in the human population. Food standards have been set to provide protection for the general population. Breaking of food standards in crops grown in or on the soil, and in animal products could result in an increased health risk for the general population.

Land use flexibility issues include the inability for farmers to convert to a more sensitive productive use e.g. conversion of pastoral land to cropping. Another issue is the conversion of rural land to urban and the inability to subdivide for residential use without rehabilitation e.g. arsenic was historically used for pest control in orchards in New Zealand and old orchard soils may have high residual arsenic levels which prevent residential subdivisions being built.

Higher levels of contaminants in soil, resulting in food standard exceedences by crops and animal products grown on that soil are likely to cause regulators to restrict the types of production to low uptake crops (which may be less valuable).

Off-site issues usually revolve around accumulation of toxic or injurious elements in other parts of the environment:

- Off-site migration and accumulation in enclosed water sediments (of lakes, estuaries, the Baltic Sea).
- Risk of long-term contamination of rural groundwater through wide-scale leaching.
- Secondary poisoning through excess accumulation in terrestrial or aquatic food chains.

Guidelines

Guidelines for trace elements in soils can (in principle) be developed to cover any of the issues identified above. For example, a given guideline might be intended to cover one or more of the following:

1. protection of soil microbial health;
2. protection of soil invertebrates (e.g. earthworms);
3. protection of plant health (phyto-toxicity);
4. prevention of micronutrient deficiency;
5. protection of higher wildlife (e.g. insectivorous birds);
6. protection of groundwater;
7. suitability for root and leafy vegetable production from a food standard point of view;
8. suitability for grain production from a food standard point of view;
9. suitability for animal production (including protection of grazing stock health);
10. protection of human health.

In practice:

Guideline types 1–3 are often combined to allow use of a single number to protect all soil organisms (ecological receptors).

Guideline types 7–9 may be developed for the purposes of protecting trade, as well as meeting food standards.

Guideline 4 may be a figure below which deficiency of an essential element will occur, or might represent the level above which a second element reduces availability of the primary element and therefore causes deficiency.

Before applying a guideline, it is important to ascertain what it is designed to protect. Very few guidelines protect all of the receptor classes listed above.

Manage the Risk

Good governance will encourage a best practice approach to managing contaminant build up in soils over the long-term through regulations and incentives. Proactive management to minimise the risk to health, land use flexibility and the environment should have an overview of the next 100 years. Such management would seek to prevent trace element losses to the environment and balancing of inputs to outputs e.g. to reduce unwanted trace element contents input in fertilisers, only apply the amount of fertiliser nutrient needed. Experience shows many farmers apply far more fertiliser than they need for “insurance”.

Dutch government departments are stimulating sustainable alternatives by the development of knowledge and information in cooperation with agricultural business. The Dutch organisation for entrepreneurs in agricultural and horticultural business (LTO) has recognised the trace element problem and is now willing to undertake activities to reduce the accumulation of heavy metals in agricultural soils (Schipper et al. 2008). However, experience suggests broad recognition by farmers will take years of effort.

One useful tool could be open documentation as it allows the consumer to choose one brand over another, e.g. the amount of contaminants in fertilisers could be reported on the product packaging. Germany has attempted to limit heavy metal contents in fertilisers by defining limit values under the German Fertiliser Ordinance. However, since most of the fertilisers sold on the German market are EU fertilisers, an EU regulation is needed for these limit values to prove more effective.

Monitoring of food products for standard exceedances, and identification of origin, is a powerful tool for public protection. However, there can be no protection if there is not a food standard (due to lack of information) e.g. no food standard is known for uranium, although there are guidelines for drinking water (such as WHO 2011) and the German Federal Environmental Agency has a guideline value in drinking water of 10 µg/L. The WHO drinking water guideline for chemical toxicity of uranium gave a Total Daily Intake value of 0.6 µg/kg of bodyweight/day. Radiation protection was based on the assumption that any exposure to radiation involved some level of risk, with no threshold value below which there is no risk. The Individual Dose Criterion was to represent a very low level of risk that was not expected to give rise to any detectable adverse health effect. The annual dose associated with the intake of radionuclides from drinking-water was set at 0.1 mSv/year (WHO 2011).

Fertiliser Management

Farmers and growers should work closely with their fertiliser representatives to determine the most cost effective, efficient and appropriate fertiliser application and land management options. This will vary depending on the type of farming, the size of operation, soil type, the need to protect vulnerable or special environments and other considerations e.g. management of the environmental effects of cadmium should be directed to minimising the occurrence of outwash (e.g. by better irrigation timing) or cadmium transfer by using less cadmium-rich phosphate fertiliser or minimising phosphate fertiliser use in areas susceptible to surface runoff (McDowell 2010).

An important aspect of this book is closing the phosphorous cycle—recycling phosphorous within the farm system, thus reducing or eliminating imports of contaminant-containing phosphate fertiliser. The imminent extraction of phosphorous from local waste products in a clean form, suitable and economical enough for farm use (Naidu et al. 2012), is a fertile area of research that is the likely future for the fertiliser industry.

In the interim, it is good agricultural practice to minimise the amounts of contaminant trace elements input to farms. Alternatives to mineral phosphate fertiliser include using local farmyard manures and slurries, which are a step to closing the phosphorous cycle.

Environmental Monitoring and Research

A well designed monitoring programme will provide information on the ongoing status of trace elements in the soil environment; if they accumulate or are being lost to another part of the environment (e.g. ground or surface water), which will enable those involved directly (farmers etc.) or indirectly (policy makers, governance) to assess and manage potential risks.

Where there are knowledge gaps, such as a food standard for uranium, these need to be addressed by researchers and funded by authorities.

Remediation

The best option is to simply not allow trace elements to accumulate to environmentally excessive levels. Unlike hydrocarbons, biodegradation into innocuous carbon dioxide and water is not possible. Irrespective of soil chemical and biochemical reactions, the same metal will ultimately be present. Where accumulation does occur it is challenging to remediate due to the permanent, diffuse and widespread nature of the contamination, and the permanent nature of elemental contamination (elements do not break down except for radioactive decay, which still can result in a daughter contaminant).

Planting buffer zones can reduce the amount of particle-born contaminants transferred to streams, rivers and lakes. Buffer zones removed 20–78 % of total P (and presumably other particle-bound contaminants) in Finland (Uusi-Kamppa et al. 1996) and wet meadows and grey alder forests in Estonia retained up to 12.8 kg P/ha/year (Mander et al. 1996).

Liming can be used to raise the pH and reduce the mobility of cations like cadmium and copper (Kabata-Pendias and Pendias 2001; Gray et al. 1999). However, calcium in the lime can compete with cadmium for binding sites on the soil (Maurer et al. 2012). Besides, liming does not remove the contaminants and they can become available again should the pH drop.

Deep ploughing can be used to dilute the contaminant within the soil profile. However, this does not remove the contaminant; but mixes it deeper into the soil and closer to groundwater.

Capping the contaminated soil with an impervious layer of clean material essentially removes trace element contaminants from contact with the rest of the environment. As rain water cannot penetrate the cap there is no drainage and leaching of contaminants to groundwater. Similarly, plant roots cannot penetrate to reach the contaminants nor can the contaminants be resuspended as dust. However, this does not remove the contaminants and they can become available again should the cap fail.

Robinson et al. (2009) give a detailed review of phytomanagement technologies. Phytomanagement is a term that describes an assemblage of related technologies, including **phytomining** (Nicks and Chambers 1994), **phytoremediation**, which is divided into phytoextraction and phytostabilization (Salt et al. 1995), as well as biofortification (King 2002). While biofortification is the improvement in trace element uptake by food or fodder crops, the other technologies may deliver a cost-effective means of mitigating environmental risks associated with a contaminated site.

Phytoremediation may aim to immobilise contaminant trace elements in the soil, so that they neither leach nor enter the plant shoots (phytostabilization). Alternatively, one may engineer or manipulate the soil-plant system so that plants take up the contaminating trace elements into the shoots, where they may be volatilised (phytovolatilization), or removed from the site when the plants are harvested (phytoextraction). Phytomining describes the phytoextraction of valuable trace elements from ore bodies or metal-contaminated environments where the target metal concentration is too low for conventional mining. In soils that are deficient in essential trace elements, the goal of phytomanagement is to increase the trace element concentration in the edible plant organs (biofortification). The boundaries between these technologies are not distinct. Phytoextraction of trace elements from a contaminated soil may produce biomass enhanced in essential trace elements (biofortification) or saleable quantities of valuable trace elements (phytomining), while reducing erosion and trace element leaching from the site (phytostabilization).

From a phytomanagement perspective, contaminated land can be utilised as a resource. Bioenergy and timber production on contaminated land has several advantages over production on non-contaminated fertile soils. Most importantly, productive agricultural land will not be taken out of food production, which, in the case of bioenergy production, has led to increased food costs (Economist 2007). Since the biomass produced either fixes carbon (timber) or produces CO_2-neutral energy (bioenergy), carbon credits may add further value to the operation. In addition to timber or bioenergy, the biomass of vegetation from a phytomanagement program may have other uses, where the possibly elevated concentrations of trace elements in the plant tissues are either unimportant (because they will not be consumed) or beneficial essential nutrients (such as selenium or zinc). This may include cut flowers, cotton, and stock fodder. In the latter case, the biomass should be carefully monitored for excessive concentrations or the presence of other contaminants.

Phytomanagement is a long-term and site-dependent technology. It is only feasible if it satisfies environmental regulation and either returns a profit or costs less than alternative technologies or inaction.

When groundwater is contaminated with metals, remediation alternatives are limited to mobilisation or immobilisation. Generally, it is easier to immobilise a metal than it is to mobilise and recover it. However, an immobilised metal may still be a source of future problems if subsurface ambient conditions change (Vance 1994). Mobilisation and pumping is limited to highly polluted situations and is not suitable for low-level, diffuse contamination (Uusi-Kamppa et al. 1996).

Concluding Remarks

Phosphate fertiliser application is essential for high production using conventional farming techniques and likely to remain so for the immediate future. However, no one type of phosphate fertiliser currently on the market is completely free of contaminants. Some trace element contaminants from phosphate fertilisers will continue to accumulate in the soil, while others will be transported to waterways. The harmful effects caused by trace elements tend to be chronic, gradually building up over continual exposures. It is easy for potential trace element induced ecological and human-health consequences may go on unnoticed. Policymakers and regulators need to be aware of the issues surrounding trace elements to be able to make policies and regulations that meet human health or water quality requirements. Policies and regulations should expect land users to account for inputs and outputs of selected trace elements on the land they farm where these trace elements can impact on the lives of others.

This chapter has identified the elements of key concern as cadmium, uranium, mercury, boron, molybdenum, fluoride, arsenic and the rare earth elements. The ranking exercise suggested further study of the fate of fertiliser-derived molybdenum, mercury and zinc, their transfer between the lithosphere, hydrosphere and the biosphere, and their ultimate repository is advisable.

The following knowledge gaps were identified: There was no data on the uptake by plants of gadolinium, samarium and thulium and limited data on the toxicity of all the rare earth elements; There was limited data on the movement of fluoride from soil to water and on its toxicity for plants and humans; There was limited data on the toxicity of uranium.

The most effective means of managing the risk is to simply not allow trace elements to accumulate to environmentally excessive levels. The future of farming is likely to involve closing the phosphorous cycle and recycling phosphorous within the farm system, thus reducing or eliminating imports of contaminant-containing phosphorous fertiliser. The fertiliser industry should develop phosphate sources other than phosphate rock, such as the extraction of phosphorous from local waste products, in a clean form, suitable and economical for farm use.

References

Abu-Qare AW, Abou-Donia MB (2002) Depleted uranium—the growing concern. J Appl Toxicol 22:149–152

Allison J, Allison T (2005) Partition coefficients for metals in surface water, soil, and waste. U.S. Environmental Protection Agency, Office of Research and Development EPA/600/R-05/074, Washington, DC

Al-Shawi A, Dahl R (1999a) The determination of cadmium and six other heavy metals in nitrate/phosphate fertilizer solution by ion chromatography. Anal Chim Acta 391:35–42

Al-Shawi A, Dahl R (1999b) Determination of total chromium in phosphate rocks by ion chromatography. J Chromatogr A 850:137–141

Andrews S, Cooke J (1989) Distribution of trace element pollutants in a contaminated ecosystem established on metalliferous fluorspar tailings. 3: Fluoride. Environ Pollut 60:165–179

Azouazi M, Ouahidi Y, Fakhi S, Andres Y, Abbe J, Benmansour M (2001) Natural radioactivity in phosphates, phosphogypsum and natural waters in Morocco. J Environ Radioact 54:231–242

Baes C, Sharp RD, Sjoreen A, Shor R (1984) A review and analysis of parameters for assessing transport of environmentally released radionuclides through agriculture. United States Department of Energy, Oak Ridge

Barisic D, Lulic S, Miletic P (1992) Radium and uranium in phosphate fertilizers and their impact on the radioactivity of waters. Water Res 26:607–611

Barry M, Meehan B (2000) The acute and chronic toxicity of lanthanum to Daphnia carinata. Chemosphere 41:1669–1674

Bech J, Suarez M, Reverter F, Tume P, Sanchez P, Bech J, Lansac A (2010) Selenium and other trace elements in phosphate rock of Bayovar-Sechura (Peru). J Geochem Explor 107:136–145

Boikat U, Fink A, Bleck-Neuhaus J (1985) Cesium and cobalt transfer from soil to vegetation on permanent pastures. Radiat Environ Biophys 24:287–301

Bramley RGV (1990) Cadmium in New Zealand agriculture. NZ J Agric Res 33:505–519

Burnett W, Schultz M, Hull C (1996) Radionuclide flow during the conversion of phosphogypsum to ammonium sulphate. J Environ Radioact 32:33–51

Burt R, Wilson M, Mays M, Lee C (2003) Major and trace elements of selected pedons in the USA. J Environ Qual 32:2109–2121

Busby C, Schnug E (2008) Advanced biochemical and biophysical aspects of uranium contamination. In: De Kok LJ, Schnug E (eds) Loads and fate of fertilizer-derived uranium. Backhys Publichers, Leiden, p 229
Cadmium Working Group (2008) Cadmium in New Zealand report one: cadmium in New Zealand agriculture. MAF Information Paper. Ministry of Agriculture and Forestry, Wellington, New Zealand. Available from: http://www.mpi.govt.nz/news-resources/publications.aspx?title=Cadmium
Camelo L, de Miguez S, Marban L (1997) Heavy metals input with phosphate fertilizers used in Argentina. Sci Total Environ 204:245–250
Cevik U, Baltas H, Tabak A, Damia N (2010) Radiological and chemical assessment of phosphate rocks in some countries. J Hazard Mater 182:531–535
Chandrajith R, Seneviratna S, Wickramaarachchi K, Attanayake T, Aturaliya T, Dissanayake C (2010) Natural radionuclides and trace elements in rice field soils in relation to fertilizer application: study of a chronic kidney disease area in Sri Lanka. Environ Earth Sci 60:193–201
Charter R, Tabatabai M, Schafer J (1995) Arsenic, molybdenum, selenium, and tungsten contents of fertilizers and phosphate rocks. Commun Soil Sci Plant Anal 26:3051–3062
Chen W, Chang A, Laosheng W (2007) Assessing long-term environmental risks of trace elements in phosphate fertilizers. Ecotoxicol Environ Saf 67:48–58
Choi A, Sun G, Zhang Y, Grandjean P (2012) Developmental fluoride neurotoxicity: a systematic review and meta-analysis. Environ Health Perspect. doi:10.1289/ehp.1104912
Chu HY, Zhu JG, Xie ZB, Zeng Q, Li ZG, Cao ZH (2003) Availability and toxicity of exogenous lanthanum in a haplic acrisols. Geoderma 115:121–128
Committee Examining Radiation Risk from Internal Emitters (2004a) Report of the Committee Examining Radiation Risks from Internal Emitters Chilton UK. National Radiological Protection Board
Committee Examining Radiation Risk from Internal Emitters (2004b) Minority report of the Committee Examining Radiation Risk from Internal Emitters (CERRIE). Bramhall R, Busby C and Dorfman P. Sosiumi Press, Aberystwyth
Cronin S, Manoharan V, Hedley M, Loganathan P (2000) Fluoride: a review of its fate, bioavailability, and risks of fluorosis in grazed-pasture systems in New Zealand. NZ J Agric Res 43:295–321
Cronin S, Neall V, Lecointre J, Hedley M, Loganathan P (2003) Environmental hazards of fluoride in volcanic ash: a case study from Ruapehu volcano, New Zealand. J Volanol Geothermal Res 121:271–291
Czarnowski W, Krechniak J (1990) Fluoride in the urine, hair, and nails of phosphate fertiliser workers. Br J Ind Med 47:349–351
De Vries W, Romkins P, Schulz G (2007) Critical soil concentrations of cadmium, lead, and mercury in view of health effects on humans and animals. Rev Environ Contam Toxicol 191:91–130
Doelsch E, Van de Kerchove V, Saint Macary H (2006) Heavy metal content in soils of reunion (Indian Ocean). Geoderma 134:119–134
Economist, April 4, 2007. Castro was right. In The Economist
Efroymson R, Will M, Suter G, Wooten A (1997a) Toxicological benchmarks for screening. Contaminants of potential concern for effects on terrestrial plants: 1997 Revision. U.S. Department of EnergyES/ER/TM-85/R3, Oak Ridge
Efroymson R, Will M, Suter G (1997b) Toxicological benchmarks for contaminants of potential concern for effects on soil and litter invertebrates and heterotrophic process: 1997 Revision. U. S. Department of Energy, Oak Ridge
Eisenbud M, Gesell T (1997) Environmental radioactivity. Academic Press, San Francisco
Erdem E, Tinkiliç N, Yilmaz VT, Uyanik A, Ölmez H (1996) Distribution of uranium in the production of triple superphosphate (TSP) fertilizer and phosphoric acid. Nutr Cycl Agroecosyst 44:129–131

Eriksson J (2001) Concentrations of 61 trace elements in sewage sludge, farmyard manure, mineral fertiliser, precipitation and in oil and crops. Swedish Environmental Protection Agency, Stockholm, Sweden

European Committee on Radiation Risk (2003) The 2003 recommendations of the European Committee on Radiation Risk. In: Busby C, Bertell R, Schmitz-Feuerhake I, Yablokov A (eds) The health effects of ionising radiation exposure at low doses for radiation protection purposes, Regulators Edition. ECRR; Aberystwyth: Green Audit, Brussels

European Committee on Radiation Risk (2010) The 2010 recommendations of the European Committee on Radiation Risk. The health effects of exposure to low doses of ionizing radiation. ECRR; Aberystwyth: Green Audit, Brussels

European Fertilizer Manufacturers Association (2000) Phosphorus: essential element for food production. European Fertilizer Manufacturers Association (EFMA), Brussels

Fakayode S, Onianwa P (2002) Heavy metal contamination of soil, and bioaccumulation in Guinea grass (Panicum maximum) around Ikeja Industrial Estate, Lagos, Nigeria. Environ Geol 43:145–150

Feldman I, Jones J, Cross R (1967) Chelation of uranyl ions by adenine nucleotides. J Am Chem Soc 89:49–53

Friday G (1998) Ecological screening values for surface water, sediment, and soil. US Department of Energy, Oak Ridge

Gargouri M, Chtara C, Charrock P, Nzihou A, El Feki H (2011) Synthesis and physicochemical characterization of pure diammonium phosphate from industrial fertilizer. Ind Eng Chem Res 50:6580–6584

Germann K, Bock W, Schroter T (1984) Facies development of upper cretareas phosphorites in Egypt: sedimentological and geochemical aspects. In: Klitzsch E, Said R, Schrank E, Hecht F (eds) Research in Egypt and Sudan: Sonderforschungsbereich 69, results of the special research project arid areas, period 1981–1984. D. Reimer, Berlin, p 457

Gerzabek M, Strebl F, Temmel B (1998) Plant uptake of radionuclides in lysimeter experiments. Environ Pollut 99:93–103

Gowariker V, Krishnamurthy V, Gowariker S, Dhanorkar M, Paranjape K (2009) The fertilizer encyclopedia. Hoboken, New Jersey. A John Wiley & Sons, Inc

Gray C, McLaren R, Roberts A, Condron L (1999) Effect of soil pH on cadmium phytoavailability in some New Zealand soils. NZ J Crop Horticultural Sci 27:169–179

Guzman E, Rios M, Garcia J, Regil E (1995) Uranium in phosphate rock and derivates. J Radioanal Nucl Chem 189:301–306

Hainfeld JF, Slatkin DN, Smilowitz HM (2004) The use of gold nanoparticles to enhance radiotherapy in mice. Phys Med Biol 49:N309–N315

Hakam O, Choukri A (2008) Determination of the uranium content in some natural formations in morocco by use of nuclear analytical methods. Spectrosc Lett 41:261–266

Handi J, Kallweit E, Henning M, Szec L (2000) On the long-term behaviour of Ag110 in the soil–plant system and its transfer from feed to pig. J Environ Radioact 48:159–170

Hassoun R (2011) A statistical evaluation of the contribution of mineral and tap water to the dietary intake of As, B, Cu, Li, Mo, Ni, Pb, U and Zn by humans. PhD thesis, Braunschweig Technical University, Braunschweig, Germany

Houston EC, Yates LD, Haunschild RL (1955) Fertilizer manufacture, diammonium phosphate fertilizer from wet-process phosphoric acid. J Agr Food Chem 3:43–48

Huxley HE, Zubay G (1961) Preferential staining of nucleic acid containing structures for electron microscopy. J Biophys Biochem Cytol 11:273

Institute for Radiological Protection and Nuclear Safety (2005) Health consequences of internal contaminations by radionuclides. Comments on the ECRR Report. The health effects of ionizing radiation exposure for radiation protection purposes. Report DPRH 2005-20. Fontenay aux Roses, IRSN

International Atomic Energy Agency (2010) Handbook of parameter values for the prediction of radionuclide transfer in terrestrial and freshwater environments. International Atomic Energy Agency, Vienna

International Commission in Radiological Protection (1996) ICRP Publication 72: age-dependent doses to the members of the public from intake of radionuclides part 5, compilation of ingestion and inhalation coefficients. Annals of the ICRP 26/1. Elsevier, Oxford

International Commission on Radiation Units and Measurements (1984) Radiation dosietry: electron beams with energies between 1 and 50 MeV. ICRU Report 35. International Commission on Radiation Units and Measurements Berthesda MD, USA

Jha S, Nayak A, Sharma Y (2008) Response of spinach (Spinacea oleracea) to the added fluoride in an alkaline soil. Food Chem Toxicol 46:2968–2971

Jha S, Nayak A, Sharma Y (2009) Fluoride toxicity effects in onion (*Allium cepa* L.) grown in contaminated soils. Chemosphere 76:353–356

Kabata-Pendias A, Mukherjee A (2007) Trace elements from soil to human. Berlin, Springer

Kabata-Pendias A, Pendias H. (2001) Trace elements in soils and plants, 3rd ed. CRC Press, Boca Raton

King JC (2002) Evaluating the impact of plant biofortification on human nutrition. J Nutr 132:511S–513S

Kopittke P, McKenna B, Blamey F, Wehr J, Menzies N (2009) Metal-induced cell rupture in elongating roots is associated with metal ion binding strengths. Plant Soil 322:303–315

Kpomblekou-A K, Tabatabai M (1994) Metal contents of phosphate rocks. Commun Soil Sci Plant Anal 25

Lambert R, Grant C, Sauvé S (2007) Cadmium and zinc in soil solution extracts following the application of phosphate fertilizers. Sci Total Environ 378:293–305

Lauria D, Riberiro F, Rodrigues Silva J (2006) Phosphate fertilizer influence on 238U content in vegetables. In : Proceedings IJM Acapulco 2006 on CDROM. Primer Congreso Americano del IRPA 2006/First American IRPA Congress 2006 XXIV Reunión Anual de la SMSR y XVII Congreso Anual de la SNM/XXIV SMSR Annual Meeting XVII Annual SNM Congress Acapulco México, del 3 al 8 de Septiembre 2006/ Acapulco Mexico, 3–8 Sept 2006'Acapulco México)

Lemarchand E, Schott J, Gaillardet J (2005) Boron isotopic fractionation related to boron sorption on humic acid and the structure of surface complexes formed. Geochim Cosmochim Acta 69:3519–3533

Loganathan P, Hedley M, Wallace G, Roberts A (2001) Fluoride accumulation in pasture forages and soils following long-term applications of phosphorus fertilisers. Environ Pollut 115:275–282

Loganathan P, Hedley M, Grace N (2003) Fertiliser contaminants in New Zealand grazed pasture with special reference to cadmium and fluorine: a review. Aust J Soil Res 41:501–532

Lottermoser B (2009) Trace metal enrichment in sugarcane soils due to the long-term application of fertilisers, North Queensland, Australia: geochemical and Pb, Sr, and U isotopic compositions. Aust J Soil Res 47:311–320

Mander U, Lohmus K, Kuusemets V, Ivask M (1996) The potential role of wet meadows and grey alder forests as buffer zones. In: Haycock N, Burt T, Goulding K, Pinay G (eds) Buffer zones: their processes and potential in water protection. Haycock Associated Ltd., St. Albans

Manoharan V, Loganathan P, Tillman R, Parfitt R (1996) Assessing aluminum phytotoxicity in long-term phosphate fertilized pasture soils. Commun Soil Sci Plant Anal 27:1731–1748

Maurer F, Christl I, Hoffmann M, Kretzschmar R (2012) Reduction and reoxidation of humic acid: Influence on speciation of cadmium and silver. Environ Sci Technol. Just Accepted Manuscript, Downloaded from http://pubs.acs.org on 2 Aug 2012

Maxwell GR (2005). Synthetic nitrogen products: a practical guide to the products and processes. Springer, Berlin, pp 285–298

Mazzilli B, Bourlegat F, Saueia C, Favaro D (2011) Rare earth elements in phosphogypsum and phosphate fertilizers in Brazil. Accessed from http://pintassilgo2.ipen.br/biblioteca/2011/eventos/17686.pdf

McBride M, Spiers G (2001) Trace element content of selected fertilizers and dairy manures as determined by ICP-MS. Soil Sci Plant Anal 32:139–156

McDowell R (2010) Is cadmium loss in surface runoff significant for soil and surface water quality: a study of flood-irrigated pastures? Water Air Soil Pollut 209:133–142
McLaughlin M, Tiller K, Naidu R, Stevens D (1996) Review: The behaviour and environmental impact of contaminants in fertilizers. Aust J Soil Res 34:1–54
Meenakshi S, Maheshwari R (2006) Fluoride in drinking water and its removal. J Hazard Mater 137:456–463
Menzel RG (1968) Uranium, radium and thorium content in phosphate rocks and their possible radiation hazard. J Agric Food Chem 16:231–234
Mermut A, Jain J, Song L, Kerrich R, Kozak L, Jana S (1996) Trace element concentrations of selected soils and fertilizers in Saskatchewan, Canada. J Environ Q 25:845–853
Mirlean N, Roisenberg A (2007) Fluoride distribution in the environment along the gradient of a phosphate-fertilizer production emission (southern Brazil). Environ Geochem Health 29:179–187
Mirlean N, Baisch P, Machado I, Shumilin E (2008) Mercury contamination of soil as the result of long-term phosphate fertilizer production. Bull Environ Contam Toxicol 81:305–308
Molina M, Aburto F, Calderon R, Cazanga M, Escudey M (2009) Trace element composition of selected fertilizers used in chile: phosphorus fertilizers as a source of long-term soil contamination. Soil Sed Contam 18:497–511
Mollah A, Begum A (2001) A study on transfer factors of 60Co and 65Zn from soil to plants in the tropical environment of Bangladesh. Environ Monit Assess 68:91–97
Naidu R, Lamb DT, Bolan NS, Gawander J (2012) Recovery and reuse of phosphorus from wastewater sources. In: Currie LD, Christensen CL (eds) Advanced nutrient management: gains from the past—goals for the future. http://flrc.massey.ac.nz/publications.html. Occasional Report No. 25. Fertilizer and Lime Research Centre, Massey University, Palmerston North, New Zealand
Nakamaru Y, Uchida S (2008) Distribution coefficients of tin in Japanese agricultural soils and the factors affecting tin sorption behavior. J Environ Radioact 99:1003–1010
National Council on Radiation Protection and Measurements (1987) Exposure of the population of the United States and Canada from Natural Background Radiation NCRP 94. NCRU, Bethesda, USA
New Zealand Ministry For The Environment (2010) Draft methodology for deriving soil guideline values protective of human health. Ministry for the Environment, Wellington
New Zealand Water and Wastes Association (2003) Guidelines for the safe application of biosolids to land in New Zealand. Water and Wastes Association, Wellington, New Zealand
Nicks LJ, Chambers MF (1994) Nickel farming. *Discover Magazine*, 19 Sept
Nielsen PE, Hiort C, Soennischsen SO, Buchardt O, Dahl O, Norden B (1992) DNA binding and photocleavage by Uranyl VI salts. J Am Chem Soc 114:4967–4975
Norden S, Avila R, de la Cruz I, Stenberg K, Grolander S (2010) Element specific and constant parameters used for dose calculations in SR-Site. Svensk Kärnbränslehantering AB, Stockholm
Notholt A, Sheldon R, Davidson D (1989) Phosphate deposits of the world, vol 2. Phosphate Rock Resources. Cambridge University Press: Cambridge
Ogunleye P, Mayaki M, Amapu I (2002) Radioactivity and heavy metal composition of Nigerian phosphate rocks: possible environmental implications. J Environ Radioact 62:39–48
Oyedele D, Asonugho C, Qworoye O (2006) Heavy metals in soil and accumulation by edible vegetables after phosphate fertilizer application. Electr J Enviro Agric Food Chem 5:1446–1453
Pantelica A, Salagean M, Georgescu I, Pincovschi E (1997) INAA of some phosphates used in fertilizer industries. J Radioanal Nucl Chem 216:261–264
Perez-Lopez R, Nieto J, Lopez-Coto I, Aguado J, Bolivar J, Sanristeban M (2010) Dynamics of contaminants in phosphogypsum of the fertilizer industry of Huelva (SW Spain): from phosphate rock ore to the environment. Appl Geochem 25:705–715
Perpich B, Soule C, Zamani S, Timchak L, Uebehoer G, Nagghappan L, Helwick R (2005) Mobile wastewater treatment helps remediate concentrated acidic process water at fertilizer plant. Florida, Water Resour, pp 24–36

Raven K, Reynolds J, Loeppert R (1997) Trace element analyses of fertilizers and soil amendments by axial-view inductively-coupled plasma atomic emission spectrophotometry. Commun Soil Sci Plant Anal 28:237–257
Righi S, Lucialli P, Bruzzi L (2005) Health and environmental impacts of a fertilizer plant—Part I: Assessment of radioactive pollution
Robinson B, Banuelos G, Conesa H, Evangelou M, Schulin R (2009) The phytomanagement of trace elements in Soil. Crit Rev Plant Sci 28:240–266
Roca M, Vallejo V, Roig M, Tent J, Vidal M, Rauret G (1997) Prediction of cesium-134 and strontium-85 crop uptake based on soil properties. J Environ Qual 26:1354–1362
Roemer W, Gruendel M, Guethoff F (2010) Concentrations of U-238, U-235, Th-232 and Ra-226 in some selected raw phosphates, phosphate fertilizers, soil and plant samples from a long term P fertilization experiment. J für Kulturpflanzen 62:200–210
Roselli C, Desideri D, Meli M (2009) Radiological characterization of phosphate fertilizers: Comparison between alpha and gamma spectrometry. Microchem J 91:181–186
Rutherford P, Dudas M, Samek R (1994) Environmental impacts of phosphogypsum. Sci Total Environ 149:1–38
Salt DE, Blaylock M, Kumar N, Dushenkov V, Ensley BD, Chet I, Raskin I (1995) Phytoremediation—a novel strategy for the removal of toxic metals from the environment using plants. Biotechnology 13:468–474
Santos AJG, Silva PSC, Mazzilli BP, Favaro DIT (2006) Radiological characterisation of disposed phosphogypsum in Brazil: evaluation of the occupational exposure and environmental impact. Radiat Prot Dosimetry 12:179–185
Saueia CHR, Mazzilli BP (2006) Distribution of natural radionuclides in the production and use of phosphate fertilizers in Brazil. J Environ Radioact 89:229–239
Saueia C, Mazzilli B, Favaro D (2005) Natural radioactivity in phosphate rock, phosphogypsum and phosphate fertilizers in Brazil. J Radioanal Nucl Chem 264:445–448
Sauvé S, Cook N, Hendershot W, McBride M (1996) Linking plant tissue concentrations and soil copper pools in urban contaminated soils. Environ Pollut 94:153–157
Sauvé S, Norvell W, McBride M, Hendershot W (2000) Speciation and complexation of cadmium in extracted soil solutions. Environ Sci Technol 34:291–296
Schipper P, Bonten L, Plette A, Moolenaar S (2008) Measures to diminish leaching of heavy metals to surface waters from agricultural soils. Desalination 226:89–96
Schnug E, Birke M, Costa N, Knolle F, Fleckenstein J, Panten K, Lilienthal H, Haneklaus S (2008) Uranium in German mineral and tap waters. In: De Kok LJ, Schnug E (eds) Loads and fate of fertilizer-derived uranium. Backhys Publichers, Leiden, p 229
Senesi G, Baldassarre G, Senesi N, Radina B (1999) Trace element inputs into soils by anthropogenic activities and implications for human health. Chemosphere 39:343–377
Serne R (2007) Kd Values for Agricultural and Surface Soils for Use in Hanford Site Farm, Residential, and River Shoreline Scenarios. US Department of Energy, Oak Ridge
Sheppard S (2011) Robust prediction of Kd from soil properties for environmental assessment. Human Ecol Risk Assess 17:263–279
Sheppard M, Vandergraaf T, Thibault D, Reid D, Keith J (1983) Technetium and uranium: sorption by and plant uptake from peat and sand. Health Pysics 44:635–643
Sheppard S, Long J, Sanipelli B, Sohlenius G (2009) Solid/liquid partition coefficients(Kd) for selected soils andsediments at Forsmark andLaxemar-Simpevarp. Svensk Kärnbränslehantering ABSKB Rapport R-09-27, Stockholm
Speirs FW (1949) The influence of energy absorption and electron range on dosage in irradiated bone. Br J Radiol 22:521–533
Syers J, Mackay A, Brown M, Currie L (1986) Chemical and physical characteristics of phosphate rock materials of varying reactivity. J Sci Food Agric 37:1057–1064
Tack FM, Verloo MG, Vanmechelen L, Van Ranst E (1997) Baseline concentration levels of trace elements as a function of clay and organic carbon contents in soils in Flanders (Belgium). Sci Total Environ. 201:113–123

Takeda A, Kimura K, Yamasak iS (2004) Analysis of 57 elements in Japanese soils, with special reference to soil group and agricultural use. Geoderma 119:291–307

Takeda A, Tsukada H, Takaku Y, Hisamatsu S, Nanzyo M (2006) Accumulation of uranium derived from long-term fertilizer applications in a cultivated Andisol. Sci Total Environ 367:924–931

Tankard A (1974) Petrology and origin of the phosphorite and aluminium phosphate rock of the Langebaanweg-Saldanha area south-western Cape Province. Ann South Afr Museum 65:217–249

Tayibi H, Choura M, Lopez F, Alguacil F, Lopez-Delgado a (2009) Environmental impact and management of phosphogypsum. J Environ Manage 90:2377–2386

Taylor MD, Gibb R, Willoughby J, Hewitt A, Arnold G (2007) Soil maps of cadmium in New Zealand. Landcare Research Contract Report: LC0607/84, Landcare Research, Lincoln, New Zealand

Torres-Dorante LO, Claassen N, Steingrobe B, Olfs HW (2005) Hydrolysis rates of inorganic polyphosphates in aqueous solution as well as in soils and effects on P availability. J Plant Nutr Soil Sci 168:352–358

Uchida S, Tagami K (2007) Soil-Crop transfer factors of naturally occurring Radionuclides and stable elements for long-term dose assessment. In WM'07: waste management symposium 2007—global accomplishments in Environmental and Radioactive Waste Management: Education and Opportunity for the Next Generation of Waste Management Professionals (WM Symposia, 1628 E. Southern Avenue, Suite 9—332, Tempe, AZ 85282 (United States), Tucson, AZ (United States), 25 Feb–1 Mar 2007)

Uprety D, Hejcman M, Szakova J, Kunzova E, Tlustos P (2009) Concentration of trace elements in arable soil after long-term application of organic and inorganic fertilizers. Nutr Cycl Agroecosyst 85:241–252

U.S. Department of Health and Human Services (2003) Toxicological profile for fluorides, hydrogen fluoride and fluorine. U.S. Department of Health and Human Services, Atlanta, Georgia, USA

U.S. Environmental Protection Agency (1999) Estimating risk from contaminants contained in agricultural fertilizers, Washington, DC

U.S. Environmental Protection Agency (2005) Applying to EPA for approval of other uses of phosphogypsum: preparing and submitting a complete petition under 40 CFR 61.206, A Workbook-61.204 Distribution and use of phosphogypsum for outdoor agricultural purposes

U.S. Environmental Protection Agency, U.S. Department of Energy (1999) Understanding variation in partition coefficient, Kd, values Volume II: review of geochemistry and available Kd values for cadmium, cesium, chromium, lead, plutonium, radon, strontium, thorium, tritium (3H), and uranium.EPA 402-R-99-004B, Washington, DC

U.S. National Research Council. Subcommittee on Mineral Toxicity in Animals (1980) Mineral tolerance of domestic animals. National Academy Press, Washington DC

Uusi-Kamppa J, Turtola E, Hartikainen H, Ylaranta T (1996) The interactions of buffer zones and phosphorus runoff. In : Haycock N, Burt T, Goulding K, Pinay G (eds) Buffer zones: their processes and potential in water protection. Haycock Associated Ltd., St. Albans, England

Vance DB (1994) In-situ bioremediation of metals contaminated groundwater. Nat Environ J 1994 (4):24–25

Vandenhove H, Van Hees M, Wouters K, Wannijn J (2007) Can we predict uranium bioavailability based on soil parameters? Part 1: effect of soil parameters on soil solution uranium concentration. Environ Pollut 145:587–595

Waheed S, Sabiha-Javied Siddique N, Shakoor R, Tufail M (2011) Rare earths elements in phosphorite and granulated single super-phosphate fertilizers of Pakistan, a study using instrumental neutron activation analysis. J Radioanal Nucl Chem 289:521–528

Warren G, Alloway B, Lepp N, Singh B, Bochereau F, Penny C (2003) Field trials to assess the uptake of arsenic by vegetables from contaminated soils and soil remediation with iron oxides. Sci Total Environ 311:19–33

Wassef A, El Shafey S, Taha F, Bahr W (1977) The geological and economic evaluation of Abu Tartur phophorite deposit, Western Desert, Egypt. Ann Geol Surv Egypt 7:130
Whicker J, Pinder J III, Ibrahim S, Stone J, Breshears D, Baker K (2007) Uranium partition coefficients (K-d) in forest surface soil reveal long equilibrium times and vary by site and soil size fraction. Health Phys 93:36–46
Williams C (1974) Heavy metals and other elements in fertilisers: environmental considerations. In: Leach D (ed) Fertilisers and the environment. NSW Branch, Australian Institute of Agricultural Science, Sydney, Australia
World Health Organization (2011) Guidelines for drinking-water quality. World Health Organization, Geneva. Accessed 25 July 2012 from http://www.who.int/water_sanitation_health/publications/2011/dwq_guidelines/en/index.html
Yamazaki I, Geraldo L (2003) Uranium content in phosphate fertilizers commercially produced in Brazil. Appl Radiat Isot 59:133–136
Zhao X, Wang D (2010) Mercury in some chemical fertilizers and the effect of calcium superphosphate on mercury uptake by corn seedlings (*Zea mays* L.). J Environ Sci 22:1184–1188
Zheng Y-M, Liu Y-R, Hu H-Q, He J-Z (2008) Mercury in soils of three agricultural experimental stations with long-term fertilization in China. Chemosphere 72:1274–1278
Zielinski R, Simmons K, Orem W (2000) Use of 234U and 238U isotopes to identify fertilizer-derived uranium in the Florida Everglades. Appl Geochem 15:369–383
Zielinski R, Orem W, Simmons K, Bohlen P (2006) Fertilizer-derived uranium and sulfur in rangeland soil and runoff: a case study in Central Florida water. Air Soil Pollut 176:163–183
Zielinski RA, Al-Hwaiti MS, Budahn JR, Ranville JF (2011) Radionuclides, trace elements, and radium residence in phosphogypsum of Jordan. Environ Geochem Hlth 33:149–165

Organic Xenobiotics

Elke Bloem and Sylvia Kratz

Abstract As well as fertilizer benefits, reutilization of phosphor (P) from secondary sources and recycling materials such as sewage sludge, farmyard manures, composts, biogas residues or meat and bone meal products on agricultural soils bears the great risk of contaminating farm sites with organic xenobiotics. Contamination of groundwater, surface and drinking water as well as of aquatic sediments and soils with pharmaceuticals has been reported (Hamscher et al. in Analytical Chemistry 74:1509–1518, 2002; Hirsch et al. in Science of the Total Environment 225:109–118, 1999; Kümmerer in Pharmaceuticals in the environment. Sources, fate, effects and risks. Springer, Heidelberg, 2001b; Ternes in Water Research 32:3245–3260, 1998). According to available data, biodegradation is more the exception than the rule, which is one reason for the widespread occurrence of these compounds in the environment (Kümmerer in Chemosphere 45:957–969, 2001a). Antibiotics and disinfectants are especially designed to have a strong impact on microorganisms, making environmental effects very likely to occur. Therefore it is necessary to consider possible risks of using secondary raw materials containing such xenobiotics despite their advantages as a fertilizer or source of organic material. For many organic contaminants no environmental risk assessment (ERA) is available. For veterinary drugs, a note of guidance on ERA was issued in January 1998 but hardly any eco-toxicological data are available for products approved before 1998 (Baguer et al. in Chemosphere 40:751–757, 2000; EU in EMEA/CVMP/055/96. EMEA, London, 1996) or for human therapeutics. For many pharmaceuticals as well as organic chemicals their environmental effects are quite unknown, particularly any interactions. New technologies enabling the detection of traces of organic xenobiotics and their metabolites even in critical matrices (slurry, sewage sludge) are partly responsible for the higher awareness of possible risks. Environmental tests on the effects of xenobiotics are still rare and have only been performed in the last decade and mainly for aquatic organisms. It is the aim of this chapter to give an overview of

E. Bloem (✉) · S. Kratz
Institute for Crop and Soil Science, Julius Kühn-Institut (JKI),
Federal Research Centre for Cultivated Plants, Bundesallee 50,
38116 Brunswick, Germany
e-mail: elke.bloem@julius-kuehn.de

E. Schnug and L.J. De Kok (eds.), *Phosphorus in Agriculture: 100 % Zero*,
DOI 10.1007/978-94-017-7612-7_13

the different classes of organic contamination that are relevant in organic P fertilizers. Moreover technical solutions to decrease or minimize the risk of soil contaminations by using organic P fertilizers are discussed and the different recycling materials are evaluated with respect to their environmental risks. It is not possible to address all different organic compounds and their impact on the environment in one chapter because of the sheer number of compounds and metabolites. The focus was put on the input of pharmaceutical contamination into agriculture because sewage sludge and farmyard manures are important organic P fertilizers, which can contain high amounts of pharmaceuticals. But even so this field is too large to cover the whole spectrum taking into account that in Germany alone 50,000 drugs are registered for human use, containing more than 3000 different active compounds (Kümmerer in Pharmaceuticals in the environment. Sources, fate, effects and risks. Springer, Heidelberg, 2001b). In order to illustrate the general problems arising from field application of environmentally relevant substances, two main classes of contaminants were discussed in detail: antibiotics and substances with an endocrine potential. For more detailed information several extensive review articles and books are available covering the different aspects of pharmaceutical drugs in the environment, their occurrence and behavior in the environment, their chemical characteristics, and their impact on soil organisms, plant growth and aquatic life (Boxall et al. in Reviews of environmental contamination and toxicology 180:1–91, 2004; Halling-Sørensen et al. in Chemosphere 36:357–393, 1998; Kümmerer in Pharmaceuticals in the environment. Sources, fate, effects and risks. Springer, Heidelberg, 2001b; Kumar et al. in Advances in Agronomy 87:1–54, 2005a; Sarmah et al. in Chemosphere 65:725–759, 2006). Pathogens, another important possible organic contamination of secondary P fertilizers, were also not addressed in detail, as the topic is comprehensive enough to fill books on its own (e.g. Bowman in Manure pathogens: manure management, regulations, and water quality protection. WEF Press, United States, 2009).

Keywords Antibiotics · Endocrine substances · Farmyard manures · Organic contamination · Organic P fertilizer · Sewage sludge

Agricultural Use of Farmyard Manure, Slurry, Organic Waste Products and Recycled Materials

Important organic P fertilizers are farmyard manures, sewage sludge (+ashes), composts, meat and bone meal (+ashes) and biogas digestates. Farmyard manures and composts have a long tradition in agriculture and they are appreciated as soil ameliorants and a source of additional organic matter. Organic waste products such as sewage sludge or recycled materials like meat and bone meal ashes or sludge ashes are also used as fertilizer materials but problems with heavy metals and organic contaminations have been reported.

Organic fertilizers are used in agriculture with changing proportions. The usage of organic fertilizers is shown for Germany in two different years (2000 and 2010) together with the associated P loads (Table 1). Farmyard manures are—on a quantity basis—the most important organic fertilizers used in German agriculture, accounting for more than 90 % of the total application. They supply about 55–63 % of the annual crop demand for P, while mineral fertilizers only account for 28–40 %. In the EU-27, over 1500 million tons of animal manures are produced yearly (Holm-Nielsen et al. 2009). In comparison, sewage sludge represents a small proportion of organic fertilizers: only 3.7 % of the total P demand was supplied in 2010 by its application. Nevertheless, more than 6.5 million tons dry weight of sewage sludge are generated annually in the EU, of which more than 37 % are utilized in agriculture (Smith 1996). Therefore organic contamination by its application is a relevant issue.

According to official data, bio-composts as well as biogas digestates represent a small fraction of the organic P fertilizers and only a small proportion of 1.4 and

Table 1 Main organic fertilizers used in German agriculture: total amounts and percentage of the total P demand covered by its application

Organic fertilizers	Total amount used in agriculture		P demand supplied by the organic fertilizers [%]			
	2000	2010	2000		2010	
			P [t]	[%]	P [t]	[%]
Farmyard manure						
Cowshed slurry [1000 m^3 a^{-1} fw]	90,825	81,128	155,775	33.8	139,029	37.1
Cowshed manure [1000 t a^{-1} fw]	29,942	26,727				
Cowshed liquid manure [1000 m^3 a^{-1} fw]	8706	7775				
Pig slurry [1000 m^3 a^{-1} fw]	99,581	118,372	65,741	14.3	66,481	17.7
Pig manure [1000 t a^{-1} fw]	495	635				
Pig liquid manure [1000 m^3 a^{-1} fw]	680	832				
Poultry dung [1000 t a^{-1} fw]	2842	2991	14,495	3.1	15,542	4.1
Other livestock [1000 t a^{-1} fw]	8169	7767	14,433	3.2	12,742	3.4
Total farmyard manure [1000 t a^{-1} fw]	~241,240	~246,227	250,444	**54.7**	233,794	**63.4**
Sewage sludge [in 1000 t a^{-1} dw]	769	566	18,579	**4.1**	13,682	**3.7**
Bio-composts [in 1000 t a^{-1} fw]	1792	2293	4142	**0.9**	5300	**1.4**
Bone meal [in 1000 t a^{-1} fw]	No data	176	No data		10,542	**2.9**
Biogas digestate [in 1000 t a^{-1} fw][a]	No data	2834	No data		2226	**0.6**
Use of mineral P fertilizers [t a^{-1} P_2O_5]	420	235	184,948	**40.4**	103,470	**28.0**

Data on livestock and P production are derived from the official nutrient balance for German agriculture calculated by JKI (Institute for Crop and Soil Science, unpublished data). Amounts of livestock manure are estimated using conversion factors published by KTBL (2009). [a]Values represent only a small number (992) of the existing biogas plants (5905) in 2010, see text. Bold numbers show the percentage of P that is supplied by the different organic sources (farmyard manures in sum, sewage sludge, bio-composts, bine meal, digestates) in comparison to mineral P fertilizer in 2000 and 2010.

0.6 %, respectively, of the total P demand was supplied by these compounds in 2010 (Table 1). Because of its high P content, bone meal delivered more P to agriculture in 2010 despite the low quantities that were applied. Data for the usage of bone meal are only available since 2001 and the amounts used in German agriculture have been more or less constant since then.

An increasing production of biogas digestates has been recorded since 2006. On-farm production of renewable energy is expanding in Europe and land spreading of biogas digestates is partly replacing the use of fresh manure which is used together with other organic products such as crop residues, wastes from food industry, municipal wastes and dedicated energy crops in the bioreactors. For Germany it was estimated that in 2011 approximately 20 % of all animal wastes were already used in biogas plants and that the total amount of digestate produced was in the range of 65.5 million m^3, containing a total amount of 74,075 Mg P (Möller and Müller 2012). Official values derived from the German P balance calculation suggest a much lower share of biogas digestates in the total amount of organic fertilizers (Table 1). However, data published by the Federal Statistical Office, which is the basis for the balance, includes only bio-wastes from private households and industries, green wastes from gardens and parks, food residues, sewage sludge and wastes from food processing and agriculture which are delivered to biogas plants. Farmyard manures and crops planted for biogas production by farmers (which are the main inputs to a rising number of rural biogas plants) are currently not accounted for in official statistics. The estimates in Table 1 for the amounts of digestates are much too low as only 992 biogas plants were considered while 5905 were working at that time. Therefore the real amount of digestates is much higher and the P demand supplied by them is presumably in the same order of magnitude as sewage sludge or even higher with the trend further increasing, as by 2011 7000 biogas plants were recorded in Germany.

Biogas digestates could become an important organic fertilizer, if the boom in biogas production continues further. Digestates are either directly applied to the soil or are treated before (solid-liquid separation, drying, dilution, filtration, composting) (Möller and Müller 2012). Many questions regarding the quality of biogas digestates as a fertilizer source are still unanswered such as pathogen transfer, organic contaminations and high ammonium availability (Goberna et al. 2011; Teglia et al. 2011). The direct application of digestates without post-treatment is often not recommended (Teglia et al. 2011). Moreover anaerobic digestion of animal manures may cause a lower P availability in the digestate (Güngör and Karthikeyan 2008).

The data in Table 1 clearly reveal that farmyard manure is by far the most important organic fertilizer used in German agriculture. Contamination of farmyard manure with organic xenobiotics are therefore of special importance, and measures to decontaminate it or prevent its contamination most probably have the highest impact on environmental safety.

Relevant Classes of Organic Contaminations in Organic Fertilizer Materials

Organic fertilizers are contaminated by many different organic xenobiotics, which differ according to the origin and treatment of the material. The most important organic xenobiotics found are pharmaceuticals used in human therapy or for veterinary purposes. Moreover cleaning agents and disinfectants for cleaning animal housing and for disinfection of hooves ($CuSO_4$-solutions), udder and slurry (cyanamid solutions), pesticides used in agriculture and a huge amount of different chemicals used in households or by industries can be detected entering the waste water treatment plant or the soil and water bodies directly via the application of manure, sewage sludge or sewage effluents. An overview over the expected occurrence of different contaminants in organic fertilizers is given in Table 2. Most pharmaceuticals used in animal husbandry or human therapy are not absorbed completely but are excreted to a high extent either in their original form or as metabolites in urine or faeces to end up in sewage sludge or farmyard manure. Once applied to the field, these pharmaceuticals can enter surface and groundwater through nonpoint source pollution.

Farmyard manures are mainly contaminated by pharmaceuticals that were used in animal husbandry as therapeutics, vaccines, and feed additives. Anti-infectives such as ivermectin are one example of veterinary drugs that directly affect soil life as the faeces of treated animals have a toxic effect on dung-degrading invertebrates and soil fauna (Wall and Strong 1987; McCracken and Foster 1993; Gunn and Sadd 1994). Sewage sludge can contain nearly all-possible organic contamination because households together with industries deliver a potpourri of very different xenobiotics. The same is true for composts. In principle composts produced from yard wastes and household bio-wastes should contain only little amounts of organic

Table 2 Expected occurrence of organic xenobiotics in farmyard manure, sewage sludge and composts (adapted from Zullei-Seibert and Skark 2002)

Class of contaminant	Expected abundance of organic xenobiotics		
	In farmyard manure	In sewage sludge	In compost
Pharmaceuticals (human therapy)	–	◆	♦
Pharmaceuticals (animal nutrition and therapy)	◆	–	–
Hormones	◆	◆	♦
Phthalates	–	◆	◆
Phenoles	◆	◆	◆
Organo-tin compounds	–	◆	♦
Polyacrylate	–	◆	♦
Polyacrylamide	–	◆	♦
Pharmaceuticals in personal care products	–	◆	♦

◆ High expected abundance, ♦ low expected abundance, –no occurrence expected

contamination, but in reality they can be contaminated with very different compounds because of careless disposal of forbidden compounds such as meat products or even worse pharmaceuticals. In addition, some composts are produced from sewage sludge, municipal solid wastes, industrial and agricultural wastes as well as biogas digestates (Deportes et al. 1995). Therefore composts can be contaminated by the same xenobiotics that are detected in sewage sludge or farmyard manures.

Human and Veterinary Pharmaceuticals

Different active agents are used in human therapy mainly for the treatment of chronic diseases such as diabetes and high blood pressure, thus analgetics, lipid lowering agents, and anticonvulsants end up in sewage sludge or the effluent from waste water treatment plants (Fig. 1). 30,000 t of about 3000 different active compounds in 9000 preparations are annually used in human therapy in Germany (Schönauer and Grommelt 2007). It was shown that farmyard manure is the most important organic fertilizer accounting for more than 90 % in German agriculture (Table 1). Veterinary drugs that are excreted to a high proportion and end up in manure are of special relevance as environmental contaminants. The most widely used veterinary drugs all belong to the class of antibiotics (tetracyclines, sulfonamides and ß-lactames).

Antibiotics

Antibiotics are the most important class of pharmaceuticals that are used in human as well as in animal therapy. Even in plant agriculture antibiotics are used mostly as spray applications in orchards against bacterial diseases such as fire blight and bacterial leaf and fruit spot (Vidaver 2002). Antibiotics are third among all pharmaceuticals used in human medicine accounting for 6 % of all prescriptions (Schwabe and Paffrath 2001) and dominate veterinary medicine accounting for 70 % of all medications used (Halling-Sørensen et al. 1998). Since 1928 when Fleming discovered the first antibiotic, penicillin, many other natural antibiotics have been discovered and synthetic as well as semi-synthetic derivatives have been developed for the treatment of infections. In human medicine antibiotics are used more or less exclusively therapeutically, whereas veterinary antibiotics are also given to healthy animals for so-called non-therapeutic uses, precautionary in prophylaxis or as feed additives for growth promotion or to improve feed efficiency (Blaha 1996).

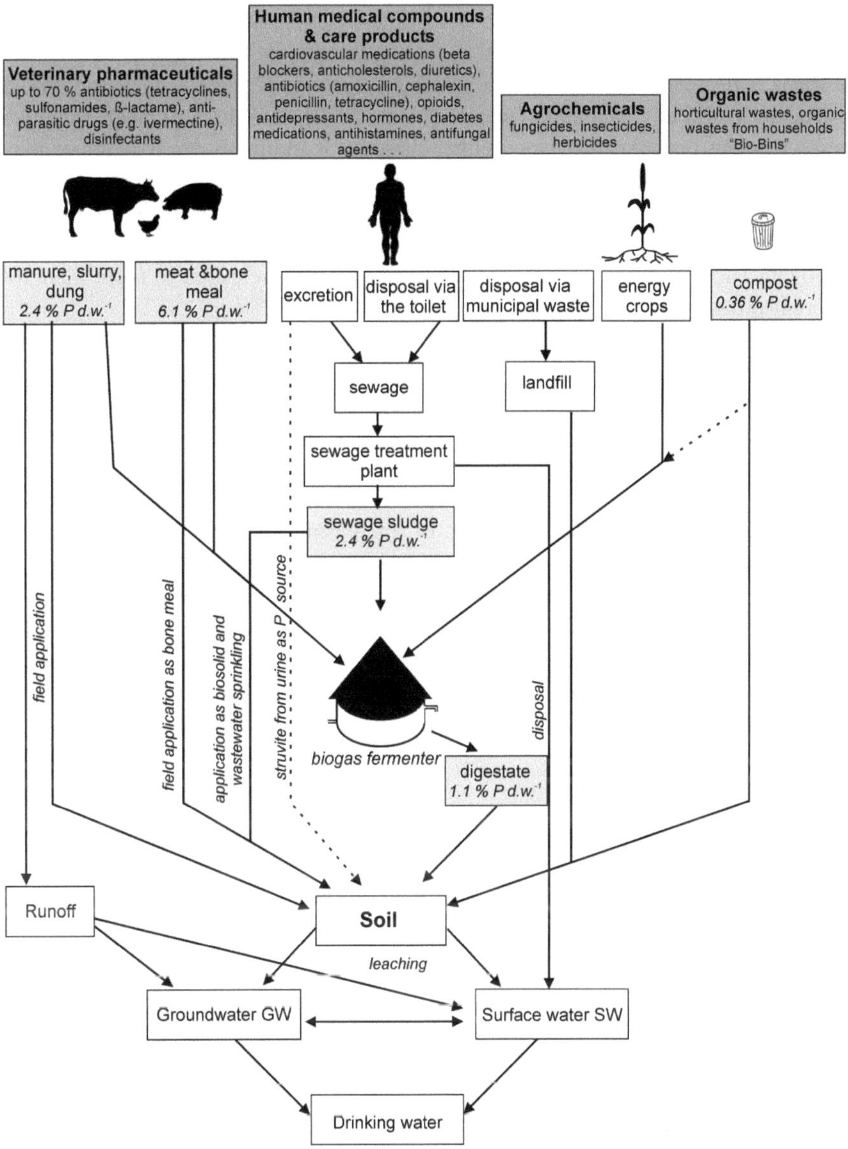

Fig. 1 Input pathways of organic xenobiotics into the soil by organic P fertilizers

Development of Antimicrobial Resistance

The application of antibiotics as feed additives in sub-therapeutic doses (<100 mg kg^{-1} feed) has been critically discussed in the past as more and more multi-resistant pathogenic organisms developed causing serious problems for public health (Richter et al. 1996). The use of antibiotics in sub-therapeutic doses for growth promotion along with casual use of antibiotics in household products such as soaps and creams are thought to be major reasons for increasing antimicrobial resistance in the environment (Chander et al. 2007). Residues of pharmaceutical antibiotics can provoke resistance either directly or indirectly by transfer of plasmids (Wegener et al. 1998). Though antibiotics used in human therapy were not allowed to be used as feed additives, cross-resistances developed, for example between avoparcin and vancomycin (Wegener et al. 1998). Some strains of human pathogens such as *Enterococcus faecalis* are meanwhile resistant to more than 100 different drugs (Schmidt 2002).

These antibiotic resistant bacteria can be spread into the environment when manure containing such resistant bacteria is applied to land or they can be enriched when residual antibiotics in manure or sewage wastes generated from households and hospitals provide an environment for the selection of antibiotic resistant bacteria in soils (Chander et al. 2005). As a consequence of rising problems with resistant or multi-resistant bacteria in hospitals, the environmental risk of veterinary medicinal products became a matter of increasing public scrutiny and legal requirements. Several antibiotics lost their permission as feed additives in the EU in 1999 such as spiramycin, zinc-bacitracin, virginiamycin, tylosin phosphate, carbadox and olaquindox. Since 2006 antibiotic feed additives for growth promotion have been completely banned in the EU (Hamscher and Mohring 2012) but official statistics for Germany show that the use of antibiotics in livestock production has increased drastically since then (see Figs. 2 and 3).

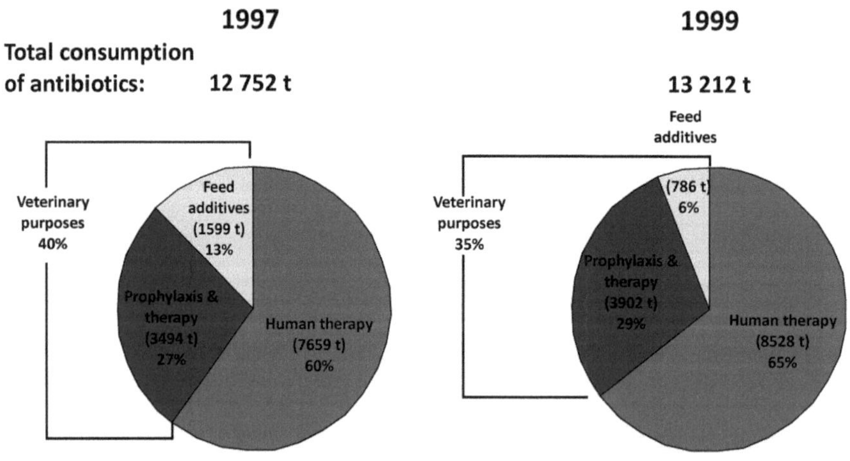

Fig. 2 Consumption of antibiotics in human therapy and for veterinary purposes in Europe in 1997 and in 1999 (FEDESA 2001)

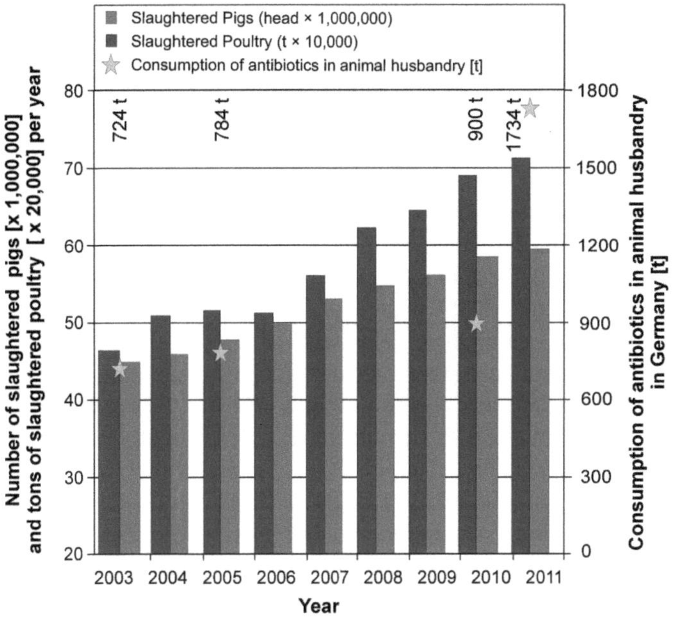

Fig. 3 Consumption of antibiotics for veterinary purposes and development of meat production (pigs and poultry) in Germany. *Data source* Germap (2008), Haunhorst (2012), BVL (2012), Destatis (2011)

Antibiotics as Feed Additives

Feed additives in animal husbandry account for a substantial proportion of total antibiotic use, being used primarily in intensive farming for pigs (38.5 %), poultry (61 %) and calves. In 1997 a total of 12,752 t of antibiotics were used in Europe (Fig. 2). From this amount 60 % were used in human therapy and around 40 % for veterinary purposes, of which 12.5 % were used as feed additives. After several antibiotics were banned in the EU, a change in the use of antibiotics was observed in 1999. The total use increased to 13,216 t (65 % in human therapy, around 35 % for veterinary purposes). Only 6 % of the total consumption of antibiotics was still used as feed additives (FEDESA 2001). The decreased use of antibiotics as feed additives was partly outweighed by an increase in the therapeutic use by 11 %. According to the Institute of Medicine 50 million pounds of antibiotics are produced in the United States annually and 32 % of these are used for non-therapeutic purposes in agriculture—or even more (78 %), as the Union of Concerned Scientists estimates (Chander et al. 2007). In total, the consumption of antibiotics in animal husbandry as well as meat production itself is still increasing as shown for Germany (Fig. 3). Since 2011 official consumption data provided by producers have been collected for Germany. The high value of 1734 t of antibiotics consumed in 2011 suggests that the consumption was underestimated in the past.

Excretion of Antibiotics

Antibiotics are only partly absorbed in the animal intestines and substantial proportions are excreted. Most antibiotics are water-soluble (e.g. tetracyclines, sulphonamides and some ß-lactam antibiotics) and are excreted via the urine as the parent compound (e.g. tetracyclines and ß-lactam antibiotics) or metabolites (e.g. sulphonamides and macrolides) while less soluble compounds such as macrolides and pleuromutilin derivatives are excreted with faeces (Halling-Sørensen 2000). As much as 40–90 % of the antibiotics added in feed and water are generally excreted unaltered (Thiele-Bruhn 2003; Kumar et al. 2005a). Thus, application of manure as fertilizer is likely to contaminate fields with antibiotics. Tetracyclines (TCs) and sulfonamides were found at a magnitude of mg kg^{-1} in manure after designated use; tylosin, salinomycin and tiamulin were found in lower concentrations of μg kg^{-1} (Hamscher and Mohring 2012). Chander et al. (2005) could demonstrate that even though antibiotics are tightly adsorbed by clay particles they are still biologically active and may influence the selection of antibiotic resistant bacteria in the terrestrial environment. Data about absorption, storage, metabolism and excretion of such compounds are important to estimate environmental inputs. Moreover residence time, degradability and accumulation behavior in the environment are important characteristics for their risk assessment (Seel 1998). Several reviews are available about factors affecting the persistence of antibiotics in the environment such as soil type, sorption and fixation, mobility, degradation and transformation, climate and class of antibiotic (e.g. Boxall et al. 2004; Thiele-Bruhn 2003).

Effects of Antibiotics in the Environment Shown Exemplary for Tetracyclines

Tetracyclines (TCs) represent the most important pharmaceutical class in veterinary medicine accounting for 58 % of the antibiotics used in Germany (Winckler and Grafe 2000) and for about 66 % of the antibiotic consumption in Europe (FEDESA 2001). Calculations for the consumption of pharmaceuticals and possible inputs by farmyard manures into the soils are very often based on estimated values because many data such as pharmaceuticals directly provided by veterinarians or administered over the drinking water, are not documented. Winckler and Grafe (2000) calculated the “Predicted Environmental Concentration” (PEC) for TC which was applied to the soils via slurry from piglets under the assumptions that piglets were treated with a regular doses of 65 mg TC kg^{-1} for 10 days, that the TCs were excreted at 100 % and that slurry was applied to the soil according to the fertilization recommendations. Under these assumptions, a TC concentration of 0.9–2.1 ppm was calculated in the top soil. Data confirmed that a high rate (42–70 %)

of the administered TCs was excreted by pigs during the first 7 days after application (Winckler and Grafe 2000). A TC concentration in the soil of 0.5–2 ppm is already a critical value if the entire amount is available, as it is in the range of the minimum inhibition concentration in humans and animals serum (Forth et al. 1997). For turkey production, an even higher PEC of 0.9–8.7 ppm in the soil was calculated (Winckler and Grafe 2000). TCs were meanwhile found in many different environmental niches and calculated soil concentrations were verified by real experimental data (Table 3). Xenobiotics that reach soil concentrations higher than 100 μg kg^{-1} which is the case for TCs, or whose concentrations are close to the effect concentrations on organisms are classified as ecotoxicologically relevant compounds that need further risk assessment (Hembrock-Heger 2007).

For the risk assessment of a compound direct effects as well as the degradation behavior are important. Compounds with slow degradation accumulate in the environment and critical concentrations can be achieved on a long-term basis. Very different degradation rates were reported in the literature for TCs. Kühne et al. (2000) observed a high degradation under laboratory conditions and determined a half life of TCs in liquid manure of 4.5 and 9 days in ventilated and non-ventilated samples, respectively. Similarly, Jagnow (1977) determined a decrease in the antimicrobial potential of poultry dung supplemented with 10 ppm TC of about 66 % in 12 weeks

Table 3 Tetracycline concentrations in environmental compartments

Environmental compartment	Maximum concentrations	Reference
Influents and effluents in sewage treatment plants (STPs)	0.34 μg l^{-1} OTC 0.28 μg l^{-1} CTC 0.45 μg l^{-1} TC	Yang and Carlson (2004)
Surface water	>0.01–1.34 μg l^{-1} OTC 0.11–0.69 μg l^{-1} CTC 0.06–0.14 μg l^{-1} TC	Calamari et al. (2003), Hamscher and Mohring (2012), Kolpin et al. (2002), Yang and Carlson (2004)
Groundwater	0.15–0.19 μg l^{-1} OTC 0.17–0.22 μg l^{-1} CTC 0.11–0.27 μg l^{-1} TC	Review by Boxall et al. (2004), Hamscher and Mohring (2012)
Farmyard manures	0.82–366 mg kg^{-1} OTC (maximum of 872 mg OTC kg^{-1} faeces after treatment) 1–203.3 mg l^{-1} CTC 0.1–66 mg l^{-1} TC	Bergmann et al. (2011), Campagnolo et al. (2002), De Liguoro et al. (2003), Höper et al. (2002)
Soils fertilized with farmyard manures	<10–322 μg OTC kg^{-1} soil 27–810 μg CTC kg^{-1} soil 43–199 μg TC kg^{-1} soil	Bergmann et al. (2011), Hamscher et al. (2002), Kues et al. (2004a, b)
Sediments of fish farms	0.1–285 mg kg^{-1} OTC	Review by Boxall et al. (2004), Coyne et al. (1994)
Runoff from fields fertilized with slurry	32–71.7 mg l^{-1} OTC	Kay et al. (2005)
Drainage water from soils fertilized with slurry	36.1 μg l^{-1} OTC	Kay et al. (2004)

under laboratory conditions while under outdoor conditions (5–10 °C) no degradation occurred. The temperature dependency of the degradation of TCs was also shown by Gavalchin and Katz (1994) who found a reduction in the activity of chlortetracyclines (CTC) after 30 days of 44 % at 30 °C, of 12 % at 20 °C and no degradation at 4 °C. In 300 l cooling tanks with pig slurry kept at 8 °C, Winkler and Grafe (2000) observed nearly no degradation of TCs during 6 weeks in the sediment where high portions of the TCs are bound, and only a slight degradation in the liquid phase. The authors concluded that even over longer periods no degradation of TCs will occur in pig slurry. TCs were found to bond strongly to the soil matrix so that leaching is not likely to occur (Rabolle and Spliid 2000). Nevertheless Hamscher and Mohring (2012) could demonstrate that—despite strong sorption—significant amounts of TCs were transferred to the groundwater with a maximum concentration of 0.13 μg l^{-1}. Even higher concentrations were detected in surface waters (CTC: 0.11–0.69 μg l^{-1}, oxytetracycline: 0.15–1.34 μg l^{-1}) contaminated by run-off from agricultural sites (Hamscher and Mohring 2012).

Antibiotics such as TCs are critical environmental contaminants as they are biologically active and have a strong impact on soil and aquatic microorganisms. Soils which were fertilized with farmyard manures from pigs containing TCs show a higher content of TC-resistant isolates compared to control plots (Huysman et al. 1993). In farmyard manure derived from piglets treated with sulfonamides, 46 % of the original application was detected of which 50 % were in a metabolized form (Berger et al. 1986). Storage of this manure caused a reactivation of metabolites back to their original form before degradation was observed.

Toxic Effects of Antibiotics on Aquatic Species

Toxic effect data (EC_{50}) of antibacterial agents on aquatic species can be found in the literature (Lansky and Halling-Sørensen 1997; Macri et al. 1988; Migliore et al. 1997) and range from mg l^{-1} for crustaceans and fish down to 2–3 orders of magnitude lower for algae and bacteria because of their designated mode of action against micro-organisms (Halling-Sørensen 2000; Lützhøft et al. 1999). Some toxic effect data are summarized in Table 4. The EC_{50} value is the statistically calculated concentration of a compound at which 50 % of the exposed algae are inhibited in their exponential growth during the experimental time. Wollenberger et al. (2000) determined EC_{50} values for TC and oxytetracycline (OTC) in the *Daphnia magna* reproduction test to be in the range of about 50 mg l^{-1}. According to the studies of Wollenberger et al. (2000) most antibiotics have no acute effects on *D. magna* at commonly found environmental concentrations but show effects in long-term exposure and on reproduction. Lower EC_{50} values were determined for algae and bacteria. Halling-Sørensen (2000) found EC_{50} values between 5–100 μg l^{-1} for the prokaryotic cyanobacteria *Microcystis aeruginosa* and 1–100 mg l^{-1} for the eukaryotic green algae *Selenastrum capricornutum* when they tested the effect of 8 different antibiotics. For TCs and CTCs respectively they determined EC_{50} values

Table 4 Toxicological effect data of tetracyclines for different organisms (EC_{10} or EC_{50}: effect concentration where 10 or 50 % of the organisms are affected; LC_{10} or LC_{50}: lethal concentration where 10 or 50 % of the organisms are dead)

Organism	Toxicological effect	Reference
Sludge bacteria	EC_{50} = 80–1200 μg l^{-1} OTC EC_{50} = 30–400 μg l^{-1} CTC EC_{50} = 80–2200 μg l^{-1} TC	Halling-Sørensen et al. (2002), review by Thiele-Bruhn (2003)
Microcystis. aeruginosa (Cyanobacteria)	EC_{50} = 207 μg l^{-1} OTC EC_{50} = 50 μg l^{-1} CTC EC_{50} = 90 μg l^{-1} TC	Boxall et al. (2004), Halling-Sørensen (2000), Lützhøft et al. (1999)
Soil microorganisms	EC_{10} = 810–930 μg kg^{-1} OTC	Kues et al. (2004a, b) in Hembrock-Heger (2007)
Selenastrum capricornutum (green algae)	EC_{50} = 3.1 mg l^{-1} CTC EC_{50} = 4.5 mg l^{-1} OTC EC_{50} = 2.2 mg l^{-1} TC	Boxall et al. (2004), Halling-Sørensen (2000), Lützhøft et al. (1999)
Nitzschia closterium	EC_{50} = 16 mg l^{-1} TC	Review by Boxall et al. (2004)
Vibrio fischeri	EC_{50} = 25 μg l^{-1}	Backhaus and Grimme (1999)
Lemna gibba (higher aquatic plant)	EC_{10} = 730–1000 μg l^{-1} OTC EC_{10} = 36–59 μg l^{-1} CTC EC_{10} = 194–230 μg l^{-1} TC	Brain et al. (2004)
Daphnia magna (freshwater water flea)	EC_{50} = 20.6–104 mg l^{-1} OTC EC_{50} = 40.3–49.8 mg l^{-1} TC	Wollenberger et al. (2000)
Folsomia fimetaria *Aporrectodea caliginosa* *Enchytraeus crypticus* (Springtails, Earthworms, Enchytreids)	EC_{10} = >5000 mg kg^{-1}OTC EC_{10} = >1954 mg kg^{-1}OTC EC_{10} = >3000 mg kg^{-1}OTC LC_{10} = for all 3 organisms >5000 mg kg^{-1}	Review by Thiele-Bruhn (2003)
Morone saxatilis (fish species)	LC_{50} = 62.5–150 mg l^{-1}	Boxall et al. (2004)
Plants (maize, wheat, beans)	Growth stimulation up to 160 mg kg^{-1} OTC or CTC; higher values are lethal for plants	Review by Halling-Sørensen et al. (1998)

of 0.09 and 0.05 mg l^{-1} for blue algae and of 2.2 and 3.1 mg l^{-1} for green algae. The higher sensitivity of cyanobacteria towards antibiotics was also shown by Harrass et al. (1985) who found a ten-time higher sensitivity of *M. aeruginosa* to streptomycin compared to *S. capricornutum*. Similarly, Lützhøft et al. (1999) determined *M. aeruginosa* to be two or three orders of magnitudes more sensitive toward different antibiotics compared to *S. capricornutum* and *Rhodomonas salina*. One of the most widely used biotests to determine the aquatic toxicity of xenobiotics toward bacteria is the acute bioluminescence inhibition assay with the marine bacterium *Vibrio fischeri* which is also sensitive to antibiotics in the μg l^{-1} range. An EC_{50} value of 25 μg l^{-1} was determined for TCs (Backhaus and Grimme 1999). Backhaus and Grimme (1999) demonstrated that environmental concentrations in

surface waters, rivers and sediments close to fish farms were already in a range "provoking direct toxic effects on the bacterium".

As antibiotics are designed to affect mainly microorganisms, much higher concentrations are necessary to show toxic effects on higher soil organisms. No observed effect concentrations (NOEC) of 2000–5000 mg TC kg^{-1} soil were determined for reproduction and mortality of earthworms, springtails and enchytreids (Baguer et al. 2000). Effects of TCs on soil community and the impact of increased antibiotic resistant genes in the gene pool of soil organisms are still discussed (Halling-Sørensen et al. 2005; Hammesfahr et al. 2011; Thiele-Bruhn 2003). Boxall et al. (2004) reviewed the toxicity of veterinary drugs on a broad range of test organisms. According to Thiele-Bruhn (2003) knowledge about the ecotoxicity of antibiotics is still scarce, mainly because of the lack of well-adapted test methods. Adoption of unsuited methods can cause erroneous results and conclusions, for example when the duration of an exposition experiment is too short (see Backhaus and Grimme 1999).

Uptake of Antibiotics into Plants

In several studies effects of antibiotics on plant growth and uptake into plant organs were demonstrated (Hembrock-Heger 2007; Boxall 2010; Grote et al. 2006; Hamscher 2008; Jjemba 2002; Kumar et al. 2005b; Migliore et al. 1995). Grote et al. (2006) observed in an experiment with winter wheat grown on soils fertilized with CTC-containing pig manure that CTCs and their degradation products were transferred from the soil into the plants where they were translocated into the grain. Greenhouse studies indicated that corn, lettuce and potato plants take up sulfamethazine with concentrations ranging from 0.1 to 1.2 mg kg^{-1} dry weight. Corn, green onions and cabbage absorb concentrations of 1–17 ng CTCs per g fresh weight but no tylosin from soils where antibiotic-containing manures were applied (Dolliver et al. 2007; Kumar et al. 2005a). Edible roots and tuber crops may be of particular concern as a higher accumulation of antibiotics was found in the peel of carrots and in the skin of potatoes (Dolliver et al. 2007; Boxall et al. 2006). It is possible that adherent soil particles contributed to these elevated levels (Dolliver et al. 2007).

Therefore low concentrations of antibiotics and their metabolites enter the food chain not only via meat or animal products but also via plants grown on manure treated fields. Antibiotics show different transfer factors as well as variance in degradation but a transfer into the plant was determined for some of them (Grote et al. 2006). Endangerment of humans via the food chain is thus possible (Kennedy et al. 2000). Hamscher et al. (2002) concluded from soil investigations that significant amounts of TCs enter the environment via repeated liquid manure application, build up persistent residues, and accumulate in soils. Risks connected to resistant bacteria are not restricted to a particular country, because animal products are traded worldwide and that way resistant bacteria are also spread (Kumar et al. 2005a).

Endocrine Disruptors

Endocrine-disrupting compounds (EDCs) are a second very important class of contamination in farmyard manures and sewage sludge. EDCs or oestrogen-mimicking chemicals are defined as substances that can mimic or antagonize the action of hormones. They either affect endocrine glands directly or their activity at the active sites. EDCs include both natural and synthetic steroid oestrogens, and a variety of oestrogen-mimicking chemicals and chemicals with androgenic or anti-androgenic effects. EDCs may alter diverse physiological functions including modified and abnormal blood hormone levels, cause reduced fertility and fecundity, masculinization of females and feminization of males in different species (Sumpter 1998). As well as natural substances there are artificial man-made contaminations that act as endocrine disruptors, too, among which synthetic oestrogens used in the formulation of contraceptive pills (17 α-ethinylestradiol, mestranol) are the best-known. Several other chemicals or degradation products interfere with the hormonal balance when released into the environment and so far more than 200 compounds have been identified (Gülden et al. 1997; Schönauer 2001). Most studies have been done for oestrogens and oestrogen-like chemicals called xenooestrogens (Table 5). It is important to mention that androgens, glucocorticoids and progestagens are released into the environment in the same order of magnitude as oestrogens and affect the hormonal balance of organisms as well (Liu et al. 2012).

Natural as well as pharmaceutical oestrogens display the highest affinity for binding to nuclear oestrogen receptors and have the highest oestrogenic potency, which highlights their importance. The relative oestrogenic activity of

Table 5 Compounds showing oestrogenic activity (Andersen et al. 1999; Combalbert and Hernandez-Raquet 2010; Daston et al. 2003; Murkies et al. 1998)

Natural compounds	Examples
Hormones produced by humans and animals	Estrone (E1), 17 ß-estradiol (E2), estriol (E3)
Phytooestrogens produced by plants	Isoflavones, coumestan, lignans
Mycooestrogens produced by molds	Resorcyclic acid lactones
Synthetic compounds—xenoesterogens	
Synthetic oestrogens	17 α-ethinylestradiol (EE2), mestranol
Pesticide ingredients	p-DDT, dieldrin, methoxychlor
Plastifiers ingredients	Bisphenol A (BPA), phthalates
Paints ingredients	Tributyltin
Detergents and degradation products	Nonylphenol, octylphenol
Industrial chemicals	Polychlorobiphenyls (PCB), polyaromatic hydrocarbons (PAH), dioxins

xenooestrogens such as bisphenol A and nonylphenol, is less than 1 % of that of 17 ß-estradiol, the most potent oestrogen (Combalbert and Hernandez-Raquet 2010). Nevertheless xenooestrogens display strong effects on wildlife despite very low doses and much lower binding affinities than endogenous hormones, which is discussed further in the section about environmental and physiological effects in this chapter.

Sources of Compounds with Oestrogenic Activity

Main oestrogen sources are the human population and livestock excreting oestrogens via the urine and faeces (D'Ascenzo et al. 2003; Hanselman et al. 2003; Liu et al. 2012), reaching the natural environment through discharge from sewage treatment plants and manure disposal. The amount varies as a function of sex, physiological and developmental stage and, for animals, their species. The highest excretion rates were observed for pregnant women as well as pregnant farm animals (Lange et al. 2002). Humans excrete large quantities of oestrogens in their glucuronide- and sulfate- conjugated forms via the urine, which can be deconjugated again during wastewater treatment. Values as high as 30–80 ng l^{-1} estrone, 80–300 ng l^{-1} estriol and 5–30 ng l^{-1} 17 ß-estradiol were measured in raw sewage effluents of different countries before treatment. Even after treatment substantial proportions of oestrogenic compounds, especially estrone, were determined in secondary effluents as well as in sewage sludge (Combalbert and Hernandez-Raquet 2010). The efficiency of different processes of wastewater treatment plants to eliminate oestrogenic compounds is reviewed by Combalbert and Hernandez-Raquet (2010).

Animal breeding represents the second major source of steroids but in contrast to wastewater treatment there are no regulations for the treatment of manure to eliminate EDCs. Steroid excretion by livestock may be even more significant than urban/human sources as certain animals such as cattle excrete 2-times more and swine even 10-times more oestrogens than humans (Johnson et al. 2006). Steroids are still used in some countries in livestock production to prevent and treat diseases, as anabolic agents for growth promotion, and to improve productivity by synchronizing reproduction cycles (Liu et al. 2012). Different animals excrete oestrogens through different routes: Ruminants excrete oestrogens to a higher proportion via faeces, while in pigs and poultry oestrogen excretion predominately occurs via the urine (Hanselman et al. 2003; Shore and Shemesh 2003). Urinary oestrogens are mostly conjugates while faecal oestrogens are excreted as unconjugated steroids (Palme et al. 1996). As well as species, age and gender are important for the route and amount of oestrogen excretion: According to Liu et al. (2012) steroid excretion by sows occur preferentially via faeces, while piglets and barrows excrete steroids mainly via the urine. Steroid excretion by sows is so high (90.7–6310 μg day^{-1}) that the number of sows in a swine farm directly affects the total excretion mass of steroids of a particular farm. The daily steroid excretion of a sow is approximately equivalent to that of 100 people (Liu et al. 2012).

Very different amounts of steroids were determined in animal wastes depending on manure management system and animals that are kept. Combalbert and Hernandez-Raquet (2010) determined the total oestrogen concentration in liquid swine waste to be in the range of 47–34,326 ng l^{-1} and that in solid wastes from 600 to 6800 μg kg^{-1} dry weight (Table 9). The overall annual hormone excretion by farm animals, which may reach the environment with organic fertilizer application has been estimated to be at least 360 t in the EU and 330 t in the USA (Lange et al. 2002). Hanselman et al. (2003) demonstrated that N-based application rates of dairy manures could potentially increase 17 ß-estradiol concentrations in run-off water to concentrations that will potentially affect the reproductive biology of aquatic wildlife by disrupting the normal function of their endocrine systems.

Environmental EDC Concentrations and Physiological Effects

Physiological effects of hormones in the environment such as on development, sexual differentiation, production of vitellogenin (an egg yolk precursor protein that is normally produced only by adult females) and reproduction have already been found at very low concentrations. Induction of vitellogenin synthesis in immature and male fish is used as a biomarker of oestrogen exposure (Harris et al. 1997). Concentrations measured in effluents of sewage treatment plants (STPs) are presented in Table 6 to demonstrate that relatively high concentrations of these compounds were released into the environment and were actually found in aquatic ecosystems. Predicted no effect concentration (PNEC) values reported in literature are 3 ng l^{-1} for estrone, and 1–1.5 ng l^{-1} for 17 ß-estradiol (Young et al. 2002; Zhao et al. 2010). Detected concentrations (Table 6) were several times higher than the PNEC values underlining the obvious hazard created for aquatic organisms.

The fact that these substances are released to water bodies is the reason why investigations focus mainly on aquatic organisms and in particular on vertebrates, since for that group, more relevant data are available. Only limited knowledge exists about the endocrine systems of invertebrates, making the evaluation of effects of EDCs a complex task as the response can differ widely (Jobling et al. 2004). Effects of environmental contamination with EDCs on reproduction and development of different species are summarized in Table 7. Inputs into the environment are already so high (Table 6) that effects on wildlife are observed regularly, especially on aquatic organisms or animals that consume great amounts of aquatic organisms such as gulls (Fry and Toone 1981). Reproductive hormone-receptor systems appear to be especially vulnerable, and EDCs have been found to disrupt the endocrine system of many species already at nanogram-per-liter levels (Table 7).

Plant growth and seed germination were also affected by steroid hormones applied to the field with manure or sludge (reviewed by Geuns 1978). For example irrigation of alfalfa (*Medicago sativa* L.) with oestrogen concentrations of 0.05–5 μg l^{-1} caused

Table 6 Maximum release of compounds showing oestrogenic activity in the effluents of German sewage treatment plants (STPs) and surface and sea water (according to Schönauer 2001)

Active compound	Utilization	Concentrations in the effluents from STPs unless otherwise stated	Reference
Naturally released oestrogens			
Estriol (E1), 17 ß-estradiol (E2), estrone (E3), 17 α-estradiol (17 α)	Excretion by humans and animals (e.g. daily oestrogen excretion by woman: 25 μg–30 mg)	1–76 ng l^{-1} (E1) 0.5–64 ng l^{-1} (E2) <0.1–47 ng l^{-1} (E3) <0.1–5.0 ng l^{-1} (17 α)	Kalbfus (1998), Fent (2000), Belfroid et al. (1999)
Non-naturally released compounds with oestrogenic potential			
17 α-ethinyl-estradiol (EE2)	Use in contraceptiva and for hormonal treatment	<0.2–62 ng l^{-1}	Fent (2000)
Mestranol	Use in contraceptiva	0.5–43 ng l^{-1}	Fent (2000)
Clofibric acid	Lipid lowering drug	4500 ng l^{-1}	Römbke et al. (1996)
Phytooestrogens (e.g. ß-Sitosterol)	Release during wood processing; use as lipid lowering drug	47–7105 ng l^{-1} ß-Sitosterol	Wenzel et al. (1998)
Phthalates (e.g. Di-n-butylphthalate, DBP)	Widely used industrial chemical	<0.5–3.5 μg l^{-1} DBP <0.5–510 μg l^{-1} DEHP	Gülden et al. (1997)
Bisphenol A (BPA)	Widely used chemical in the production of epoxy resins and polycarbonate plastics	up to 330 ng l^{-1} in surface water (max. value of 21 μg l^{-1})	Belfroid et al. (2002)
DDT	Insecticide	<0.1–1400 ng l^{-1} in surface waters	Gülden et al. (1997)
Polychlorobiphenyls (PCB)	Widely used chemical in plasticizers, cooling agent, hydraulic fluents	<0.2–80 ng l^{-1} PCB-52 in surface waters	Gülden et al. (1997)
Nonylphenole	Detergent	<0.2–44 μg l^{-1}	Gülden et al. (1997)
Non-naturally released compounds with anti-oestrogenic potential			
Polychlorinated dibenzo-dioxins/–furans (PCDD/PCDF)	Chemical reaction products released from industries	1.2–3.1 pg I-TEQ l^{-1} in surface waters	Gülden et al. (1997)
Non-naturally released compounds with androgenic potential			
Tributyltin (TBT)	Antifouling pesticide in marine paints	7 μg l^{-1} (up to 200 ng l^{-1} in North and Baltic Sea)	Schönauer (2001), Kalbfus (1998)
Non-naturally released compounds with anti-androgenic potential			
Linuron, Diuron	Herbicide	0.47 μg l^{-1} Linuron, 11.3 μg l^{-1} Diuron	Gülden et al. (1997)
Vinclozolin	Common fungicide	0.2 μg l^{-1}	Gülden et al. (1997)

DBP Di-n-butylphthalate; *DEHP* Di(2-ethylhexyl)phthalate; *DDT* Dichloro-diphenyle-trichloro-ethane; *I-TEQ*: International toxic equivalent to report the toxicity-weighted masses of mixtures of dioxins

Table 7 Effects of endocrine disruptors on invertebrate and vertebrate species

Phyla/species	Exposure conditions	Observed effects	Reference
Mollusca			
Potamopyrgus antipo-darum (fresh water snail)	1 ng l^{-1} EE2; 1 μg l^{-1} BPA; 5 μg l^{-1} OP	Significantly increased embryo production after 21 d	Jobling et al. (2004)
Marisa cornuarietis (tropical fresh water snail)	1–100 μg l^{-1} BPA or OP for 5 month	Complex alterations at concentrations as low as 1 μg l^{-1} BPA/OP: formation of additional female organs, enlarged accessory pallial sex glands, stimulation of oocyte and spawning mass production etc.	Oehlmann et al. (2000)
Nucella lapillus (marine species)	1–100 μg l^{-1} BPA or OP for 3 month	Enlarged accessory pallial sex glands, enhanced oocyte production	Oehlmann et al. (2000)
Arthropoda			
Balanus amphitrite	0.01–1 μg l^{-1} 4-NP, 1 μg l^{-1} E2	Altered larval development, increased cypris storage protein production	Billinghurst et al. (2000)
Elminius modestus (larvae)	10 μg l^{-1} E2 or 4-NP	Induction of DNA damages	Atienzar et al. (2002)
Gammarus pulex	0.1–10 μg l^{-1} EE2 for 100 d	Modifications in population size and sex ratio (feminization)	Watts et al. (2002)
Acartia tonsa (marine copepod)	23 μg l^{-1} E2 or 20 μg l^{-1} BPA	Stimulated maturation of ovaries, higher egg production rates	Andersen et al. (1999)
Annelida			
Dinophilus gyrociliatus (polychaete worm)	Environmentally concentration of 4-NP	Increase in egg production and reduction in egg viability	Depledge and Billinghurst (1999)
Echinodermata			
Asterias rubens	fed with mussels containing 0.6 μg g^{-1} wet wt PCBs	Aberrations in early development; only 30 % of the embryos developed to normal bipinnaria larvae	Den Besten et al. (1989)
Chordata			
Amphibians			
Xenopus laevis	10^{-10} to 10^{-6} M Vinclo-zolin	Altered mate calling behavior that caused reduced reproductive success	Hoffmann and Kloas (2010)
Fish			
Rainbow trout (*Oncorhynchus mykiss*)	Fish held in undiluted or slightly diluted sewage effluent	Rapid and pronounced increase in plasma VTG at EE2 concentration of 1–10 ng l^{-1}	Harris et al. (1997), Purdom et al. (1994)
Carp (*Cyprinus carpio*)	Fish held in sewage effluent of different dilution	Dose-dependent induction of VTG production	Jobling et al. (2004)
Fathead minnow (*Pimephales promelas*)	0.1–100 ng l^{-1} EE2	Dose—dependent induction of VTG at 1–100 ng l^{-1} EE2 and production of eggs at 0.1–1 ng l^{-1} EE2	Jobling et al. (2004)

(continued)

Table 7 (continued)

Phyla/species	Exposure conditions	Observed effects	Reference
Birds			
California and western gull (*Larus californicus*; *L. occidentalis*)	0.5, 2 and 5 ppm estradiol injected into the egg yolk	Male embryos at all doses of estradiol were feminized	Fry and Toone (1981)
	2, 5, 20, 50, 100 ppm o, p′-DDT	7 of 8 embryos were feminized	Fry and Toone (1981)

E2 (17 *β*-estradiol); *EE2* (synthetic 17 *α*-ethinylestradiol); *BPA* (bisphenol A); *OP* (4-*tert* octylphenol); *DDT* (dichlorodiphenyltrichloroethane); *4-NP* (4-n-nonylphenol); *VTG* (vitellogenin)

increased shoot and root growth by at maximum 30 and 54 %, respectively compared to the control, while higher oestrogen concentrations of 50 μg l^{-1} decreased vegetative plant growth by 5 % and root growth by 8 % (Shore et al. 1992).

Observations on European otters and common seals, where a malfunctioning reproduction was observed because of high DDE (dichlorodiphenyldichloroethylene) and PCB (polychlorinated biphenyls) concentrations in their environment and in their main food, fish (Fent 2000), underline that even mammalians are already affected by high environmental EDC concentrations. Trends in reproductive health, increases in testicular and breast cancer as well as decreases in sperm quantity and quality give room to speculate about effects of EDCs in the range of ng l^{-1} even on humans (Brody and Rudel 2003; Sharpe and Irvine 2004).

Degradation of Compounds with Oestrogenic Activity

Degradation of oestrogens in waste products has mainly been investigated in sewage sludge while data related to animal wastes are still scarce. Aerobic biodegradation appears to be an important process involved in hormone elimination in marine sediment, seawater, soil, aquifer material, and groundwater while longer persistence can be expected under anoxic or anaerobic conditions (Combalbert and Hernandez-Raquet 2010; Ying and Kookana 2003, 2005). Therefore, if animal wastes are stored in lagoons or tanks under partly anaerobic conditions, steroids will persist to a greater percentage and be released later into the environment. Degradation of oestrogens highly depends on temperature and pH: Oestrogens in cattle manure were not reduced at 5 °C during 12 weeks, while they were reduced to non-detect levels within 3 weeks when the temperature was raised to 30 °C (Schlenker et al. 1998). Shore et al. (1993) found that a reduction in oestrogen concentration in broiler litter occurred at pH 5 and 7 whereas at pH 8 and 12 no reduction was observed. Conventional wastewater treatment is only efficient in the removal of 17 ß-estradiol (85–99 %), estrone removal was shown to be poor (25–80 %; Khanal et al. 2006). Oestrogens are a serious problem in sewage sludge and effluents as well as in animal wastes. Field experiments with poultry litter applied to pasture demonstrated its contribution to 17 ß-estradiol in run-off and showed that this hormone can persist for at least 7 days under field conditions (Nichols et al. 1997)

ECDs are critical environmental pollutants with more or less unknown consequences for ecosystems, animals and humans. Typical sewage effluents and sewage sludge contain many thousands of chemicals, not all of them already identified. Wildlife is not exposed to a single contaminant as in many trials, but to a mixture of very different xenobiotics. For oestrogenic compounds, an additive mode of action was shown and some chemicals accelerate the action of one another, making negligible concentrations toxic (Harris et al. 1997; Silva et al. 2002). Therefore it is important to minimize environmental contaminations with EDCs as much as possible and to improve technologies to remove them from organic fertilizer products. For different compound classes the mode of action in mixture is even more complicated to assess. Experiments have mainly been performed for aquatic environments but seldom in soils which are much more complex. Yang et al. (2012) could show in soil experiments that the toxicity of low-level pyrene on earthworms was prolonged by the co-presence of cadmium. Toxicity studies with single pure chemicals are not suitable to assess the chemical hazards of a compound in the environment, as there are always very different contaminants present, which can interact in different ways. "Effects of mixtures usually exceed those of the most active constituents alone" (Faust et al. 2003).

Contamination of Organic P Fertilizers by Antibiotics and EDCs

Many different organic contaminants are applied to the field when organic fertilizers are used in agriculture. To give an indication about loads of organic contaminants that are transferred via organic fertilizers to soils, literature data are summarized for antibiotics (Table 8) and EDCs (Table 9). A lot of data are available about heavy metal contamination in sewage sludge and contamination with many organic xenobiotics, but information on pharmaceuticals is limited. Also, extensive literature is available about farmyard manures but few studies about organic pollutants in digestates and composts have been published so far. Some data can be found about the occurrence of organic xenobiotics such as polycyclic aromatic hydrocarbons (PAHs), PCBs or pesticides (Kupper et al. 2008) but hardly any information about pharmaceutical residues. From the results summarized in Tables 8 and 9 it is hard to draw a conclusions about the question of which organic P fertilizer is favorable with respect to low contamination with antibiotics or EDCs. Data on digestates and composts are more or less missing. Furthermore, composts are a very heterogenic class of fertilizer, as they include not only organic wastes, but also digestates, farmyard manures and sewage sludge composted to improve the quality of these fertilizers.

Table 8 Contamination of organic fertilizers with antibiotics

Organic P fertilizer	Origin	Organic xenobiotics and pharmaceuticals	Applied amount	Medium excretion/medium concentration in the organic fertilizer	Reference
Manure	Chicken	Nicarbacine	125 mg kg^{-1} feed	20–38 mg kg^{-1}	Kan et al. (1996)
	Poultry	CTC	11 µg CTC g^{-1} feed	22.5 mg kg^{-1} CTC in manure	Warman and Thomas (1981)
	Calves	OTC Tylosin (TYL)	60 mg OTC kg^{-1} d^{-1} for 5 d 20 mg TYL kg^{-1} d^{-1} for 5 d	2–19 mg OTC kg^{-1} and <0.1–0.1 mg TYL kg^{-1} in 30–135 d old manure; 0.82 mg OTC kg^{-1} and no TYL after 5 month of maturation	De Liguoro et al. (2003)
	Calves	TC; Sulfadiazine (SFD)	[a]	1.1–45.3 mg kg^{-1} fw TC; 0.9–1.1 mg kg^{-1} fw SFD	Höper et al. (2002)
	Pigs	TC; CTC; Sulfadiazine (SFD)	[a]	0.7–4.0 mg kg^{-1} fw TC; 0.1–2.7 mg kg^{-1} fw CTC 0.3–0.6 mg kg^{-1} fw SFD	Höper et al. (2002)
	Pig	Sulfadiazine (SFD) Trimethoprim (TMP)	Unknown amount of SFD + TMP applied 2 month before	30 µg kg^{-1} SFD; 230 µg kg^{-1} N^4 acetyl sulfadiazine; 15 µg kg^{-1} sulfadimidine; TMP was not detected	Pfeifer et al. (2002)
	Pig	TC	Treatment for 10 d	23–52 mg kg^{-1} fw TC	Hamscher (2008)
	Pig	TC and CTC	Unknown	4 mg kg^{-1} fw TC; 0.1 mg kg^{-1} fw CTC	Hamscher et al. (2002)
	Pig	CTC, Tylosin (TYL)	Unknown	3.5–7.9 mg l^{-1} CTC; 3.3–4.0 mg l^{-1} TYL	Kumar et al. (2004)

(continued)

Table 8 (continued)

Organic P fertilizer	Origin	Organic xenobiotics and pharmaceuticals	Applied amount	Medium excretion/medium concentration in the organic fertilizer	Reference
Liquid manure	Pig	CTC, TC, OTC, Sulfamethazine (SMZ), Sulfadimethoxine (SDX), Lincomycin (LIN), Trimethoprim (TMP), Erythromycin (ERY)	[b]	68–1000 $\mu g\ l^{-1}$ CTC; 25–410 $\mu g\ l^{-1}$ TC + OTC 2.5–400 $\mu g\ l^{-1}$ SMZ; 2.5 $\mu g\ l^{-1}$ SDX 2.5–240 $\mu g\ l^{-1}$ LIN 2.5 $\mu g\ l^{-1}$ TMP; 2.5 $\mu g\ l^{-1}$ ERY	Campagnolo et al. (2002)
	Pig	Sulfonamides, Trimethoprim (TMP)	[c]	Up to 20 mg sulfonamides kg^{-1} fw, <0.1 mg TMP kg^{-1}	Haller et al. (2002)
	Cattle	TC	Unknown[d]	0.0005–200 mg l^{-1}	Aga et al. (2003)
	Cattle	Monensin-Natrium	85 g t^{-1} feed	4.7 mg l^{-1} after 2 weeks	Donoho (1984)
Slurry	Pig	Sulfadiazine (SFD), Sulfamethazine (SMZ), OTC, CTC	Screening of 344 samples of pig slurry	Median values: 3.5 mg kg^{-1} fw SFD, 4.0 mg kg^{-1} fw SMZ, 3.4 mg kg^{-1} fw CTC; max. 235 mg kg^{-1} fw SFD, 167 mg kg^{-1} fw SMZ, 331 mg kg^{-1} fw OTC	Hamscher and Mohring (2012)
	Pig	TC, CTC, OTC, Sulfadiazine (SFD), Sulfamethazine (SMZ)	Screening of 168 (1999/2000) and 176 samples (2001/2002)	Median values of positive samples [mg kg^{-1} fw]: 5.5 TC, 4.8 OTC, 4.4 CTC, 2.9 SFD, 4.4 SMZ (1999/2000); 7.3 TC, 5.1 OTC, 5.6 CTC, 3.4 SFD, 3.8 SMZ (2001/2002)	Engels (2004)
	Pig	TC	Screening of 181 samples	24 % contained >0.6 mg kg^{-1} TC; median value 11.4 mg kg^{-1} (0.6–66 mg kg^{-1})	Grafe (2001)
Slurry	Pig	TC	2.9–7.9 g (for 5 d)	380–830 ppm after 4 days of treatment (total excretion: 42–70 % of the applied amount)	Winckler and Grafe (2000)
Slurry	Pig	Sulfadiamine	Prophylactic dose for 34 d	253 mg kg^{-1} (after 51 d 82 mg kg^{-1}); 46 % of applied amount	Berger et al. (1986)

(continued)

Table 8 (continued)

Organic P fertilizer	Origin	Organic xenobiotics and pharmaceuticals	Applied amount	Medium excretion/medium concentration in the organic fertilizer	Reference
Slurry	Pig	TC	Unknown	0.04–0.7 mg l^{-1} TC	Sengeløv et al. (2003)
	Pig	CTC; Sulfadiazine (SFD), Trimethoprim (TMP)	45.0 g CTC, 112.6 g SFD and 22.5 g TMP for 10 d	87.5 (active) + 27.1 (non-active) mg kg^{-1} CTC residues 198.5 mg kg^{-1} SFD, 303.7 mg kg^{-1} N^4-SFD, 15.8 mg kg^{-1} TMP	Grote et al. (2004)
	Pig	Sulfadimidine (SDM), sulfathiazole (STZ), TMP	37.5 g Vetoprim 2400® (40 g kg^{-1} SDM, 40 g kg^{-1} STZ, 16 g kg^{-1} TMP) for 8 d	Up to 39 mg SMD and STZ l^{-1} slurry; degradation of 40–60 % within 5 weeks	Langhammer et al. (1988)
Dung	Beef cattle and calves	CTC; OTC	70 mg $head^{-1}$ d^{-1} for 10 d	5.3 μg g^{-1} dw faeces CTC; 11.3 μg g^{-1} dw faeces OTC	Patten et al. (1980)
Faeces	Pig	TC-HCl + 4-epi-TC	2.5 g TC kg^{-1} feed (=33 mg kg^{-1} body weight)	Up to 8000 mg kg^{-1} dw	Grafe (2001)
	Pig	CTC	800 mg CTC kg^{-1} feed for 3 d	112 mg CTC kg^{-1} faeces one day after application	Hansen et al. (2002)
Urine	Cattle	Ceftiofur sodium	2.2 mg kg^{-1} body weight	11.3–216 mg l^{-1} Ceftiofur equivalents in 24 h samples	Gilbertson et al. (1990)
Carcass	Pig-bones	CTC	45.0 and 56.1 g CTC[e]	56.2 mg kg^{-1} bones (directly after medication) 22.3 mg kg^{-1} bones (14 d after medication)	Grote et al. (2004)
	Pig-muscle	CTC	45.0 and 56.1 g CTC[e]	210.4 μg kg^{-1} (directly after medication) 4.0 μg kg^{-1} (14 d after medication)	Grote et al. (2004)
	Pig-liver	CTC	45.0 and 56.1 g CTC[e]	392.2 μg kg^{-1} (directly after medication) 7.4 μg kg^{-1} (14 d after medication)	Grote et al. (2004)
	Pig-kidney	CTC	45.0 and 56.1 g CTC[e]	659.0 μg kg^{-1} (directly after medication) 19.7 μg kg^{-1} (14 d after medication)	Grote et al. (2004)

(continued)

Table 8 (continued)

Organic P fertilizer	Origin	Organic xenobiotics and pharmaceuticals	Applied amount	Medium excretion/medium concentration in the organic fertilizer	Reference
Compost	Mixed compost[f]	TC; OTC 4-epimer of OTC		2.95 ± 0.42 μg g^{-1} dw TC; 6.51 ± 0.52 μg g^{-1} dw OTC 1.35 ± 0.24 μg g^{-1} dw 4-epimer of OTC	Shafrir and Avisar (2012)
	Manure-based compost	Total tetracyclines and degradation products		88.3 ± 1.7 up to 850.8 ± 746.8 μg kg^{-1} total TC and degradation products	Qiao et al. (2012)
Sewage sludge	Activated sludge	Sulfamethoxazole (SMX), Sulfa-pyridine (SPY), Clarithromycin (CLA). Azithromycin (AZI), TMP		68 ± 20 μg SMZ kg^{-1} dw, 28 ± 3 μg SPY kg^{-1} dw, 67 ± 27 μg CLA kg^{-1} dw, 64 ± 30 μg AZI kg^{-1} dw, 41 ± 15 μg TMP kg^{-1} dw	Göbel et al. (2005)

[a]Samples of liquid manure were investigated which were derived from the long term soil monitoring program in Lower Saxony; [b]Nine commercial swine farms were investigated in that study including birthing facilities, facilities where pigs were weaned, and where pigs are raised until reaching market weight. No data are available about administration of pharmaceuticals. Manure samples were derived from manure waste lagoons. [c]Samples were taken from 6 different pits containing manure slurry from cattle (1) and pigs (5). [d]Manure samples were collected from hog lagoons and cattle feedlots. [e]Piglets were medicated twice for 10 days; the period was interrupted by 11 days. [f]Mixed compost comprised of de-watered secondary sludge from 14 different STPs in Israel and shredded municipal green waste in a ratio of approximately 3:1 by weight

Table 9 Contamination of organic fertilizers with endocrine-disrupting compounds (EDCs)

Organic P fertilizer	Origin	Organic xenobiotics/pharmaceutical	Medium concentrations of contaminants in organic fertilizers	Reference
Liquid manure	Swine Cattle Poultry	Total oestrogens	47–75,127 ng l^{-1} swine manure 46–8960 ng l^{-1} cattle manure 41–7817 ng l^{-1} poultry manure	Reviewed by Combalbert and Hernandez-Raquet (2010)
	Swine Cattle Poultry	E1	27–74,700 ng E1 l^{-1} swine manure 146–4500 ng E1 l^{-1} cattle manure 41–5910 ng E1 l^{-1} poultry manure	Reviewed by Combalbert and Hernandez-Raquet (2010)
Slurry	Cattle	E1, 17α-estradiol (17α-E2), E2	41–3057 μg E1 kg^{-1}; 19–1028 μg 17α-E2 kg^{-1}; 29–1229 μg E2 kg^{-1}	Reviewed by Limpiyakorn et al. (2009)
	Swine	E1, 17α-estradiol (17α-E2), E2, E3	4728 μg E1 kg^{-1}; 890 μg 17α-E2 kg^{-1}; 1251 μg E2 kg^{-1}, 402 μg E3 kg^{-1}	
Litter	Broiler Poultry	E2	904 μg E2 kg^{-1} in broiler litter 533 μg E2 kg^{-1} in poultry litter	Reviewed by Limpiyakorn et al. (2009)
Solid manure	Swine Cattle	Total oestrogens	600–6800 μg kg^{-1} dw swine manure 130–2104 μg kg^{-1} dw cattle manure	Reviewed by Combalbert and Hernandez-Raquet (2010)
Faeces	Sows Piglets	Total steroids (androgens, oestrogens, glucocorticoids, progestagens)	>16,703 μg kg^{-1} faeces of sows >116 μg kg^{-1} faeces of piglets	Liu et al. (2012)
Flush water (residual faeces, urine, wastes)	Sows Piglets	Total steroids	>17,115 ng l^{-1} flush water sows >3564 ng l^{-1} flush water piglets	Liu et al. (2012)
Manure	Swine Cattle Chicken	Estradiol (E2) and testosterone (measured as hormone equivalents in hormone receptor binding assays)	up to 6000 ng g^{-1} dw E2 + up to 580 ng g^{-1} dw testosterone in swine manure up to 5000 ng g^{-1} dw E2 + up to 2000 ng g^{-1} dw testosterone in cattle manure 60–400 ng g^{-1} dw E2 + 10–30 ng g^{-1} dw testosterone in chicken litter	Lorenzen et al. (2004)
Manure	Chicken	Testrosterone and oestrogen (used as growth promoters)	1 μmol g^{-1} manure	Shore et al. (1988)
Sewage sludge		Natural and synthetic oestrogens	<2–63 μg kg^{-1} dw E1; 0.1–49 μg kg^{-1} dw E2; 5.3–406 μg kg^{-1} dw E3; 0.4–18 μg kg^{-1} dw EE2	Reviewed by Combalbert and Hernandez-Raquet (2010)

(continued)

Table 9 (continued)

Organic P fertilizer	Origin	Organic xenobiotics/pharmaceutical	Medium concentrations of contaminants in organic fertilizers	Reference
Sewage effluent		Natural and synthetic oestrogens	<0.3–100 ng l^{-1} E1; <0.2–30 ng l^{-1} E2; <0.5–275 ng l^{-1} E3; <0.3–9 ng l^{-1} EE2	Reviewed by Combalbert and Hernandez-Raquet (2010)
Sewage sludge	Municipal biosolids from sewage	Hormone equivalents for estradiol	11.3 ng g^{-1} dw after aerobic treatment 1233 ng g^{-1} dw after anaerobic treatment	Lorenzen et al. (2004)
Sewage sludge	Activated and digested sludge	Natural and synthetic oestrogens	<2–37 ng g^{-1} dw E1; 5–49 ng g^{-1} dw E2 <2–17 ng g^{-1} dw EE2	Ternes et al. (2002)
Sewage sludge	Secondary excess sludge	Natural and synthetic oestrogens	7 ng g^{-1} dw E1; 1.7 ng g^{-1} dw E2; 3.0 ng g^{-1} dw EE2	Andersen et al. (2003)
	Excess sludge after anaerobic digestion		25.2 ng g^{-1} dw E1; 5.1 ng g^{-1} dw E2; <1.5 ng g^{-1} dw EE2	Andersen et al. (2003)
	Activated and dewatered sludge	Natural and synthetic oestrogens	12–14 ng g^{-1} dw E1; <1 ng g^{-1} dw E2 and EE2	Braga et al. (2005)
	Excess and dewatered sludge	Natural and synthetic oestrogens	2–8 ng g^{-1} dw E1; 1–10 ng g^{-1} dw E2; 1–18 ng g^{-1} dw EE2; <1 ng g^{-1} dw E3	Muller et al. (2008)
	Primary sludge		8 ng g^{-1} dw E1; 10–13 ng g^{-1} dw E2; 2–3 ng g^{-1} dw EE2; <3 ng g^{-1} dw E3	Muller et al. (2010)
	Secondary sludge		43 ng g^{-1} dw E1; 10–13 ng g^{-1} dw E2 2–3 ng g^{-1} dw EE2; <3 ng g^{-1} dw E3	Muller et al. (2010)
	Activated sludge	Testosterone and oestrogen	10–234 pmol l^{-1} oestrogen, 8–600 pmol l^{-1} testosterone	Shore et al. (1993)
Sewage sludge		Natural and synthetic oestrogens	3.3–328.2 E1, 4.2–111.4 E2, 18.1–31.4 E3; 1.2–4.5 mestranol; 415.3–268,702.7 ß-sitosterol [in ng kg^{-1} dw]	Wenzel et al. (1998)
Compost	Mixed compost[a]		160 ± 65 ng g^{-1} dw E1 21 ± 3 ng g^{-1} dw E2	Shafrir and Avisar (2012)
Silage	Chicken manure silage used as feed for cattle	Estradiol (E2) + estrone (E1)	317 ± 36 ng g^{-1} (E1 + E2)	Shore et al. (1988)

[a]Mixed compost comprised of the dewatered secondary sludge from 14 different STPs in Israel and municipal green waste in a ratio of approximately 3:1 by weight

Results of Shafrir and Avisar (2012) indicate that the aerobic composting process does not sufficiently remove all hazardous contaminants in the secondary sludge of sewage water treatment plants (STPs). Significant concentrations of TCs, OTCs, 4-epimer OTC, E1 and E2 were detected in composts (Tables 8 and 9) while sulfamethoxazole and sulfademethoxine were below the detection limit. Degradation and transformation processes during composting affect the organic matter and organic contaminants in the bio-wastes. Hartlieb et al. (2003) demonstrated that a substantial fraction of the contaminants is locked up within the organic matrix by formation of non-extractable residues (NER). First results indicate that the anaerobic digestion process is also not able to degrade antibiotics completely, and the degradation efficiency may vary from very low elimination to up to 99 % (Hamscher and Mohring 2012). The process temperature is of high importance for the reduction of antibiotics and coliforms: Varel et al. (2012) found higher CTC degradation and removement of coliforms during anaerobic digestion when the process temperature was raised from 22 to 38 or 55 °C. However, the process at higher temperatures was not effective in reducing monensin in cattle manure. Like composts and digestates, bone meal products are rarely analyzed for pharmaceuticals. Studies by Grote et al. (2004) show that antibiotics were enriched in bones of farm animals treated with antibiotics. Therefore bone meal products can contain very high loads of antibiotics (Table 6) as well as high P contents and should not be applied to the field without prior treatment.

Several studies are available about contaminations of farmyard manures with antibiotics (Table 8), which are very often in the mg kg^{-1} range. Highest values were detected shortly after administration of antibiotics but distinct concentrations of certain antibiotics were also found weeks after their administration or in screening trials of hundreds of samples. Inputs higher than 800 g ha^{-1} TC can result from manure application containing up to 52 mg TC kg^{-1} fw (Table 8) assuming that farmyard manure was applied to the soil according to the German fertilization recommendations for P (22 kg ha^{-1} × a P). But even on basis of a median value of 5.5 mg TC kg^{-1} fw in pig slurry as determined by Engels (2004), an input of 92 g TC ha^{-1} would result. Engels (2004) calculated the soil application of TC to be in the range of 108–128 g TC ha^{-1} if 100 kg N was fertilized via the investigated pig slurry. Kumar et al. (2005a) calculated CTC and tylosin inputs into the soil with manure application to be as high as 387 g ha^{-1} CTC and 202 g ha^{-1} TYL. Therefore substantial proportions of antibiotics are distributed when manure containing such compounds is applied to the field.

For a risk assessment of antibiotics it is necessary to calculate realistic worst-case concentrations in slurry and soils. Following this approach, Montforts (2006) calculated a soil concentration of 207 μg OTC per kg dry weight in the upper 30 soil cm when pig slurry was applied. This value was confirmed by field measurements (Hamscher et al. 2002). High concentrations of antibiotics were also found months after slurry application as in the study of Christian et al. (2003) who determined a soil concentration of 15 μg kg^{-1} soil of sulfadimidine 7 months after pig slurry had been applied, indicating a very high stability of these compounds in slurry as well as in soils.

Toxicity studies show that prokaryotic cyanobacteria and bacteria are most sensitive to antibiotics with EC_{50} values of about 5–100 μg l^{-1} in liquid media. Effects of antibiotics on soil micro-organisms and the impact of increased antibiotic resistant genes in the gene pool are still discussed but may be most hazardous because there is no definite way to predict the consequences.

EDCs were also found in very high concentrations in farmyard manures as well as in sewage sludge (Table 9). Farm animals excrete large amounts of hormones, which for some species are in the range of milligrams per animal and day (Hanselman et al. 2003). Especially high concentrations of total oestrogens were found in liquid swine manure with values of up to 75,127 ng l^{-1}, followed by cattle and poultry manure with values of up to 8960 and 7817 ng l^{-1}, respectively. Lorenzen et al. (2004) determined values of up to 6 mg oestrogenicity as estradiol equivalent per kg dry weight in swine manure. Similarly, flush water used to clean animal housing was very high in total steroids (17,115 ng l^{-1}; Liu et al. 2012). Such high EDC concentrations in farmyard manures present a very critical risk, considering that it is known that many organisms react to very low concentrations (Table 7). Fish species such as fathead minnow and rainbow trout show effects on vitellogenin and egg production at concentrations as low as 0.1–100 ng l^{-1} EE2 (Jobling et al. 2004; Harris et al. 1997). Even after treatment of sewage sludge substantial concentrations of oestrogens remain in the sludge (Table 9). A comparison of aerobic and anaerobic treatment of municipal biosolids revealed that a high reduction in oestrogenicity was only observed after aerobic treatment (Table 9; Lorenzen et al. 2004).

All available organic fertilizers carry the risk of contaminating farm sites and adjacent water bodies with pharmaceuticals at concentrations where effects on soil community and wildlife can be expected. Therefore it is not possible to recommend the preferable usage of one of these fertilizers without prior treatment. First of all an extensive screening of the contamination of all organic fertilizers with pharmaceuticals is necessary as well as an evaluation of the procedures to improve product quality, such as composting. After that, it would be possible to compare the organic fertilizers with respect to their potential to contaminate soils with pharmaceuticals or their degradation products. Rather, it must be the target of future efforts to reduce, or preferably eliminate contamination of such fertilizers with pharmaceuticals. Technical solutions for an effective elimination as well as general reduction by reduced application of pharmaceuticals should be major goals.

Evaluation of Technical Procedures to Remove Organic Contaminants from Fertilizer Materials

The stability of organic contaminants mainly depends on structure, matrix, temperature, pH, lighting conditions, oxygen supply and the presence of specific degrading bacteria (Lange et al. 2002). An effective possibility to reduce antibiotic concentrations in manure is a longer maturation period before field application: De

Liguoro et al. (2003) determined an OTC concentration of 820 μg kg^{-1} in manure after 5 months maturation while the constituent in faeces and bedding before maturation showed concentrations of 871,700 and 366,820 μg kg^{-1}, respectively. Degradation rates of veterinary medicines in soils vary, with half-lives ranging from days to years (Boxall et al. 2004). In sewage treatment plants (STPs) different parameters such as the hydraulic and sludge retention times are important.

As mentioned before, a substantial proportion of contaminants remain in the sewage sludge as well as in the effluent after wastewater treatment. Mainly acidic drugs such as lipid regulators (bezafibrate), antiphlogistics (diclofenac, ibuprofen) and the metabolites clofibric acid and salicylic acid as well as neutral or weak basic drugs were found ubiquitously in rivers and streams, mostly in the ng l^{-1} range. Elimination rates of these chemicals in STPs vary from 7 % (carbamazepine) to more than 99 % (salicylic acid) but generally, more than 60 % of the total drug residues are removed (Ternes 1998). Hormone elimination rates vary from 0 to 90 % in different STPs, also showing a high variability (Combalbert and Hernandez-Raquet 2010). Because of the huge range of different contaminants with different chemical characteristics, it is hard to develop processes, which eliminate all possible contaminations at once.

For sewage sludge, a complex technical incineration procedure was developed. The main target was to get rid of heavy metals and increase P availability, but at the same time it was assumed that all organic contamination would be destroyed in the process as well (Adam et al. 2008; Prinzhorn 2005). Theoretically, at temperatures above 800 °C, all organic compounds remaining after the first sludge incineration process would be transformed to CO_2 and a clean ash product rich in plant-available nutrients should arise. However, recent investigations of the ash products revealed that as a result of incomplete incineration, new organic compounds were formed and some of the existing ones (up to 3 %) were still left in the ash. Further investigations about the nature of these persistent organic compounds and possible improvements of the technical procedure in order to reach full incineration are currently under way. Only then, the aim to produce a true "100 % zero P fertilizer" as postulated by the title of this book, i.e. a fully available P source free of xenobiotics, will be fulfilled.

A very different approach to get a clean fertilizer product from human wastes is the separation of urine and faeces and the precipitation of magnesium ammonium phosphate ($MgNH_4PO_4 \cdot 6H_2O$), also known as struvite, from the urine. Urine source separation seems to be an efficient tool in wastewater management, because urine contains most of the nutrients as well as pharmaceutical contaminants but constitutes less than 1 % of the total wastewater volume (Larsen et al. 2001, 2004). Therefore a much smaller volume has to be treated and precipitation of struvite turns two dominant wastewater nutrients, nitrogen and phosphorus, into a slow release fertilizer (Escher et al. 2006). Experiments with spiked urine showed that more than 98 % of the pharmaceuticals and hormones remained in solution during the precipitation process so that the product is almost free of most organic micro-pollutants (Ronteltap et al. 2007; Escher et al. 2006).

Future Prospects and Research Needs

First of all, precise data on pharmaceutical applications in livestock production are required to allow for a more reliable estimate of their environmental inputs. Recently released data on the consumption of antibiotics in animal production in Germany (see Fig. 3), provided by manufacturers for 2011, reveal that the amounts were most likely underestimated in the past. Moreover, an environmental risk assessment (ERA) should be the rule for all xenobiotics potentially entering the environment, including human and veterinary therapeutics as well as their metabolites. Only this way will it be possible to select between two (or more) drugs with the same therapeutic effect the one with the lower environmental impact and to phase out especially dangerous compounds (Jørgensen and Halling-Sørensen 2000). According to Jørgensen and Halling-Sørensen (2000), ecotoxicological test methods for the evaluation of drugs are missing and need to be developed. Many test methods are developed for aquatic organisms. However, there are hardly any eco-toxicological test methods for soil organisms despite the fact that highly contaminated organic fertilizers are directly applied to the soil. Increasing problems with antibiotic resistance are certainly exacerbated by overuse in human medicine and misuse by consumers. Use of antibiotics in livestock farming contributes significantly toward the problem. Though the prophylactic administration for improving animal productivity is legally forbidden, the consumption rate is still increasing (Fig. 3).

To reduce the risk of environmental contamination and to maintain antibiotics as effective drugs against infectious diseases, a responsible use and general reduction of consumption is needed. "The single greatest factor driving resistance to a given antibiotic is simply use of the drug" (Schmidt 2002). Experts recommend using antibiotics sparingly, but at dose levels intended to kill all or as many of the bacteria causing an infection as possible. Use of antibiotics around the world is characterized by widespread chronic under-treatment, leaving a hardy group of survivors that grow and multiply into resistant strains (Schmidt 2002). Therefore public education is important to reduce misuse of antibiotics in daily life, in usage as well as in disposal. An unknown but significant portion of human drugs ends up unused in the sewer system, for example 40 % of Germans dispose of liquid drugs via the toilet (Schönauer and Grommelt 2007). In animal production it is important to develop production practices that reduce dependence on antibiotics such as species appropriate animal husbandry and sufficient sanitation in housing. Most problems arise from industrial livestock farming where generally the highest amounts of pharmaceuticals are used. Moreover manure disposal practices need to be developed that minimize the spreading of residual antibiotics and antibiotic resistant bacteria into the environment.

Contamination of aquatic ecosystems and agricultural sites with EDCs are of special concern as such compounds may affect the endocrine system of wildlife and humans at very low concentrations, which have already been reached at many natural sites. As Jobling et al. (2004) stressed, there is only little information on the relative sensitivities of different wildlife groups to EDCs and/or mixtures of them.

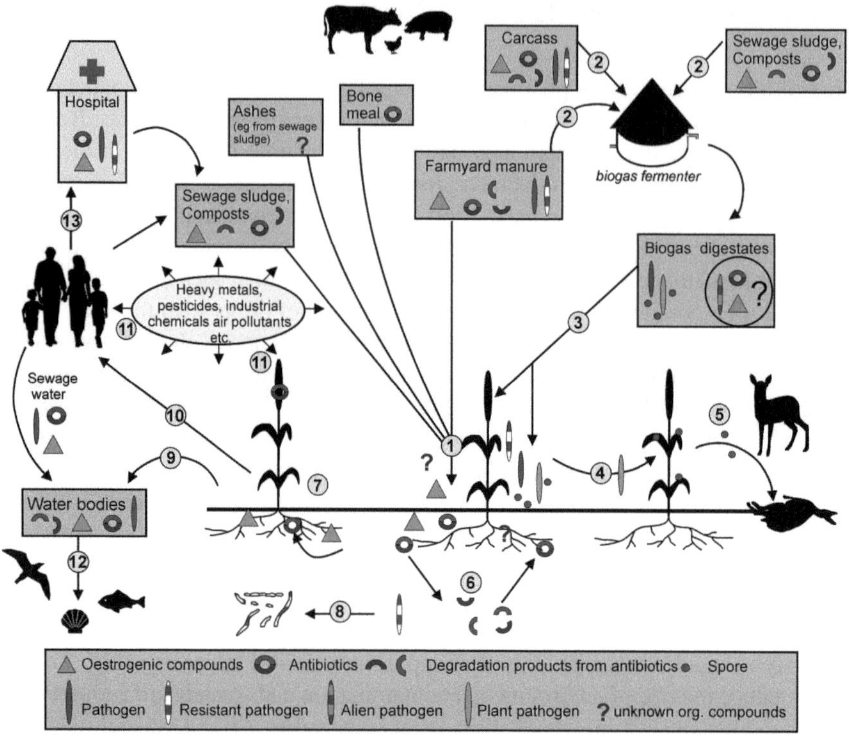

1 Pharmaceuticals, un-known organic compounds, pathogens and antibiotic resistant pathogens transfered via farmyard manures, bone meal, composts, sewage sludge and ashes (e.g. from sewage sludge) to soils.
2 Pharmaceuticals, pathogens and antibiotic resistant pathogens transfered via organic fertilizers into biogas fermenters.
3 Digestates contain pathogens (phytopathogens, human/animal pathogens and spores) and pharmaceuticals. It is unclear if new pathogens formate during fermentation and if pharmaceuticals are degraded [1].
4 Plant pathogens can be spread and can infect plants (e.g. *Synchytrium endobioticum* is still infective after anaerobic digestion [2]).
5 Pathogens and spores can be distributed via digestates on plants and soils. It was suspected by veterinarians and farmers that wild and grazing animals were infected by digestates with spores from *Clostridium botulinum* when disease incidence increased in areas where digestates were used as fertilizers [3].
6 Degradation of antibiotics and reactivation of parent compounds [4].
7 Oestrogenic compounds affect plant growth either positively at low or negatively at higher concentrations [5]. Antibiotics are taken up by plants and end up in the food chain [6].
8 Antibiotic resistant bacteria can be selected in soils containing low concentrations of antibiotics. Transfer of resistance genes to soil microorganisms is possible [7].
9 Leaching and erosion of pharmaceuticals into water bodies [8].
10 Humans and animals take up antibiotics via contaminated plant material. Resistance against antibiotics is triggered by low concentrations of antibiotics.
11 Combination toxicity was shown for different compounds such as oestrogenic compounds. Mixture effects usually exceed those of added single compounds [9].
12 Aquatic organisms and animals that feed on aquatic organisms are strongly affected by oestrogenic compounds and there is a high risk of species extinction from habitats [10].
13 Increasing problems with (multi-)resistant bacteria in hospitals jeopardizing especially babies and older people [7].

Fig. 4 Risks of applying pharmaceuticals via organic fertilizers to agricultural sites. ([1] Bagge et al. 2010; [2] Schleusner et al. 2010; [3] Boll 2011; [4] Bagge et al. 1986; [5] Shore et al. 1992; [6] Grote et al. 2006; [7] Chander et al. 2005; [8] Hamscher and Mohring 2012; [9] Faust et al. 2003; [10] Fent 2000)

Wildlife and humans are always exposed to a high number of very different chemicals and it is currently impossible to predict additive or synergistic effects.

Changes in agriculture such as the boom in biogas production and the concomitant increased usage of digestates as fertilizers bear several new risks such as the enrichment of pathogens like *Clostridium botulinum* (Boll 2011) or the development of new pathogens during the fermentation process (Bagge et al. 2010). Moreover, plant pathogens can also survive the anaerobic digestion process such as *Synchytrium endobioticum* (Schleusner et al. 2010) or the *Tobacco Mosaic* virus (Böhm et al. 2001). Use of digestates as fertilizer bears the great risk of a widespread contamination of agricultural sites with such pathogens. The environmental benefits and risks of an increasing use of digestates as fertilizer materials still need to be defined. High loads of pharmaceuticals such as EDCs in digestates cannot be excluded as EDCs mainly degrade under aerobic conditions. Additionally, the fermentation process can be affected negatively by pharmaceuticals in the substrate; for example low concentrations of Monensin-Na in sewage sludge (1–10 ppm) or pig slurry (8.6–17.2 mg l^{-1}) may reduce the biogas production significantly (Hilpert et al. 1984; Winterhalder 1985). Figure 4 summarizes the risks, which derive from soil application of different sources of organic fertilizers containing various organic contaminations. The environmental compartments are interrelated, thus contaminants most likely end up in the food chain, affecting humans as well as wildlife.

The most important target in producing high-grade organic fertilizers is the prevention of contamination by reducing the consumption of pharmaceuticals that are not necessary and by avoiding incorrect methods of disposal. In addition, highly contaminated sources should be treated. In this respect, the urine source separation is a promising possibility for separating effluent sources which are highly contaminated with xenobiotics from sources with much lower organic contaminations, e.g. those originating from washing. Most likely, treatment of these lower amounts of highly contaminated effluents will be more effective as well as economic.

It is concluded that a lot of effort is still necessary in the technical as well as in the educational and agricultural area to come close to the target of producing efficient organic P fertilizers that are almost free of organic contaminants.

Acknowledgment The authors wish to express their sincere thanks to Dr. Kerr Walker (SAC, Aberdeen) for the linguistic revision of the manuscript.

References

Adam C, Peplinski B, Kley G, Kratz S, Schick J, Schnug E (2008) Phosphorrückgewinnung aus Klärschlammaschen—Ergebnisse aus dem EU-Projekt SUSAN. Österreichische Wasser—und Abfallwirtschaft 60:55–64

Aga DS, Goldfish R, Kulshrestha P (2003) Application of ELISA in determining the fate of tetracyclines in land-applied livestock wastes. Analyst 128:658–662

Andersen HR, Halling-Sørensen B, Kusk KO (1999) A parameter for detecting estrogenic exposure in the copepod *Acartia tonsa*. Ecotoxicol Environ Saf 44:56–61

Andersen H, Siegrist H, Halling-Sørensen B, Ternes TA (2003) Fate of estrogens in a municipal sewage treatment plant. Environ Sci Technol 37:4021–4026

Atienzar FA, Billinghurst Z, Depledge MH (2002) 4-n-Nonylphenol and 17-beta estradiol may induce common DNA effects in developing barnacle larvae. Environ Pollut 120:735–738

Backhaus T, Grimme LH (1999) The toxicity of antibiotic agents to the luminescent bacterium *Vibrio fischeri*. Chemosphere 38:3291–3301

Bagge E, Persson M, Johansson K-E (2010) Diversity of spore-forming bacteria in cattle manure, slaughterhouse waste and samples from biogas plants. J Appl Microbiol 109:1549–1565

Baguer AJ, Jensen J, Krogh PH (2000) Effects of the antibiotics oxytetracycline and tylosin on soil fauna. Chemosphere 40:751–757

Belfroid AC, Van der Horst A, Vethaak AD, Schafer AJ, Rijs GBJ, Wegener J, Cofino WP (1999) Analysis and occurrence of estrogenic hormones and their glucoronides in surface water and waste water in The Netherlands. Sci Total Environ 225:101–108

Belfroid A, van Velzen M, Van der Horst B, Vethaak D (2002) Occurrence of bisphenol A in surface water and uptake in fish: evaluation of field measurements. Chemosphere 49:97–103

Berger K, Petersen B, Brüning-Pfaue H (1986) Persistenz von Gülle-Arzneistoffen in der Nahrungskette. Arch Lebensmittelhyg 37:99–102

Bergmann A, Fohrmann R, Weber FA (2011) Zusammenstellung von Monitoringdaten zu Umweltkonzentrationen von Arzneimitteln. UBA Texte 2011, 66/2011, Berlin, Umweltbundesamt

Billinghurst Z, Clare AS, Matsumura K, Depledge MH (2000) Induction of cypris major protein in barnacle larvae by exposure to 4-n-nonylphenol and 17 beta-oestradiol. Aquat Toxicol 47:203–212

Blaha T (1996) Gesundheits—und Umweltrisiken nach der Anwendung von Antiinfektiva und Antiparasitika in der Nutztierhaltung—Vermeidungsstrategien und Auswege. Dtsch Tierarztl Wochenschr 103:278–280

Böhm R, Buchenauer H, Knie A, Lorenz H, Hellwald KH, Philipp W (2001) Untersuchungen zur Seuchen—und Phytohygiene in Anaerobanlagen. Zwischenbericht. Statusseminar des BWPLUS, 28.02–01.03.2001, Karlsruhe

Boll C (2011) Chronischer Botulismus—Tod aus der Biogasanlage. Wild und Hund 10:14–19

Bowman DD (2009) Manure pathogens: manure management, regulations, and water quality protection. WEF Press, USA

Boxall ABA (2010) Veterinary medicines and the environment. In: Cuningham F, Elliott J, Lees P (eds) Comparative and veterinary pharmacology. Handbook of experimental pharmacology, vol 199. Springer, Berlin, pp 291–314

Boxall ABA, Fogg LA, Blackwell PA, Kay P, Pemberton EJ, Croxford A (2004) Veterinary medicines in the environment. Rev Environ Contam Toxicol 180:1–91

Boxall ABA, Johnson P, Smith EJ, Sinclair CJ, Stutt E, Levy LS (2006) Uptake of veterinary medicines from soils into plants. J Agric Food Chem 54:2288–2297

Braga O, Smythe GA, Schafer AI, Feitz AJ (2005) Fate of steroid estrogens in Australian inland and costal wastewater treatment plants. Environ Sci Technol 39:3351–3358

Brain RA, Johnson DJ, Richards SM, Sanderson H, Sibley PK, Solomon KR (2004) Effects of 25 pharmaceutical compounds to *Lemna gibba* using a seven-day static-renewal test. Environ Toxicol Chem 23:371–382

Brody JG, Rudel RA (2003) Environmental pollutants and breast cancer. Environ Health Perspect 111:1007–1019

BVL (2012) Erstmals Zahlen über die Antibiotikaabgabe in der Tiermedizin erfasst. Press release 12 Sept 2012. http://idw-online.de/pages/de/news495558

Calamari D, Zuccato E, Castiglioni S, Fanelli R (2003) Therapeutic drugs in the rivers Po and Lambro in Northern Italy—A strategic survey. Environ Sci Technol 37:1241–1248

Campagnolo ER, Johnson KR, Karpati A, Rubin CS, Kolpin DW, Meyer MT, Esteban JE, Currier RW, Smith K, Thug KM, McGeehin M (2002) Antimicrobial residues in animal waste and water resources proximal to large-scale swine and poultry feeding operations. Sci Total Environ 299:89–95

Chander Y, Kumar K, Goyal SM, Gupta SC (2005) Antibacterial activity of soil-bound antibiotics. J Environ Qual 34:1952–1957
Chander Y, Gupta SC, Goyal SM, Kumar K (2007) Perspective antibiotics: has the magic gone? J Sci Food Agric 87:739–742
Christian T, Schneider RJ, Färber HA, Skutlarek D, Meyer MT, Goldbach HE (2003) Determination of antibiotic residues in manure, soil, and surface waters. Acta Hydroch Hydrobiol 31:36–44
Combalbert S, Hernandez-Raquet G (2010) Occurrence, fate, and biodegradation of estrogens in sewage and manure. Appl Microbiol Biotechnol 86:1671–1692
Coyne R, Hiney M, O'Connor B, Kerry J, Cazabon D, Smith P (1994) Concentration and persistence of oxytetracycline in sediments under a marine salmon farm. Aquaculture 123:31–42
D'Ascenzo G, Di Corcia A, Gentili A, Mancini R, Mastropasqua R, Nazzari M, Samperi R (2003) Fate and natural estrogen conjugates in municipal sewage transport and treatment facilities. Sci Total Environ 302:199–209
Daston GP, Cook JC, Kavlock RJ (2003) Uncertainties for endocrine disrupters: our view on progress. Toxicol Sci 74:245–252
De Liguoro M, Cibin V, Capolongo F, Halling-Sørensen B, Montesissa C (2003) Use of oxytetracycline and tylosin in intensive calf farming: evaluation of transfer to manure and soil. Chemosphere 52:203–212
Den Besten PJ, Herwig HJ, Zandee DI, Voogt PA (1989) Effects of cadmium and PCBs on reproduction of the sea star *Asterias rubens*: aberrations in the early development. Ecotoxicol Environ Safe 18:173–180
Depledge MH, Billinghurst Z (1999) Ecological significance of endocrine disruption in marine invertebrates. Mar Pollut Bull 39:32–38
Deportes I, Benoit-Guyod JL, Zmirou D (1995) Hazard to man and the environment posed by the use of urban waste compost: a review. Sci Total Environ 172:197–222
Destatis (2011) Statistical yearbook 2011 for the Federal Republic of Germany, including international tables. Statistisches Bundesamt, Wiesbaden. ISBN 978-3-8246-0913-0. https://www.destatis.de
Dolliver H, Kumar K, Gupta S (2007) Sulfamethazine uptake by plants from manure-amended soil. J Environ Qual 36:1224–1230
Donoho AL (1984) Biochemical studies on the fate of monensin in animals and in the environment. J Anim Sci 58:1528–1539
Engels H (2004) Verhalten von ausgewählten Tetrazyklinen und Sulfonamiden in Wirtschaftsdünger und in Böden. Dissertation, Universität Göttingen, pp 1–139
Escher BI, Pronk W, Suter MJF, Maurer M (2006) Monitoring the removal efficiency of pharmaceuticals and hormones in different treatment processes of source-separated urine with bioassays. Environ Sci Technol 40:5095–5101
EU (European Union) (1996) Note for guidance: environmental risk assessment of veterinary medical products other than GMO-containing and immunological products. EMEA/CVMP/055/96. EMEA, London
Faust M, Altenburg R, Backhaus T, Blanck H, Boedeker W, Gramatica P, Hamer V, Scholze M, Vighi M, Grimme LH (2003) Joint algal toxicity of 16 dissimilarly acting chemicals is predictable by the concept of independent action. Aquat Toxicol 63:43–63
FEDESA (2001) Antibiotic use in farm animals does not threaten human health, a new study suggest. http://www.simv.org/communiqués/2001/Visby.pdf, 26 July 2012
Fent K (2000) Hormonaktive Stoffe in Gewässern: Auch eine Gefahr fürs Trinkwasser? Mitt Lebensm Hyg 91:11–25
Forth W, Henschler D, Rummel W, Starke K (1997) Allgemeine und spezielle Pharmakologie und Toxikologie. 7. Aufl. Spektrum, Akad. Verl
Fry DM, Toone CK (1981) DDT-induced feminization of gull embryos. Science 213:922–924
Gavalchin J, Katz SE (1994) The persistence of fecal-borne antibiotics in soil. J AOAC Int 77:481–485

Germap (2008) Antibiotika-Resistenz und Verbrauch. BVL Schriften. ISBN 978-3-00-025097-2
Geuns JMC (1978) Steroid-hormones and plant growth and development. Phytochemistry 17:1–14
Gilbertson TJ, Hornish RE, Jaglan PS, Koshy T, Nappier JL, Stahl GL, Cazers AR, Nappier JM, Kubicek MF, Hoffman GA, Hamlow PJ (1990) Environmental fate of Ceftiofur Sodium, a cephalosporin antibiotic. Role of animal excreta in its decomposition. J Agric Food Chem 38:890–894
Göbel A, Thomsen A, Mcardell CS, Joss A, Giger W (2005) Occurrence and sorption behavior of sulfonamides, macrolides, and trimethoprim in activated sludge treatment. Environ Sci Technol 39:3981–3989
Goberna N, Podmirseg SM, Waldhuber S, Knapp BA, Garcia C, Insam H (2011) Pathogenic bacteria and mineral N in soils following the land spreading of biogas digestates and fresh manure. Appl Soil Ecol 49:18–25
Grafe A (2001) Untersuchungen zum Einsatz pharmakologisch wirksamer Stoffe in der Veredelungswirtschaft unter besonderer Berücksichtigung der Tetracykline. Dissertation, Universität Göttingen, Cuvillier Verlag Göttingen, pp 1–157
Grote M, Vockel A, Schwarze D, Mehlich A, Freitag M (2004) Fate of antibiotics in food chain and environment originating from pigfattening (Part 1). Fresenius Environ Bull 13:1216–1224
Grote M, Schwake-Anduschus C, Stevens H, Michel R, Betsche T, Freitag M (2006) Antibiotika —Aufnahme von Nutzpflanzen aus Gülle-gedüngten Böden Ergebnisse eines Modellversuchs. J Verbr Lebensm 1:40–52
Gülden M, Turan A, Seibert H (1997) Substanzen mit endokriner Wirkung in Oberflächengewässern. UBA-Texte 1997, 46/97, Berlin, Umweltbundesamt
Güngör K, Karthikeyan KG (2008) Phosphorus forms and extractability in dairy manures: a case study for Wisconsin onfarm anaerobic digesters. Bioresour Technol 99:425–436
Gunn A, Sadd JW (1994) The effect of ivermectin on the survival, behavior and cocoon production of the earthworm *Eisenia fetida*. Pedobiologia 38:327–333
Haller MY, Müller SR, McArdell CS, Alder AC, Suter MJF (2002) Quantification of veterinary antibiotics (sulfonamides and trimethoprim) in animal manure by liquid chromatography-mass spectrometry. J Chromatogr A 952:111–120
Halling-Sørensen B (2000) Algal toxicity of antibacterial agents used in intensive farming. Chemosphere 40:731–739
Halling-Sørensen B, Nielsen SN, Lanzky PF, Ingerslev F, Lützhøft HCH, Jørgensen SE (1998) Occurrence, fate and effects of pharmaceutical substances in the environment- a review. Chemosphere 36:357–393
Halling-Sørensen B, Sengelov G, Tjörnelund J (2002) Toxicity of tetracyclines and tetracycline degradation products to environmentally relevant bacteria, including selected tetracycline-resistant bacteria. Arch Environ Contam Toxicol 42:263–271
Halling-Sørensen B, Jacobsen AM, Jensen J, Sengelov G, Vaclavik E, Ingerslev F (2005) Dissipation and effects of chlortetracycline and tylosin in two agricultural soils: a field-scale study in southern Denmark. Environ Toxicol Chem 24:802–810
Hammesfahr U, Bierl R, Thiele-Bruhn S (2011) Combined effects of the antibiotic sulfadiazine and liquid manure on the soil microbial-community structure and functions. J Plant Nutr Soil Sci 174:614–623
Hamscher G (2008) Review: Tierarzneimittel in der Umwelt unter besonderer Berücksichtigung von Stallstäuben. J Verbr Lebensm 3:165–173
Hamscher G, Mohring S (2012) Tierarzneimittel in Böden und in der aquatischen Umwelt. Chemie Ingenieur Technik 84:1052–1061
Hamscher G, Sczesny S, Höper H, Nau H (2002) Determination of persistent tetracycline residues in soil fertilized with liquid manure by high-performance liquid chromatography with electrospray ionization tandem mass spectrometry. Anal Chem 74:1509–1518
Hanselman TA, Graetz DA, Wilkie AC (2003) Manure-borne estrogens as potential environmental contaminants: a review. Environ Sci Technol 37:5471–5478
Hansen LH, Aarestrup F, Sørensen SJ (2002) Quantification of bioavailable chlortetracycline in pig feces using a bacterial whole-all biosensor. Vet Microbiol 87:51–57

Harrass MC, Kindig AC, Taub FB (1985) Responses of blue-green and green algae to streptomycin in unialgal and paired culture. Aquat Toxicol 6:1–11

Harris JE, Sheahan DA, Jobling S, Matthiessen P, Neall P, Sumpter JP, Tylor T, Zaman N (1997) Estrogenic activity in five United Kingdom rivers detected by measurement of vitellogenesis in caged male trout. Environ Toxicol Chem 16:534–542

Hartlieb N, Ertunc T, Schaeffer A, Klein W (2003) Mineralization, metabolism and formation of non-extractable residues of ^{14}C-labelled organic contaminants during pilot-scale composting of municipal biowaste. Environ Pollut 126:83–91

Haunhorst E (2012) Antibiotikaeinsatz in der Tierproduktion—ein Risiko? 5. Niedersächsisches Forum Gesundheitlicher Verbraucherschutz, Hannover. http://www.laves.niedersachsen.de

Hembrock-Heger A (2007) Eintrag von Arzneimitteln und deren Verhalten und Verbleib in der Umwelt—Literaturstudie, LANUV-Fachbericht 2, Landesamt für Natur, Umwelt und Verbraucherschutz Nordrhein-Westfalen, Recklinghausen

Hilpert R, Winter J, Kandler O (1984) Agricultural feed additives and disinfectants as inhibitory factors in anaerobic digestion. Agric Wastes 10:103–116

Hirsch R, Ternes T, Haberer K, Kratz KL (1999) Occurrence of antibiotics in the aquatic environment. Sci Total Environ 225:109–118

Hoffmann F, Kloas W (2010) An environmentally relevant endocrine-disrupting antiandrogen, vinclozolin, affects calling behavior of male *Xenopus laevis*. Hormon Behav 58:653–659

Holm-Nielsen JB, Al Seadi T, Oleskowicz-Popiel P (2009) The future of anaerobic digestion and biogas utilization. Bioresour Technol 100:5478–5484

Höper H, Kues J, Nau H, Hamscher G (2002) Eintrag und Verbleib von Tierarzneimittelwirkstoffen in Böden. Bodenschutz 4:141–148

Huysman F, van Renterghem B, Verstraete (1993) Antibiotic resistant sulphite-reducing clostridia in soil and groundwater as indicator of manuring practices. Water Air Soil Pollut 69:243–255

Jagnow G (1977) Mikrobieller Abbau der Futtermittelantibiotika Zink-Bacitracin, Flavophospholipol, Spiramycin und von Tetracyclin in feucht gelagertem und in mit Boden vermischtem Hühnerkot. Landwirtsch Forsch Kongreßband, Teil 1:227–234

Jjemba PK (2002) The potential impact of veterinary and human therapeutic agents in manure and biosolids on plant growth on arable land. A review. Agric Ecosyst Environ 93:267–278

Jobling S, Casey D, Rodgers-Gray T, Oehlmann J, Schulte-Oehlmann U, Pawlowski S, Baunbeck T, Turner AP, Tyler CR (2004) Comparative responses of mollusks and fish to environmental estrogens and an estrogenic effluent. Aquat Toxicol 66:207–222

Johnson AC, Williams RJ, Matthiessen P (2006) The potential steroid hormone contribution of farm animals to freshwaters, the United Kingdom as a case study. Sci Total Environ 362:166–178

Jørgensen SE, Halling-Sørensen B (2000) Drugs in the environment. Chemosphere 40:691–699

Kalbfus W (1998) Exposition und Wirkung endokriner Substanzen im aquatischen System. Wiener Mitteilungen 153:33–44

Kan CA, Keukens HJ, Boers E (1996) Nicarbazin residues in broiler muscle, liver and litter/faeces: early exposure or recirculation. In: EuroResidue III, conference on residue of veterinary drugs in food, Veldhoven, The Netherlands, pp 591–594, 6–8 May 1996

Kay P, Blackwell PA, Boxall ABA (2004) Fate of veterinary antibiotics in a macroporous tile drained clay soil. Environ Toxicol Chem 23:1136–1144

Kay P, Blackwell PA, Boxall ABA (2005) Transport of veterinary antibiotics in overland flow following the application of slurry to arable land. Chemosphere 59:951–959

Kennedy DG, Cannavan A, McCracken RJ (2000) Regulatory problems caused by contamination, a frequently overlooked cause of veterinary drug residues. J Chromatogr A 882:37–52

Khanal SK, Xie B, Thompson ML, Sung S, Ong SK, Van Leeuwen JH (2006) Fate, transport, and biodegradation of natural estrogens in the environment and engineered systems. Environ Sci Technol 40:6537–6546

Kolpin DW, Furlong ET, Meyer MT, Thurman EM, Zaugg SD, Barber LB, Buxton HT (2002) Pharmaceuticals, hormones, and other organic wastewater contaminants in US streams 1999–2000: a national reconnaissance. Environ Sci Technol 36:1202–1211

KTBL (2009) Faustzahlen für die Landwirtschaft (14. Aufl.). ISBN 978-3-939371-91-5

Kues J, Höper H, Sczesny S, Nau H, Hamscher G (2004a) Gehalte an Tierarzneimitteln in Wirtschaftsdüngern, Eintrag in Böden und Abbauverhalten. In: Landwirtschaftliche Verwertung von Klärschlamm, Gülle und anderen Düngern unter Berücksichtigung des Umwelt—und Verbraucherschutzes. BMU/BMVEL Wissenschaftliche Anhörung, 25.–26. Oktober 2001, Bonn, KTBL-Schrift 404, Darmstadt, pp 317–322

Kues J, Höper H, Pawelzick HT, Pluquet E, Hamscher G (2004b) Einträge potenzieller Schadstoffe in den Boden über Wirtschaftsdünger. Wirkung auf Bodenorganismen und Verlagerung. In: Ministerium für Umwelt und Naturschutz, Landwirtschaft und Verbraucherschutz des Landes Nordrhein-Westfalen, Referat IV. 6 Bodenschutz und Bundesverband Boden (Hrsg.) Bodenschutz—europäisch und lokal: Publikationen zur gemeinsamen Fachtagung, Soil protection: publication by the joint specialist conference (Umwelt); Berlin, 01.04.2004; Düsseldorf, S. 55–62

Kühne M, Ihnen D, Möller G, Agthe O (2000) Stability of tetracycline in water and liquid manure. J Vet Med A 47:379–384

Kumar K, Thompson A, Singh AK, Chander Y, Gupta SC (2004) Enzyme-linked immunosorbent assay for ultratrace determination of antibiotics in aqueous samples. J Environ Qual 33:250–256

Kumar K, Gupta SC, Chander Y, Singh AK (2005a) Antibiotic use in agriculture and its impact on the terrestrial environment. Adv Agron 87:1–54

Kumar K, Gupta SC, Baidoo SK, Chander Y, Rosen CJ (2005b) Antibiotic uptake by plants from soil fertilized with animal manure. J Environ Qual 34:2082–2085

Kümmerer K (2001a) Drugs in the environment: emission of drugs, diagnostic aids and disinfectants into wastewater by hospitals in relation to other sources—a review. Chemosphere 45:957–969

Kümmerer K (2001b) Pharmaceuticals in the environment. Sources, fate, effects and risks. Springer, Heidelberg

Kupper T, Brandli RC, Bucheli TD, Stampfli C, Zennegg M, Berger U, Edder P, Pohl M, Niang F, Iozza S, Muller J, Schaffner C, Schmid P, Huber S, Ortelli D, Becker-van Slooten K, Mayer J, Bachmann HJ, Stadelmann FX, Tarradellas J (2008) Organic pollutants in compost and digestate: occurrence, fat sustainability, benefits, impacts for the environment and for plant production. In: Proceedings of the international Congress CODIS 2008, Solothurn, Switzerland, pp 27–34, 27–29 Feb 2008

Lange IG, Daxenberger A, Schiffer B, Witters H, Ibarreta D, Meyer HHD (2002) Sex hormones originating from different livestock production systems: fate and potential disrupting activity in the environment. Anal Chim Acta 473:27–37

Langhammer JP, Büning-Pfaue H, Winkelmann J, Körner E (1988) Chemotherapeutika—Rückstände und Resistenzverhalten bei der Bestandsbehandlung von Sauen post partum. Tierärztl Umschau 43:375–382

Lansky P, Halling-Sørensen B (1997) The toxic effect of the antibiotic metronidazol on aquatic organisms. Chemosphere 35:2553–2561

Larsen TA, Peters I, Alder A, Eggen RIL, Maurer M, Muncke J (2001) Re-engineering the toilet for sustainable waste water management. Environ Sci Technol 35:192A–197A

Larsen TA, Lienert J, Joss A, Siegrist H (2004) How to avoid pharmaceuticals in the aquatic environment. J Biotechnol 113:295–304

Limpiyakorn T, Homklin S, Ong SK (2009) Hormones In: Bhandari A, Surampalli RY, Adams CD, Champagne P, Ong SK, Tyagi RD, Zhang TC (eds) Contaminants of emerging environmental concern. ASCE American Society of Civil Engineers, Reston, Virginia, pp 186–234

Liu S, Ying GG, Zhou LJ, Zhang RQ, Chen ZF, Lai HJ (2012) Steroids in a typical swine farm and their release into the environment. Water Res 46:3754–3768

Lorenzen A, Hendel JG, Conn KL, Bittman S, Kwabiah AB, Lazarovitz G, Masse D, McAllister TA, Topp E (2004) Survey of hormone activities in municipal biosolids and animal manures. Environ Toxicol 19:216–225

Lützhøft HCH, Halling-Sørensen B, Jørgensen SE (1999) Algal toxicity of antibacterial agents applied in Danish fish farming. Arch Environ Contam Toxicol 36:1–6

Macri A, Staza AV, Delupis GD (1988) Acute toxicity of furazolidone on *Artemia salina*, *Daphnia magna* and *Culex pipiens molestus* larvae. Ecotoxicol Environ Saf 8:101–105

McCracken DI, Foster GN (1993) The effect of ivermectin on the invertebrate fauna associated with cow dung. Environ Toxicol Chem 12:73–84

Migliore L, Brambilla G, Cozzolino S, Gaudio L (1995) Effect on plants of sulphadimethoxine used in intensive farming (*Panicum miliaceum*, *Pisum sativum* and *Zea mays*). Agric Ecosyst Environ 52:103–110

Migliore L, Civitareale C, Brambilla G, Delupis GD (1997) Toxicity of several important agricultural antibiotics to artemia. Water Res 31:1801–1806

Möller K, Müller T (2012) Effects of anaerobic digestion on digestate nutrient availability and crop growth: a review. Eng Life Sci 12:242–257

Montforts MHMM (2006) Validation of the exposure assessment for veterinary medicinal products. Sci Total Environ 358:121–136

Muller M, Rabenoelina F, Balaguer P, Patureau D, Lemenach K, Budzinski H, Barcelo D, De Alda ML, Kuster M, Delgenes JP, Hernandez-Raquet G (2008) Chemical and biological analysis of endocrine-disrupting hormones and estrogenic activity in an advanced sewage treatment plant. Environ Toxicol Chem 27:1649–1658

Muller M, Combalbert S, Delgenes N, Bergheaud V, Rocher V, Benoit P, Delgenes JP, Patureau D, Hernandez-Raquet G (2010) Occurrence of estrogens in sewage sludge and their fate during plant-scale anaerobic digestion. Chemosphere 81:65–71

Murkies AL, Wilcox G, Davis SR (1998) Clinical review 92—phytoestrogens. J Clin Endocrinol Metabol 83:297–303

Nichols DJ, Daniel TC, Moore PA, Edwards DR, Pote DH (1997) Runoff of estrogens hormone 17ß-estradiol from poultry litter applied to pasture. J Environ Qual 26:1002–1006

Oehlmann J, Schulte-Oehlmann U, Tillmann M, Markert B (2000) Effects of endocrine disruptors on prosobranch snails (*Mollusca gastropoda*) in the laboratory. Part 1: bisphenol-A and octylphenol as xeno-estrogens. Ecotoxicology 9:383–397

Palme R, Fischer P, Schildorfer H, Ismail MN (1996) Excretion of infused ^{14}C-steroid hormones via faeces and urine in domestic livestock. Anim Reprod Sci 43:43–63

Patten DK, Wolf DC, Kunkle WE, Douglass LW (1980) Effect of antibiotics in beef cattle feces on nitrogen and carbon mineralization in soil and on plant growth and composition. J Environ Qual 9:167–172

Pfeifer T, Tuerk J, Bester K, Spiteller M (2002) Determination of selected sulfonamide antibiotics and trimethoprim in manure by electrospray and atmospheric pressure chemical ionization tandem mass spectrometry. Rapid Commun Mass Spectrom 16:663 669

Prinzhorn P (2005) Phosphordünger aus Klärschlammaschen mit thermischer Schwermetallentfrachtung. In: WAR (Hrsg.) Rückgewinnung von Phosphor aus Abwasser und Klärschlamm. 75. Darmstädter Seminar—Abwassertechnik—am 12./13. Nov 2005 in Darmstadt. Schriftenreihe WAR 167:239–247

Purdom CE, Hardiman PA, Bye VJ, Eno NC, Tyler CR, Sumpter JP (1994) Estrogenic effects of effluent from sewage treatment works. Chem Ecol 8:275–285

Qiao M, Chen W, Su J, Zhang B, Zhang C (2012) Fate of tetracyclines in swine manure of three selected swine farms in China. J Environ Sci 24:1047–1052

Rabolle M, Spliid NH (2000) Sorption and mobility of metronidazole, olaquindox, oxytetracycline and tylosin in soil. Chemosphere 40:715–720

Richter A, Löscher W, Witte W, (1996) Leistungsförderer mit antibakterieller Wirkung: Probleme aus pharmakologischer, toxikologischer und mikrobiologischer Sicht. Prakt Tierarzt 7:603–625

Römbke J, Knacker T, Stahlschmidt-Allner P (1996) Umweltprobleme durch Arzneimittel. UBA-Texte 60/96, Berlin

Ronteltap M, Maurer M, Gujer W (2007) The behavior of pharmaceuticals and heavy metals during struvite precipitation in urine. Water Res 41:1859–1868

Sarmah AK, Meyer MT, Boxall ABA (2006) A global perspective on the use, sales, exposure pathways, occurrence, fate and effects of veterinary antibiotics (VAs) in the environment. Chemosphere 65:725–759
Schlenker G, Muller W, Glatzel P (1998) Analysis of the stability of sexual steroids in feces of cows over 12 weeks. Berl Muench Tieraerztl Wochenschr 111:248–252
Schleusner Y, Müller J, Goßmann M, Rodemann B, Pietsch M, Heiermann M, Plöchl M, Bandte M, Büttner C (2010) Phytosanitäre Risiken durch die Verwendung von infiziertem Pflanzenmaterial zur Vergärung in Biogasanlagen. Julius-Kühn-Archiv 428:289
Schmidt CW (2002) Antibiotic resistance in livestock: more a stake than steak. Environ Health Perspect 110:A396–A402
Schönauer S (2001) Hormonaktive Substanzen im Wasser—Gefahr für Gewässer und Mensch. BUND Hintergrund, BUND Arbeitskreis Wasser
Schönauer S, Grommelt HJ (2007) Hormonaktive Substanzen und Arzneimittel—„Zusammenhänge, politische Bewertung, Handlungsstrategien". Beitrag zur 7. Umweltmedizinischen Tagung Berlin (5.–6. Oktober 2007)
Schwabe U, Paffrath D (2001) Arzneiverordnungs-Report 2001. Springer, Berlin, p 982
Seel P (1998) Arzneimittel in Gewässern—neue Umweltchemikalien In: Tagungsband: Arzneimittel in Gewässern—Risiko für Mensch, Tier und Umwelt? Hessisches Landesamt für Umwelt 4.06.1998 S. 3
Sengeløv G, Agersø Y, Halling-Sørensen B, Baloda SB, Andersen JS, Jensen LB (2003) Bacterial antibiotic resistance levels in Danish farmland as a result of treatment with pig manure slurry. Environ Int 28:587–595
Shafrir M, Avisar D (2012) Development method for extracting and analyzing antibiotic and hormone residues from treated wastewater sludge and composted biosolids. Water Air Soil Pollut 223:2571–2587
Sharpe RM, Irvine DS (2004) How strong is the evidence of a link between environmental chemicals and adverse effects on human reproductive health? Br Med J 328:447–451
Shore LS, Shemesh M (2003) Naturally produced steroid hormones and their release into the environment. Pure Appl Chem 75:1859–1871
Shore LS, Shemesh M, Cohen R (1988) The role of oestradiol and oestrone in chicken manure silage in hyperoestrogenism in cattle. Aust Vet J 65:68
Shore LS, Kapulnik Y, Ben-Dor B, Fridman Y, Wininger S, Shemesh M (1992) Effects of estrone and 17 β-estradiol on vegetative growth of *Medicago sativa*. Physiol Plant 84:217–222
Shore LS, Gurevitz M, Shemesh M (1993) Estrogen as an environmental pollutant. Bull Environ Contam Toxicol 51:361–366
Silva E, Rajapakse N, Kortenkamp A (2002) Something from "nothing"—eight weak estrogenic chemicals combined at concentrations below NOECs produce significant mixture effects. Environ Sci Technol 36:1751–1756
Smith SR (1996) Agricultural recycling of sewage sludge and the environment. CAB International, Wallingford
Sumpter JP (1998) Reproductive effects from oestrogen activity in polluted water. Arch Toxicol Suppl 20:143–150
Teglia C, Tremier A, Martel JL (2011) Characterization of solid digestates: Part 2, assessment of the quality and suitability for composting of six digested products. Waste Biomass Valor 2:113–126
Ternes TA (1998) Occurrence of drugs in German sewage treatment plants and rivers. Water Res 32:3245–3260
Ternes TA, Andersen H, Gilberg D, Bonerz M (2002) Determination of estrogens in sludge and sediments by liquid extraction and GC/MS/MS. Anal Chem 74:3498–3504
Thiele-Bruhn S (2003) Pharmaceutical antibiotic compounds in soils—a review. J Plant Nutr Soil Sci 166:145–167
Varel VH, Wells JE, Shelver WL, Rice CP, Armstrong DL, Parker DB (2012) Effect of anaerobic digestion temperature on odour, coliforms and chlortetracycline in swine manure or monensin in cattle manure. J Appl Microbiol 112:705–715

Vidaver AK (2002) Uses of antimicrobials in plant agriculture. Clin Infect Dis 34:S107–S110

Wall R, Strong L (1987) Environmental consequences of treating cattle with the antiparasitic drug ivermectin. Nature 327:418–421

Warman PR, Thomas RL (1981) Chlortetracycline in soil amended with poultry manure. Can J Soil Sci 61:161–163

Watts MM, Pascoe D, Carroll K (2002) Population response of the freshwater amphipod *Gammarus pulex* (L.) to an environmental estrogen, 17 α-ethinylestradiol. Environ Toxicol Chem 21:445–450

Wegener HC, Aarestrup FM, Jensen LB, Hammerum AM, Bager F (1998) The association between the use of antimicrobial growth promoters and development of resistance in pathogenic bacteria toward growth promoting and therapeutic antimicrobials. J Anim Feed Sci 7:7–14

Wenzel A, Küchler T, Henschel K-P, Schnaak W, Diedrich M, Mueller J (1998) Konzentrationen östrogen wirkender Substanzen in Umweltmedien. Report, Project sponsored by the German Environmental Protection Agency, (UBA) Project No 216 02 011/11

Winckler C, Grafe A (2000) Charakterisierung und Verwertung von Abfällen aus der Massentierhaltung unter Berücksichtigung verschiedener Böden. Texte des Umweltbundesamtes 44/00, ISSN 0722-186X

Winterhalder K (1985) Untersuchungen über den Einfluß von Desinfektionsmitteln, Futterzusatzstoffen und Antibiotika auf die Biogasgewinnung aus Schweinegülle. Dissertation, Universität Hohenheim

Wollenberger L, Halling-Sørensen B, Kusk KO (2000) Acute and chronic toxicity of veterinary antibiotics to *Daphnia magna*. Chemosphere 40:723–730

Yang S, Carlson K (2004) Routine monitoring of antibiotics in water and wastewater with a radioimmunoassay technique. Water Res 38:3155–3166

Yang X, Song Y, Kai J, Cao X (2012) Enzymatic biomarkers of earthworms *Eisenia fetida* in response to individual and combined cadmium and pyrene. Ecotoxicol Environ Saf 86:162–167

Ying GG, Kookana RS (2003) Degradation of five selected endocrine-disrupting chemicals in seawater and marine sediment. Environ Sci Technol 37:1256–1260

Ying GG, Kookana RS (2005) Sorption and degradation of estrogen-like-endocrine disrupting chemicals in soil. Environ Toxicol Chem 24:2640–2645

Young WF, Whitehouse P, Johnson I, Sorokin N, WRc-NSF (2002) Proposed predicted-no-effect-concentrations (PNECs) for natural and synthetic steroid oestrogens in surface waters. England and Wales Environment Agency Bristol, Environment Agency R&D Technical Report P2-T04/1, p 172

Zhao JL, Ying GG, Chen F, Liu YS, Wang L, Yang B, Liu S, Tao R (2010) Estrogenic activity profiles and risks in surface waters and sediments of the Pearl River system in South China assessed by chemical analysis and in vitro bioassay. J Environ Monit 13:813–821

Zullei-Seibert N, Skark C (2002) Gehalte an Pharmaka, Hormonen und endokrin wirksamen Substanzen in Klärschlamm, Kompost und Wirtschaftsdüngern. In: Landwirtschaftliche Verwertung von Klärschlamm, Gülle und anderen Düngern unter Berücksichtigung des Umwelt- und Verbraucherschutzes. BMU/BMVEL Wissenschaftliche Anhörung, 25.–26. Oktober 2001, Bonn, KTBL-Schrift 404, Darmstadt, pp 105–116

Energy Neutral Phosphate Fertilizer Production Using High Temperature Reactors

Nils Haneklaus and Ewald Schnug

Abstract Increasing global phosphorus (P) needs and energy demands makes an efficient utilization of finite resources such as phosphate rock (PR) obligatory. The anticipated growth in nuclear energy may require the use of uranium (U) from unconventional resources such as PR as future raw material for nuclear reactor fuel. In parallel, increasing amounts of energy will be needed to process lower grade PR resources in order to satisfy rising phosphate fertilizer requirements for global food production. This work assesses the possibility of using the new generation of high temperature reactors (HTRs) for future energy neutral phosphate fertilizer production. Although advantages from an environmental and resource conserving point of view are obvious, challenges associated with the technical implementation and economic feasibility need to be addressed before implementation.

Keywords Beneficiation · Elemental phosphorus · High temperature gas-cooled reactor (HTGR) · Phosphate rock · Yellow cake

Introduction

U is an accompanying element in PR. Worldwide U concentrations in PR range from 25 to 50 mg/kg with local deposits showing concentrations as high as 600 mg/kg (IAEA 2009). US PR shows mean U concentrations of 20–300 mg/kg (EPA 2012). The International Atomic Energy Agency (IAEA) recognizes ores with minimum concentrations of 300 mg/kg as U resources (IAEA 2009). U in PR can be extracted during phosphate fertilizer production using established

N. Haneklaus (✉)
Department of Nuclear Engineering, University of California, Berkeley,
4118 Etcheverry Hall, MC 1730, Berkeley, CA 94720-1730, USA
e-mail: n.haneklaus@berkeley.edu

E. Schnug
Faculty 2 Life Sciences, Technical University Braunschweig,
Pockelsstraße 14, 38106 Brunswick, Germany

E. Schnug and L.J. De Kok (eds.), *Phosphorus in Agriculture: 100 % Zero*,
DOI 10.1007/978-94-017-7612-7_14

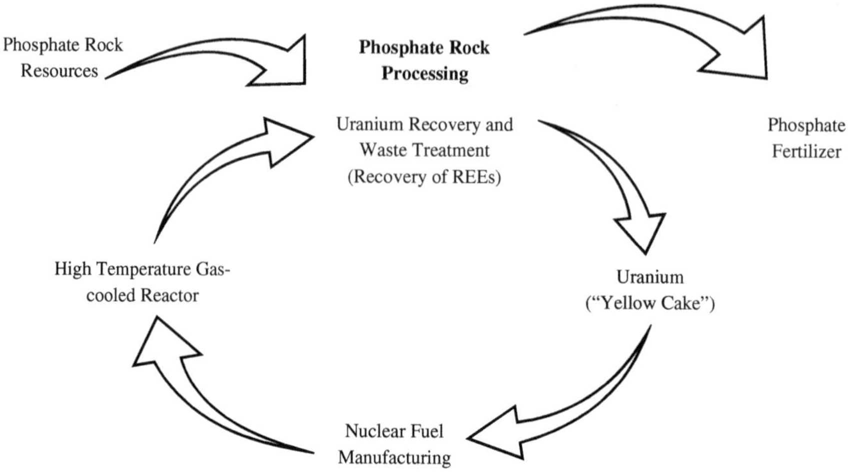

Fig. 1 Overview of the proposed processing of PR (modified after Haneklaus et al. 2015a)

technologies such as solvent extraction. After enrichment the extracted U can be used as raw material for nuclear reactor fuel. If used to fuel HTRs heat and/or electric power is generated that can be used to process PR and to extract U. In addition, surplus energy is available for additional steps during fertilizer production such as ammonia production, which is an energy intensive process within fertilizer manufacturing, or the extraction of rare earth elements (REEs) from presently unprocessed byproducts (phosphogypsum). Technical innovations will depend in future highly on a sufficient and reliable supply with REEs (Erdmann and Graedel 2011). The price for lanthanides increased from 2002 to 2011 about 60-fold for La, Ce, Nd and Pr (Anonymous 2011). The risk of supply disruption on a short to medium term basis has been addressed by several authors (De Boer and Lammertsma 2012; Erdmann and Graedel 2011) so that beneficiation of PR will provide a high-priced commodity. The idea to use HTRs as a greenhouse gas lean energy source for energy neutral phosphate fertilizer processing has first been proposed by Haneklaus et al. (2015a). Conservative estimates indicate that U concentrations as high as 80 mg/kg U for wet acid and 110 mg/kg U for thermal PR processing are sufficient for energy neutral phosphate fertilizer production. Figure 1 provides a brief overview of the proposed process (Haneklaus et al. 2015a).

The raw material for nuclear fuel manufacturing (U_3O_8/"Yellow Cake") will most likely be accumulated on site. The raw material can be sold and/or send to an enrichment plant for further processing the same way a U mine sells its product. Mines that extract and sell U in addition to other primary ores are presently operated successfully at different places worldwide (e.g. Olympic Dam Copper/U mine in Australia; Kable 2014). Finally, the enriched material could be sent to a reactor fuel manufacturer who produces the HTR (or other) nuclear reactor fuel from the recovered U.

High Temperature Gas-Cooled Reactors (HTGRs)

High temperature gas-cooled reactors (HTGRs) are a representative group of HTRs. Several HTGR designs with pebble bed or block type reactor core, using U and/or thorium (Th) as nuclear reactor fuel are available today. Numerous research reactors as well as several demonstration plants have been operated worldwide: UK (Dragon 1973–1976), USA (Peach Bottom 1967–74, Fort St. Vrain 1976–89), Germany (AVR 1976–88, THTR 1986–89), Japan (HTTR since 1998), China (HTR-10 since 2000 and HTR-PM presently under construction). Other HTR concepts such as the Fluoride-salt-cooled High-temperature Reactor (FHR), the Compact High Temperature Reactor (CHTR) or several fast reactor designs that reach equally high outlet temperatures (at minimum 600 °C) are presently under development and may be recognized suitable for PR processing in the near future.

The design of HTGRs is different from present Light Water Reactors (LWRs) and Gas-cooled Reactors (GCRs) resulting in different possible applications. HTGRs are thermal reactors that use graphite as a moderator and helium (He) as an inert coolant. To generate electricity HTGRs can be operated in a direct or indirect cycle. In the direct cycle heated He from the core is forthwith fed to a He turbine (Brayton Cycle). In the indirect cycle which has been used in previous HTGRs and which is used in current HTGR projects such as the HTR-PM in China heated He from the core is employed in a primary cycle (primary loop) in order to convert water to steam in a secondary cycle (secondary loop). The steam is fed to a commercial steam turbine (Rankine Cycle) to generate electricity. The HTGR fuel consists of small-coated particles that are about 1 mm in diameter (IAEA 2012). A few thousand of these fuel particles are compressed with graphite to either compacts which are arranged in prismatic blocks forming a block type reactor core or pebbles resulting in a loose pebble bed reactor core.

Using these designs HTGRs can reach core outlet temperatures of up to 1000 °C (GIF 2014). Figure 2 shows a simplified schematic cross section of an exemplary HTGR core. Relatively cold He (approximately 250 °C) passes the core through the graphite side reflector entering it at the top. Helium flows through the core reaching a core outlet temperature of up to 1000 °C. The core can either be a pebble bed (left side) or a block type (right side). For simplification the cross section shows both core types together (right: pebble, left: block type). Heated He leaving the core is cooled and cleaned from impurities such as graphite dust before being used again.

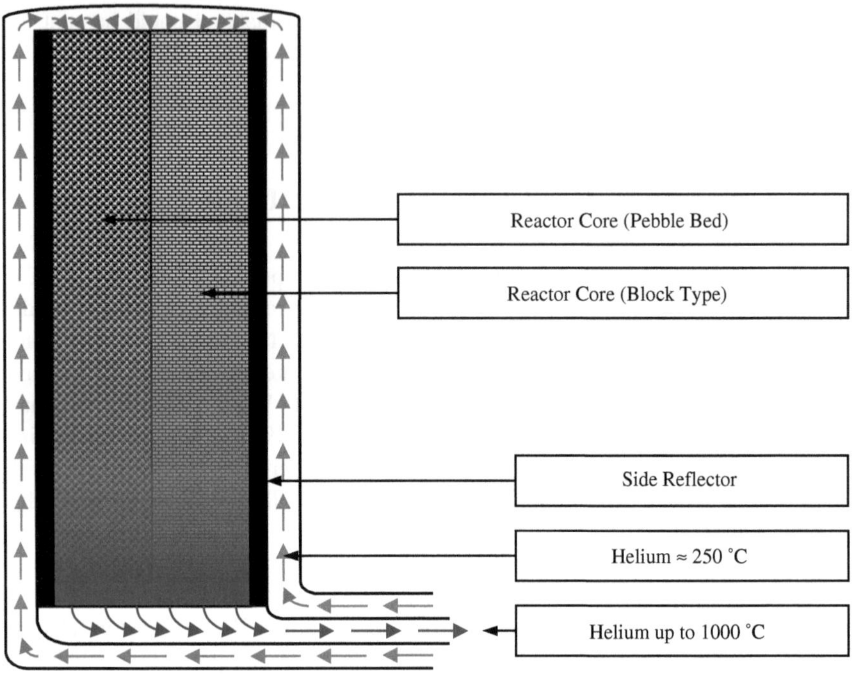

Fig. 2 Simplified exemplary cross section of a high temperature gas-cooled reactor (modified after Haneklaus et al. 2015a)

The possibility to realize temperatures as high as 1000 °C makes HTGRs unlike current LWRs feasible for high temperature (>600 °C) heat applications beyond electricity production (e.g. steam generation, steam methane reforming, hydrogen production) (GIF 2014). Using HTGRs for nuclear heat applications makes power conversion obsolete so that higher efficiency factors of the entire plant may be reached. In addition to high temperature heat applications, considerably high efficiency of electricity generation (HTGR 46 % with gas turbine, LWR 33 % with steam turbine) (Takei et al. 2006) can be achieved. Furthermore, HTGRs have desirable safety characteristics ("walk away safety") if designed as a modular reactor with relatively low power density (e.g. AVR power density 2.6 MW_{th}/m^3 compared to 92 MW_{th}/m^3 pressurized water reactor) (Voss 2010).

The convincing safety features may favor the use of HTGRs next to industrial facilities in increasingly critical societies. The modular design offers flexibility in siting and addresses deployment needs for smaller grids while allowing factory fabrication and short construction times.

Energy Neutral Phosphate Fertilizer Processing Employing High Temperature Reactors

Assuming a worldwide increasing demand for PR (USGS 2014), it will become necessary to process low-grade resources which show not only a lower P content and a higher concentration of valuable impurities such as U, Th and REEs, but also undesired gangue materials such as cadmium. The quality of the fertilizer products needs to comply with international environmental standards. Currently an upper limit value of 50 mg Cd per kg P_2O_5 (if the P_2O_5 content of the fertilizer product is >5 exists in Germany). Schnug and Haneklaus (2015) suggest an upper limit of 50 mg per kg P_2O_5 for U, too. At present the technology to process such ores is considered as being too costly because of extraordinarily high energy demands. Currently none of the previously mentioned heavy metals is extracted so that significant amounts of U are dispersed on soils by P fertilization (Schnug and Haneklaus 2015). On arable soils U is mobile and prone to leaching into water bodies. In intensive agricultural areas Kratz et al. (2006) and Knolle et al. (2008) observed elevated levels of U in surface and tap water.

The amount of energy needed for processing a certain type of phosphate containing ore is closely related to the P concentration in the ore (Gupta 2003). Ultimately, the grade of the ore determines the employed processing technique, which is associated with defined energy requirements. Alternatively, accompanying Th may also be considered as a raw material for nuclear reactor fuel.

It is not possible to process lower grade PR by the currently predominant wet acid processing techniques so that these methods require additional energy intensive beneficiation (e.g. calcination) if lower grade PR is processed. An alternative is energy intensive thermal processing currently used to produce elemental P instead of phosphoric acid (Haneklaus et al. 2014b). Elemental P can relatively easy be processed to high-grade phosphoric acid. In both cases HTGRs could provide the required electricity or process heat for energy intensive beneficiation prior to wet acid processing or direct thermal processing. Figure 3 briefly illustrates the proposed energy neutral cycle for both PR processing alternatives. It is unusual that one facility uses both wet acid and thermal processing in parallel. However, for simplicity both routes are shown in one diagram here (Haneklaus et al. 2015a).

The overview of a self-sustaining HTGR which uses U from PR to provide the nuclear fuel demonstrates the advantages of this innovative concept which comprise safe technology, greenhouse gas lean energy production, beneficiation of hazardous wastes (phosphogypsum), extraction of finite resources (U, REEs), marketing of high-grade P fertilizers, prevention of dispersion of U in agro-ecosystems, particularly water bodies thus minimizing health risks by intake of fertilizer-derived U from drinking water.

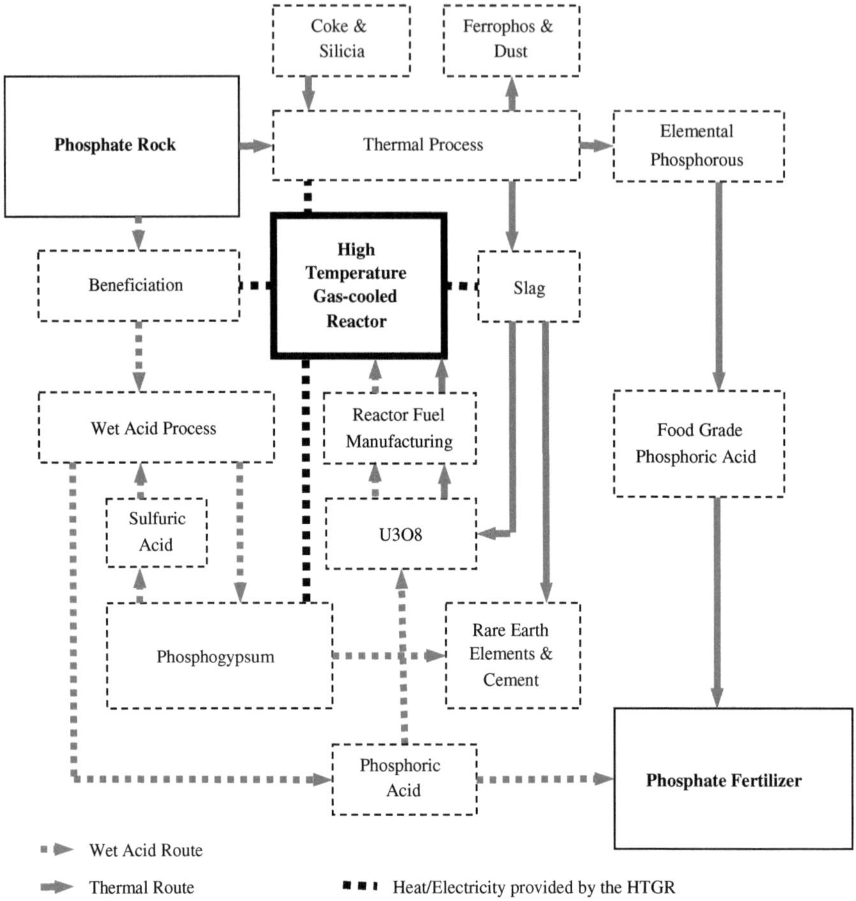

Fig. 3 Intermediate and end products in the course of energy neutral PR processing featuring wet acid and thermal processing (modified after Haneklaus et al. 2015a)

Discussion

The technical and economic feasibility of the proposed energy neutral process will be part of a coordinated research project proposed by IAEA in 2015 (Haneklaus et al. 2014a, 2016). A first case study (Haneklaus et al. 2015b) that considers U concentrations in phosphoric acid from a wet acid phosphate fertilizer plant currently operated in the Philippines and reactor designs developed by the Japan Atomic Energy Agency (JAEA) provides a first glance on the technical and economic challenges associated with the concept discussed here. It was found that the quantity of U extracted during fertilizer processing (40.48 t natural U/year) is (after enrichment and fuel manufacture) sufficient to operate three HTR50S units that require the equivalent of 23.71 t natural U/year and provide the electricity needed for PR processing and U extraction. Fuel costs account only for a relatively small portion (<10 %) of the costs associated with HTRs. In addition to costs saved for natural U higher prices for higher quality fertilizer products or selling extracted REEs from presently unprocessed byproduct phosphogypsum may be an economic incentive. Phosphogypsum samples analyzed from the respective plant in the Philippines suggest that REEs worth up to 27.5 million USD (assuming a recovery rate of 50 %) could be extracted annually (Haneklaus et al. 2015b).

Conclusions

Various environmental benefits such as reduced greenhouse gas emissions during fertilizer processing as well as higher quality fertilizers can be associated with the outlined idea. Although desirable in many ways, further investigations with view to the technical implementation and the economic feasibility are needed to evaluate the overall viability of the proposed conception.

References

Anonymous (2011) Rare earth elements. British Geological Survey. www.mineralsUK.com. Accessed 08 Aug 2013

De Boer MA, Lammertsma K (2012) Scarcity of rare earth elements. Royal Netherlands Chemical Society (KNCV), pp 1–25

EPA (2012) Fertilizer and fertilizer production waste. http://www.epa.gov/radiation/tenorm/fertilizer.html. Accessed 23 Oct 2014

Erdmann L, Graedel TE (2011) Criticality of non-fuel minerals: a review of major approaches and analyses. Environ Sci Technol 45:7620–7630

GIF (2014) Generation IV International Forum—VHTR. https://www.gen-4.org/gif/jcms/c_9362/vhtr. Accessed 23 Oct 2014

Gupta CK (2003) Chemical metallurgy—principles and practice. Wiley, Weinheim, p 741f

Haneklaus N, Reitsma F, Tulsidas H, Tyobeka B, Schnug E, Allelein HJ, Birky BK, Peterson PF, Dyck G, Koshy T (2014a) Using High Temperature Reactors for Energy Neutral Mineral Development Processes—a proposed IAEA Coordinated Research Project. In: Uranium Raw Material for the Nuclear Fuel Cycle (URAM 2014), Vienna, Austria, June 2014

Haneklaus H, Schnug E, Tulsidas H, Reitsma F (2014b) Using high temperature gas-cooled reactors for low grade phosphate rock processing. In: Sustainable Phosphorous Summit (SPS 2014), Montpellier, France, September 2014

Haneklaus N, Schnug E, Tulsidas H, Tyobeka B (2015a) Using high temperature gas-cooled reactors for greenhouse gas reduction and energy neutral production of phosphate fertilizers. Ann Nucl Energy 75:275–282

Haneklaus N, Reyes R, Lim WG, Tabora EU, Palattao BL, Petrache CP, Vargas EP, Kunitomi K, Ohashi H, Sakaba N, Sato H, Goto M, Yan X, Nishihara T, Tulsidas H, Reitsma F, Tarjan S, Sathrugnan K, Jacimovic R, Al Khaledi N, Birky BK, Schnug E (2015b) Energy neutral phosphate fertilizer production using high temperature reactors: a Philippine case study. Philippine J Sci 144(2):69–79

Haneklaus N, Reitsma F, Tulsidas H (2016) High Temperature Reactors for a new IAEA Coordinated Research Project on energy neutral mineral development processes. Nucl Eng Des. http://dx.doi.org/10.1016/j.nucengdes.2016.01.013

IAEA (2009) World distribution of uranium deposits (UDEPO) with uranium deposit classification. IAEA-TECDOC-1629

IAEA (2012) Advances in high temperature gas cooled reactor fuel technology. IAEA-TECDOC-1674

Kable (2014) Olympic dam copper–uranium mine, Adelaide, Australia. Mining Technology. http://www.mining-technology.com/projects/olympic-dam/. Accessed 23 Oct 2014

Knolle F, Birke M, Hassoun R, Jacobs F, Schnug E (2008) Uranium in German mineral water—occurrence and origins. In: De Kok L J and Schnug E (eds) Loads and fate of fertilizer-derived uranium. Backhys Publishers, Leiden, pp 127–134

Kratz S, Schnug E (2006) Rock phosphates and P fertilizers as sources of U contamination in agricultural soils. In: Merkel B and Hasche-Berger A (eds) Uranium in the Environment, Springer, Berlin, pp 57–67

Schnug E, Haneklaus N (2015) Uranium in phosphate fertilizers—review and outlook. In: Merkel B, Arab A (eds) Uranium—past and future challenges. Springer, Heidelberg, pp 123–130

US Geological Survey (2014) Minerals information, mineral commodity summaries. http://minerals.usgs.gov/minerals/pubs/commodity/phosphate_rock/mcs-2014-phosp.pdf. Accessed 19 Sept 2014

Takei M, Kosugiyama S, Mouri T, Katanishi S, Kunitomi K (2006) 高温ガス炉ガスタービン発電システム(GTHTR300)の経済性評価 - Economical Evaluation on Gas-Turbine High Temperature Reactor 300 (GTHTR300) (in Japanese)." 日本原子力学会和文論文誌 5(2):109–117

Voss K (2010) Universität Stuttgart, IER, Vorlesungsmanuskript, Energiewirtschaft und Energieversorgung Band 2, Prof. Dr.-Ing. A. Voß, Dezember 2010

Justice and Sustainability: Normative Criteria for the Use of Phosphorus

Felix Ekardt

Abstract This article will provide normative criteria for the use of phosphorus as a scarce natural resource. It gives some theoretical background on the relevance of an ethical and legal approach to environmental issues, especially with regard to environmental human rights, taking into account both transnational and national perspectives. A special focus lies on the intergenerational and global relevance of human rights.

Keywords Democracy · Ethics · Fundamental rights · Human rights · Justice · Law · Sustainability · Sustainable development

Theoretical Background—Ethical and Legal

Under what circumstances can we call social life "just", or the law "right"? This is the ultimate question of all thinking about politics, morals, and the law. This question is also relevant when it comes to the question how we use phosphorus, how we balance the colliding interests (for instance between contemporary and future generations) Conceptually, the term justice is concerned with the normative validity of a society's basic order. Thus, a normative *theory of justice* (or ethics) answers the question: How shall humans behave or what shall the founding order look like? This must strictly be distinguished from the question of how humans factually do act and what the factual reasons for this action are (and what humans factually "deem right")—this is a question of the descriptive action theory or *anthropology* respectively theory of society.[1] A link between the theory of justice

[1]This distinction is not clear, e.g., in Habermas (1985). Many readers, and probably the author himself, seem to attach a normative meaning to this book; the actual topic, however, is anthropology respectively descriptive theory of societies.

F. Ekardt (✉)
Research Unit Sustainability and Climate Policy, Leipzig, Berlin, Germany
e-mail: mail@sustainability-justice-climate.eu

F. Ekardt
Faculty of Law, University of Rostock, Rostock, Germany

E. Schnug and L.J. De Kok (eds.), *Phosphorus in Agriculture: 100 % Zero*,
DOI 10.1007/978-94-017-7612-7_15

and the action theory is the equally empirical governance theory or *control theory*, i.e., the doctrine of the choice of means to effectively and factually enforce previously defined normative aims (e.g., the right to freedom from impairments to life and health), possibly after a normative balancing with other conflicting objectives (e.g., economic freedom). Such means or instruments could be for instance taxes, cap and trade systems, voluntary commitments, or regulatory law.

A volume on a resource like phosphorus needs normative visions and principles to provide orientation and to line up normative requirements. This may enable this book to go beyond access and supply and to provide a comprehensive view on resources such as phosphorus and their use in societies today as well as for future generations. This is relevant for both the conditions of resource exploitation and for the problem of resource overuse. In the perspective of both ethics and constitutions (in international, European, and national law), the resource topic is characterized by colliding human rights: On the one hand, the freedom rights of consumers and companies—on the other hand, rights to the elementary preconditions of freedom such as food, water, climate stability, security, energy access, a basic supply of essential resources, an absence of wars and civil wars, etc. Generally speaking, any normative conflict can be regarded as a conflict of competing interests and thus as a balancing problem. It refers to the fundamental phenomenon of law: to find a just balance of conflicting interests. In this contribution, phosphorus problems will be the example for this. Since politics allows an industrial society, industrial facilities, approve traffic permits, etc., it knowingly accepts statistical deaths in some decades, i.e. impairment of the right to the elementary conditions of freedom as a result of emissions of air pollutants, etc., especially of young people. This is done balancing those interests with our freedom to consume and the economic freedom of the contemporary consumers. The framework for the legislative balancing is usually referred to as proportionality test. The administration is mainly determined by legislative acts and its balancing authority is initially (mostly) limited to the interpretation of the factual requirement of the standards which the legislature has enacted as an expression of its balancing (if those standards leave room for interpretation).

This contribution, while dealing with justice, gives a perspective from ethics respectively from a (re-)interpretation of national constitutions, the EU Charter of fundamental rights and the European convention on human rights in the light of sustainability. Sustainability has ever more often been named a key objective of policy for 20 years, whether by the UN, the EU or the German Government. It is however not always taken very seriously. The intention of sustainability is to extend justice (respectively law/morals/politics) in an intergenerational and global respect (cf. for this understanding of the principle of sustainability (and with references to opposing views) Ekardt 2011; with a similar result (but somewhat differing arguments) cf. Appel 2005; Ott and Döring 2004). In contrast, a common understanding is that sustainability is simply a balanced pursuit of the three pillars of environmental, economic and social issues, if necessary even without a time- or space-spanning aspect (Steinberg 1998; Beaucamp 2002). It was the topic elsewhere that this is at least misleading, that it sticks to the demand for, in the full sense of the word, eternal (!) growth which—in a physically finite world—cannot be fulfilled, and that this "pillar-perspective" is also incompatible with international law's founding documents of sustainability (Ekardt 2011).

The subject of this contribution takes us to national, European, and international human rights as the basic norm of any liberal democratic constitution (on national and transnational level). Human rights also form the typical core of any modern ethics. Environmental protection and intergenerational and global justice, however, is rarely addressed as guaranteed by fundamental rights in the existing legal and ethical discourse, but is rather assigned to the category of "national objectives," thus based on Article 20a of the German Constitution (Grundgesetz, GG) or, in European law, on Article 191 TFEU. Nevertheless, it seems essential to consider fundamental rights. The interpretation of fundamental rights, unlike state goals, does not only generate power but also legally enforceable obligations of the government. Furthermore, fundamental rights are the strongest element of a liberal-democratic constitution. Moreover, on a constitutional level, overcoming the economically oriented understanding of freedom could also be the essential desideratum of a more future and globally oriented (thus: sustainable) legal interpretation. By the way, restrictions in favor of environmental or for instance resource protection "for the sake of real people's (conditions of) freedom" (as embodied in fundamental rights) might also be motivationally much more plausible than the usual, fairly misleading antagonism of "self-development versus environmental protection," as latently affirmed by national objective provisions.

Accordingly, earlier—and even today in international law—there was often, or is respectively, a discussion about environmental fundamental rights (not only with regard to future generations, of course), as environmental fundamental rights would mean a break with those traditional views diagnosed above. In the academic international law debate (unlike the practice of international law), the idea of strong or even absolute, i.e. not subject to any balancing, environmental fundamental rights seems to gain support (for an outline of the common discussion, cf. Schmidt-Radefeldt 2000; Steinberg 1998 (explicitly criticizing "environmental fundamental rights"); Hattenberger 1993; Gibson 1990; Nickel 1993; Donnelly 1993). In national debates, however, environmental fundamental rights are considered non-specifiable and subject to balancing; therefore ultimately not helpful. Of course, the vague content of an "environmental fundamental right" would only result if one generally introduced a fundamental right "to environmental protection". This, however, is not my intention here. I am only concerned with the question, whether a correct interpretation of fundamental and human rights (nationally or transnationally) results in greater levels of sustainability—and for instance resource protection with regard to phosphorus—than is often assumed. Such an interpretation would build fundamental rights in the way they already exist in all western countries as well as in the international declarations on human rights signed by almost every state of the world, with the consequence that current policy might be in conflict with fundamental respectively human rights (both concepts mean basically the same). Of course, even if the issue is within the scope of a fundamental right, the problem of necessary balancing cannot be avoided. But this problem applies in precisely the same way to other fundamental rights as well (balancing is commonly called "proportionality test"). Therefore, the subject of the following analysis will not be true fundamental rights "to environmental protection." At the same time, we will not limit ourselves to accepting the common assumption that basically all aspects of fundamental rights which concern environmental issues are covered by the right to life and health, which then (a) included no provision for preventive aspects, (b) de facto prefers the defensive aspect of the fundamental right to its "protection obligation" (supposedly because of further needs for balancing, separation of powers, etc.), and (c) for the rest fails to concretize environmental protection which would be required to render it practically relevant. It is precisely this approach toward "protection obligations" (including its administrative consequences) that will be subject to criticism in the course of the following analysis.

Human Rights—Only Subordinate and Vague "Protection Obligations" with Regard to Sustainability? The Traditional Legal Point of View in the EU and Germany

It is well known that for instance the German constitutional and administrative courts are very reluctant to recognize environmental positions based on fundamental rights and previously rejected corresponding claims for violations of fundamental rights on environmental protection issues (see in detail Ekardt 2011). They already avoid the term "protection *rights*" which would clarify that subjective, individual rights are concerned (even if they are subject to balancing with conflicting legal positions). Especially (but not only) in constitutional law cases there is often not clear distinction between the tests of admissibility and substantive foundation of the claim. Thus, eventually—camouflaging the question whether a subjective, individual right exists—it remains unclear, what the respective issue is: whether the claimant has an own right that allows him to bring an action, or whether the underlying action is within the scope of the respective fundamental right or it is an issue of restrictions of the respective fundamental right. In spite of the different results (compared to actions in the area of environmental issues of fundamental rights) this mainly applies even to abortion decisions. The basis for all this is the already mentioned idea that protection rights only describe a goal, but no exact scope of protection, and that one only has to examine whether the protective measures taken are obviously insufficient. However, the latter will always be denied, since in Germany some legislative efforts can be found for every subject, which then qualify as per se "not evidently insufficient." It will be elaborated later that both this result and its reasoning (which is in fact rather proclaimed and reasoned) might deserve criticism.

From the outset, the European Court of Justice of the EU (ECJ) case law is hardly devoted to the issue of protection rights as such—European fundamental rights are included in the (since the Lisbon Treaty binding) Charter of Fundamental Rights (ECFR) and in Article 6, paragraph 1–3 of the EU treaty (see in detail Ekardt 2011). The ECJ has not even specifically addressed fundamental protection rights against the community. Within the Member States, it recognizes the possibility of those rights. Of course, to exaggerate only slightly, the ECJ structurally fails to do almost anything which could bind the EU in any way. It rather seems to be driven by the unspoken intention to give the EU Commission and Council plenty of rope in the determination of their policies. Thus the existing case law lacks any real reference points for the issues discussed in this article. Though the ECJ regularly requires Member States to comply with certain environmental requirements, this has nothing to do with the recognition of protection rights. It only refers to the fact that the Member States are obliged to effectively implement certain environmental decisions of the EU Commission, the Council and the Parliament. Thus, at its core, it is just an issue of enforcement of simple (not constitutional) European law; and it also completely unrelated to the precise content of that law. Protection rights, however, would seek to oblige the EU legislative bodies against their will to something. There is, however, no example apparent for such right. And because of the indicated intentions of the ECJ, it seems likely that this is not going to change significantly.[2] Though Article 37 ECFR, which formally has entered into force at the end of 2009, does contain a commitment to environmental protection–as did the previous EU and EC Treaties—it is not designed as a fundamental right.

Regarding the European Court of Human Rights (ECtHR), which is responsible for the interpretation of the European Convention of Human Rights (ECHR—with is valid for all geographically European countries and is extremely similar to other international human rights treaties), the situation is basically similar. Like the German Federal Constitutional Court, the ECtHR has in fact recognized obligations of the states to undertake protective actions in non-environmental cases based on fundamental rights, though not often. Furthermore, the ECtHR has already granted information rights concerning environmental damages—though confusingly not based on the right to life and health, but on the right to privacy under Article 8 ECHR. However, all environmental cases of the ECHR are ultimately limited to ensuring that in the course of administrative decisions, the concerns of individuals are adequately considered and, for example, the facts are raised carefully. This was expressed most recently in a case of mobile communications. It appears that the obligation to adopt other, more effective laws on the basis of protection rights, which would trigger a reorientation of the whole society and would not just keep my privacy somehow "free from pollutants and noise," has not been a subject of an affirmative ECHR judgments, so far.

In any case, the mere factual existence of case law does not per se mean that it is right. And it does not simply apply because judgments only decide a specific case,

[2]Of course, there are cases, though they are not numerous, in which the ECJ has declared EU legal acts void for formal reasons, e.g. due to a lack of legislative competence. But there does not appear to be any case in which the ECJ has ever required the EU to enact legal provisions against their will.

but do not determine an abstract and general norm.[3] Thus, in the following we will test and analyze a somewhat altered interpretation of existing law (based on judicial interpretation, i.e. by interpreting fundamental rights, not on policy considerations à la "suggesting a legislative change of the catalog of fundamental rights"). But what could an intergenerationally and globally extended, i.e. better complying with the requirements of sustainability, interpretation of freedom and fundamental rights look like to be more precise than the rather vague discussion of an environmental fundamental right? In deviation from the probably prevailing view in Germany, on closer examination we can notice that the wording and the systematic position of the fundamental concept of freedom, which is implied in the fundamental rights, in the German Basic Law and in the ECFR—as well as ultimately also in the ECHR—suggest a more complex interpretation than previously, which has important implications in the intergenerational context.[4] Therefore, the resulting findings can ultimately be applied to any national or transnational human rights protection—for instance with regard to phosphorus.

Intergenerational and Global Scope of Human Rights, Protecting the Conditions of Freedom, and Multipolarity of Freedom[5]

The starting point for our considerations is the idea of freedom rights as classical-liberal guarantees of self-development. So far there is no need to criticize the prevailing view. In addition, freedom also has an intergenerational (and global) dimension (Ekardt 2011). Why? In a nutshell (in more details on the three main arguments, cf. Ekardt 2011; partly cf. also Unnerstall 1999 and Calliess 2001): At their point in life, young and future people are of course people and therefore are protected by human rights–today this already applies to people in other countries. And the right to equal freedom must be directed precisely in that direction where it is threatened—in a technological, globalized world freedom is increasingly threatened across generations and across national borders. Therefore it is clear that fundamental rights also apply intergenerationally and globally, i.e. in favor of the likely main victims of environmental damages.

[3]Laws, regulations, constitutions, etc. remain the only abstract and general norms. Nevertheless it is acceptable that the practice often turns to existing judgments, because (and only) in the event that no substantial grounds be argued in favor of a change of legal opinion, the burden of argumentation bears on the party challenging the existing legal opinion from previous case law (inter alia for reasons of legal certainty), cf. Alexy (1991); on the rationality of the application of the law and the methods of legal interpretation, see also Ekardt (2011).

[4]The issue here is thus an interpretation of all fundamental rights. The rights of equality which do not seem to fit are ultimately special protections of the same freedom and thus do not contradict the following considerations.

[5]For more details and references on this subject see Ekardt (2011), Sects. 4, 5.

The classical-liberal understanding of freedom, which is mainly focused on the economic freedom of those living here and now, must be supplemented in other points, too. E.g. liberties must be interpreted unambiguously in a way as to include the elementary physical freedom conditions—thus not only a right to social welfare, as it was recently acknowledged by the German Federal Constitutional Court, but also to the existence of a relatively stable resource base and a corresponding global climate. For without such a subsistence level and without life and health, there is no freedom.[6] This fundamental right to the elementary conditions of freedom is explicitly provided to the extent life and health are concerned (see Articles 2 paragraph 2 of the German constitution; articles 2, 3 ECFR; articles 2, 8 ECHR). In all other cases it must be based on the interpretation of the general right to freedom. Contrary to the prevailing view I argue that the German Article 2 paragraph 1 of the German constitution has a counterpart in Article 6 ECFR as a general EU right to freedom (using a interpretation in accordance with its wording). The same is true for Article 5 ECHR and other similarly structured bills of rights. At least parts of a general right to freedom are also indisputably included in the right to privacy under Article 8 ECHR.—Based on what has been said so far, this right to life, health and subsistence also applies intergenerationally and globally and is the subject of human rights protection e.g. against environmental damages.

"Protection of freedom where it is endangered" also means that freedom also includes a right to protection (by the state) against fellow citizens (and not only in exceptional circumstances)—not only for, but also for future generations. This is a protection for example against environmental destruction, which is threatening my freedom and its conditions, such as climate change, *by the state against my fellow citizens*. Without that point there would be no human rights protection against intergenerational damages such as climate change since states are not the primary emitters of greenhouse gases. The problem rather lies in the fact that states tolerate or approve e.g. greenhouse gas emissions by private actors. This particular idea need be explained in detail since it is not commonly articulated, as has been indicated above. But if fundamental rights *equally* included a protection of freedom against the state, but also by the state against fellow citizens and, therefore, conflicts of interest of any kind must regularly be understood as multipolar (not bipolar) conflicts of freedoms (*multipolarity*), then this would rebut the traditional, more objective, status of fundamental rights protection (protection obligations instead of protection rights, thus non-actionable duties!) and the traditional imbalance between defensive and protective side of fundamental rights, i.e. the regular elimination of protection obligations, unless there is a case of "evident insufficiency" (understood as something which realistically never occurs, namely the complete absence of

[6]The international trend toward "social" fundamental rights to the various facets of the minimum subsistence thus has a theoretical justification. Such a "constitution of international law" can be derived from the legal source of the "general principles of law" (cf. Article 38 of the Statute of the International Court of Justice) without recourse to, e.g., the International Covenant on Economic Social and Cultural Rights; cf. Ekardt (2011), Sect. 7.

regulation in an area of law). Multipolarity would equally refute the assumption, that the protective side of the fundamental rights is almost entirely taken up with administrative norms, which are supposedly subject to wide legislative discretion, and is not of significant importance either with regard to standing in administrative cases nor regarding the application of substantive law.

What are the arguments for multipolarity and how can we respond to certain typical counter-arguments? In the following I will discuss whether protection rights exist regarding only the scope of fundamental rights (which would trigger standing in administrative and constitutional law cases). The details of necessary balancing (which will e.g. determine how much weight fundamental rights will have when interpreting substantive administrative law, e.g. discretion, in light of those rights) will be analyzed later on. This clear distinction between scope of fundamental rights and balancing differs significantly from case law which rarely clarifies whether its skepticism about protection (fundamental) rights refers to issues of standing, scope or restrictions of fundamental rights (this remains unclear even in the—ephemeral—recourse to protection rights in cases of administrative law).

First, the multipolarity of fundamental rights follows from the very idea of freedom, which is the center of liberal-democratic constitutions—and, as indicated in a footnote, as a philosophical necessity. Fundamental rights as elementary rights are intended to give firm protection against typical hazards for freedom. For hereby they realize the necessary *autonomy of the individual* which is embodied in the principle of dignity. This autonomy is not only threatened directly by the state, but also by private actors, whose actions are "only" approved or tolerated by the state. To dispute this statement, one would have to argue, e.g., that the construction of an industrial plant is relevant to the freedom of the operator but not to the residents' freedom. The classical-liberal thinking, in fact, tends to such an assumption. This view has also been adopted by the current case law. But the very purpose of a liberal state is to allow a balance of conflicts as *impartial* as possible, i.e. independent of special perspectives, and not to prefer a specific (e.g. more economically oriented) life plan. All this shows that protection rights do exist, that defense and protection are equally important–and that we should speak of protection rights, not obligations, since otherwise the equality would just not be recognized.[7]

Second, the multipolarity of fundamental rights appears in limitation or balancing provisions such as Article 2 paragraph 1 of the German constitution or Article 52 ECFR which are also presumed at several instances in the ECHR: As paradigmatic defining principles of liberal-democratic bills of rights these norms also, more practically, prescribe that the freedom of action is limited by "the rights and freedoms of others." The European "constitution" (here) in the form of the ECFR and the ECHR as well as the German Basic Law thus assumes that if the

[7]Incidentally, "protection" as defined in this argument can also consist in granting a benefit to an individual, such as a monetary payment to secure a minimum level of subsistence.

state resolves specific conflicts, not only different interests but explicitly different fundamental *rights* clash.[8]

The preceding tried to show (I) that, and why, there must be protection rights as aspects of fundamental rights and (II) that they are subjective, individual rights. And not only this: The arguments—especially that defense and protection are mentioned side by side—also point out that (III) defense must be on an equal footing with protection.[9]

One objection against that will be: The whole re-interpretation of human rights in the light of sustainability overthrew democratic parliaments, and in "protection rights" cases there was per se larger leeway than in "defensive rights" cases. So, does my re-interpretation of human rights damage democracy? This raises the old question of the relationship between freedom and democracy. Not only some lawyers, but also some philosophers think (partly implicitly) that democracy even has latent priority over freedom. It is initially correct that freedom and democracy contribute to each other. A democracy which is based on certain principles, e.g., a separation of powers, however, promises greater freedom, rationality and impartiality than a "radical" democracy. That is precisely why constitutions just like the German Base Law are based on a separation of powers and are not structured as radical democracies. Particularly justice between generations and global justice, i.e., the freedom of young people and those living after us, are arguments against radical democracy. Since for future and young people and those living geographically far away democracy is not an act of self-determination but of heteronomy. For today

[8]The *third* argument is the wording of provisions such as Article 1 paragraph 1 sentence 2 of the German constitution or Article 1 ECFR, which have been briefly referred to above. Public authorities shall "respect" and "protect" human dignity and also the liberties, which under Article 1 paragraph 2 GG ("therefore") exist *for dignity's sake*, and thus must be interpreted according to its structure. This relation ("therefore") can also be found in the materials of the ECFR. In addition, the double dimension ("respect/protection") of human dignity and *therefore also of the fundamental rights*–given the function of dignity as a reason for all human rights which was just described–shows that freedom can be impaired by threats from various sides and that, therefore, it implies defense and protection. But most of all, the word "protect" would lose its linguistic sense if it only meant that the state shall not exercise direct coercion against the citizens (otherwise the state could simply retreat to not acting at all instead of "protecting"). Hence norms such as Article 1 paragraph 1 of the German constitution and Article 1 ECFR also imply a protection against fellow citizens. And defense and protection are linguistically on equal footing there. All this implies again that there are fundamental rights of defense and protection and that protection and defensive rights must be equally strong—and that we should speak of protection rights, not of somewhat less strong mere protection obligations. This holds true even though (in the interests of an institutional system based on democracy and a separation of powers, which is indeed the most effective protection of freedom) this "protection" cannot be understood as a direct effect of fundamental rights among citizens, but as a claim against the state for protection (see, specifically Article 1 paragraph 3 of the German constitution and Article 51 ECFR).—Article 1 paragraph 2 of the German Constitution as well as the title of this section-and also the materials on the ECFR-talk about "human rights." Thus not only "some" rights are based on dignity, as one might respond, but all of them. Therefore, the structure of human rights, i.e., "equal respect and protection" applies to all and not just some human rights.

[9]In favor of an equal footing cf. already (but without comprehensive reasoning) Calliess (2001).

they are not participants in this democracy. This then leads to a democracy, which is not a principle opposing freedom, but a principle resolving conflict *between* freedoms. This function makes it reasonable to have further conflict resolving institutions, e.g., courts. All this is particularly true if it can be shown that freedom may only be restricted to enhance freedom or freedom conditions—of which the elementary above that were proven just as in our context relevant, may be subjectivized, the other conditions which only support freedom (such as supporting the arts or kindergartens) is not.

The legislature may make different choices, and the task of constitutional courts is (only) to control the framework of those decisions based on a set of balancing rules, which can be derived from the very liberties. The issue is always that some institution of control such as a constitutional court reviews the adherence to rules of balancing. Afterwards, the legislature may react by (partly) altering the constitution. Or the issue is that another institution of control such as a non-constitutional court assesses compliance with the legislative will by the administration or compliance with rules of balancing when such balancing has been passed on to the administration, etc. The aim must be a ping pong, which multipolarity supports freedom (one the one hand preventing abuses of power, on the other hand regarding democracy as a shield for freedom) and is also adequate in terms of impartiality, with a "multiple-level discourse," which in turn supports rationality since it mobilizes a maximum of good reasons, among the state powers.[10]

The Example of Phosphorus

Now, we can draw some conclusions with regard to phosphorus (for more details on this chapter see Ekardt 2011). By those means, it can also be pointed out how balancing rules derived from human rights can work in practice:

---

[10]First, a constitutional court may never order a judgment against a parliament stating "You have to do precisely this." Contrary, it must always limit its decisions to saying "At least you must not continue doing this." For instance, the Constitutional Court may not demand from the Parliament: "Phase out the use of coal power within four and a half years." It may say: "The previous phasing out is too slow; take a new decision on the issue until XX.YY.2010, taking into account the following fact situations, normative concerns, as well as procedural and balancing rules." Conversely, a constitutional court could rule on an action brought by an energy company: "Of course, the legislature may phase out nuclear power generation—but it must remain within a certain limit which it has crossed unfortunately, as it has demanded phasing out the use of nuclear energy within three days." This is all the more true as the ping pong also includes the administration and the lower courts, as just outlined by the brief introductory note on the "passing on" of balancing by the legislature. It allows authorities to respond to a court decision with new decisions, which then in turn are subject to judicial control. The same is true with respect to the legislator and the constitutional jurisdiction. And the legislature may also react on decisions of lower courts with legislative changes, etc.

- As we have seen, Freedom also has an intergenerational and global dimension, since at their point in life, young and future people are of course people and therefore are protected by human rights—today this already applies to people in other countries. Fundamental rights also apply intergenerationally and globally, i.e. in favor of the likely main victims of the current handling and overuse of resources such as phosphorus.
- Freedom rights must be interpreted unambiguously in a way as to include the above-mentioned elementary *preconditions of freedom*—thus not only a right to social welfare in general, but also to the provision and maintenance of a relatively stable resource base, food supply, security, water supply, life-supporting functions and ecosystem services.[11] This implies with regard to phosphorus: a guarantee for a proper food supply on a worldwide and intergenerational scale; a principle of materials stewardship of all actors that deal with natural resources management; a life-cycle perspective of natural resources; responsibility for maintaining life-supporting functions and services of ecosystems; and a general priority in favour of resource savings.
- "Protection of freedom where it is endangered" also implies that freedom also includes a right to protection (by the public power) against fellow citizens (and not only in exceptional circumstances). This implies a protection for example against environmental or social destruction, which is threatening freedom and its conditions, such as overuse of resources, *by any public power against my fellow citizens including companies or farmers using phosphorus.*
- Protection rights in the environmental context are not excluded despite the fact that for instance many resource problems concern—for instance with regard to phosphorus—only *forthcoming* hazards of fundamental rights. By the same token, the scope of protection rights is indeed affected by such hazards (and not only by definite encroachments). Undoubtedly, future trends are not per se exactly predictable and therefore "uncertain". However, such an objection would fail, because impairments of fundamental rights which are "only possible" are *not* irrelevant with respect to fundamental rights, especially under the threat of irreversibility of the "possible" infringement. Otherwise, fundamental rights would no longer serve the very purpose of legal fundamental rights: to guarantee the protection of autonomy exactly where autonomy is threatened with impairment.

The necessary balancing between all the above-mentioned aspects of sustainability-oriented human rights and the classical liberal guarantees of freedom

[11]In liberal democracies, there are also "further" (in contrast to "elementary") preconditions of freedom such as macroeconomic stabilization, biodiversity, etc., which are extremely helpful, but not absolutely necessary to constitute freedom. Therefore, such "further" preconditions of freedom are usually seen not as human rights but as mere obligations of the public powers (without corresponding rights of individuals). This does not mean at all that these "further" conditions are less important.

for consumers and enterprises offers some leeway. Nevertheless, especially with regard to overuse of resources, some definite conclusions can be derived:

- A very often overseen aspect of freedom is the polluter pays principle, which in turn follows from the principle of freedom itself. For freedom must include responsibility for the foreseeable (including environmental or social) consequences of one's own actions—even in other countries and in the future, and also for the unpleasant consequences of one's own life plan. The negative consequences of an action which otherwise benefit me (for instance, of cheap free resources today) must always fall back on me, if only by way of cost recovery for the damage created by that action. This justifies limitations of phosphorus use and instruments that try to avoid the harmful consequences of overuse.
- Another balancing rule is that the assumptions of underlying facts must be correct. Every decision must, for instance, be based on the latest climate research in order to understand what dangers threaten the freedom of future generations. In situations of uncertain facts such as climate change, there is also a duty to make preliminary decisions and to review them later. The current phosphorus policy already disregards the balancing rule that its decisions shall be based on a correct factual basis: In particular, existing actions are probably erroneously deemed suitable to avoid the looming drastic problems in the future.
- Furthermore, politics has not yet taken into account in its decision making that the fundamental right of freedom has also an intergenerational and global cross-border dimension and that, therefore, legal positions of future generations and the proverbial Bangladeshis need be considered in parliamentary/legal decisions.
- The task of politics is to solve the constant conflicts between the one's and another one's freedom and, in addition, to guarantee the availability of external freedom preconditions. But generally, this does not mean that the political and democratic process has to provide an equal distribution in the sense that certain material goods like phosphorus would necessarily have to be equally distributed. Consequently, the details of social distribution are subject to political discretion. However, with respect to elementary preconditions of freedom an equal treatment, as for liberties themselves (i.e. unlike for "further" freedom-promoting conditions), is necessary to provide that everyone gets a particular absolute minimum of something. For without these basic requirements like food, water, clothing, basal energy access there can be no freedom from the outset. With regard to food, this has direct implications for phosphorus. The "equal distribution principle" in this context is supported by two arguments:

  - Without a right to an equal absolute minimum level of elementary freedom preconditions freedom would be of no value for the poor—and liberal constitutions respectively human rights *guarantee* equal liberties. This "equal subsistence" means specifically two things: everyone must have a minimum

level of resources, energy, etc. available—however, all must be (because this is also basic) protected from disastrous encroachments such as climate change as far as possible. Resource exploitation and damages caused by western lifestyle must be reduced absolutely, while every man (worldwide and also in the future) needs at least a certain minimum of phosphorus—and many people worldwide do not nearly reach their "equal" per capita share so far. This makes it rather obvious to be cautious about inequalities with regard to the subject of this paper.
  - If a collective good such as resources like phosphorus is monetised, it seems plausible to turn the usage rights or the "proceeds" of an unequal distribution (the atmosphere use) in equal parts for all persons as far as possible—for no one can claim for themselves that she had accomplished a special "performance" to produce that good. This second argument can also be seen as an argument e contrario of the polluter pays principle (which also follows from the principle of freedom). Not generally "equal wealth" (nationally or worldwide) but very probably a basic resource supply and equal greenhouse gas emission rights for all—worldwide and intergenerationally—appear reasonable. This leads to a theoretical justification of a principle of common heritage of mankind applied to geological *and* anthropogenic stocks. This can also mean a participation of rents deriving from resource exploitation.
  - Sometimes a minimum use of natural resources can hardly be defined per se—but freedom of mankind in general may be endangered by overuse of resources. Therefore international resource politics should contribute to a globally fair distribution of resource use and an adequate burden sharing.
  - A "fair" distribution of resources could seek to minimise risks for security and economic turmoil due to dependence on resources.

- On a preliminary basis, a higher access rate to resources (such as phosphorus) for developing countries could be justifiable for their fight against poverty.
- Another important consequence of the principles justified above is: The colliding human rights call for distinct rules of public authorities. Purely voluntary solutions will probably not be enough (see also Ekardt et al. this volume).
- On a procedural basis, the colliding human rights imply a broad participation of all stakeholders in all legislative and administrative decisions with relevance to resources such as phosphorus.

The implications of all this might be: absolute reduction of phosphorus use in industrialized countries; relative decoupling for developing countries including newly industrialising countries; minimising problem shifting between environmental media, types of resources, economic sectors, regions and generations; driving resource productivity (total material productivity) at a rate higher than GDP growth.

References

Alexy R (1991) Theorie der juristischen Argumentation, 2d edn. Suhrkamp, Frankfurt am Mainz
Appel I (2005) Staatliche Zukunfts- und Entwicklungsvorsorge. Mohr Sies, Tübingen
Beaucamp G (2002) Das Konzept einer zukunftsfähigen Entwicklung im Recht. Mohr Siebeck, Tübingen
Calliess C (2001) Rechtsstaat und Umweltstaat. Mohr Siebeck, Tübingen
Donnelly J (1993) Third generation rights. In: Brölmann C, Lefeber R, Zieck M (eds) Peoples and minorities in international law. Nijhoff, Dordrecht, pp 119–139
Ekardt F (2011) Theorie der Nachhaltigkeit: Rechtliche, ethische und politische Zugänge – am Beispiel von Klimawandel. Ressourcenknappheit und Welthandel, Nomos, Baden-Baden
Gibson N (1990) The right to a clean environment. Saskatchewan Law Rev 1:5–23
Habermas J (1985) The theory of communicative action, 2 vol. Beacon Press, London
Hattenberger H (1993) Der Umweltschutz als Staatsaufgabe, München. In: Beck CH, Kromarek P (ed) Environnement et droits de l'homme, Bruxelles 1987
Nickel J (1993) The right to a safe enviroment. Yale Law J 3:281–300
Ott K, Döring R (2004) Theorie und Praxis starker Nachhaltigkeit. Metropolis, Marburg
Schmidt-Radefeldt R (2000) Ökologische Menschenrechte. Nomos, Baden-Baden
Steinberg R (1998) Der ökologische Verfassungsstaat. Suhrkamp, Frankfurt am Mainz
Unnerstall H (1999) Rechte zukünftiger Generationen. Königshausen & Neumann, Würzburg

Governance Instruments for Phosphorus Supply Security

Felix Ekardt, Beatrice Garske, Jessica Stubenrauch and Jutta Wieding

Abstract This article will show which policy measures could lead to more sound use of phosphorus, and which might not. From the perspective of natural science, there are varying accounts as to how long phosphorus will still be available. It is, however, undisputed that phosphorus is a finite resource with vital importance to humankind. It is also certain that overusing phosphorus causes environmental damage further increased by permanent application. In this context, the usage of phosphorus from stocks contaminated by heavy metals might increasingly become an issue. Regulatory policy incentives are required towards establishing more sound phosphorus use with increased focus on recycling. So far, phosphorus has involved tentative proclamations rather than legally binding measures. However, classical regulatory measures directed at single actions—the command-and-control approach—will not suffice in addressing resource and environmental problems, as phosphorus problems are more appropriately viewed as quantity problems. Additional economic instruments are needed in order to avoid enforcement deficits, rebound and shifting effects, and to promote recycling while decreasing phosphorus consumption. These economic instruments would also help in addressing several environmental problems regarding land use, resources, biodiversity loss and climate change all at once.

Keywords Certificate markets · Economic instruments · European Union · Governance · Regulatory law · Subsidies · Taxes

---

F. Ekardt (✉) · B. Garske · J. Stubenrauch · J. Wieding
Research Unit Sustainability and Climate Policy, Leipzig, Berlin, Germany
e-mail: mail@sustainability-justice-climate.eu

F. Ekardt · B. Garske · J. Stubenrauch · J. Wieding
Faculty of Law, University of Rostock, Rostock, Germany

E. Schnug and L.J. De Kok (eds.), *Phosphorus in Agriculture: 100 % Zero*,
DOI 10.1007/978-94-017-7612-7_16

Phosphorus as a Scarce and Environmentally Relevant Resource: Call for Action?

The principle of sustainability forms the basis of modern environmental policy. By definition, sustainability is a way of life and economy which will—globally and for generations—allow for continued life on earth (Ekardt 2011). This requires a focus on using renewable resources within the limits of their regeneration and sparing use of non-renewable resources. One of those non-renewable resources is phosphorus (P), a key nutrient vital for plants, humans and animals. Moreover, P inputs into soil and waters potentially cause considerable environmental damage. This article will first give an overview of the difficult debate on these problems. Secondly, we will address the question of effective political options for action, which will be our main focus. We will demonstrate how the issue of P is interconnected to other environmental issues such as land use, resource scarcity, biodiversity loss and climate change. Likewise, questions of social distribution will be addressed.

P is a finite, non-substitutable resource whose long-term availability is absolutely necessary in securing the global food supply. The last several years have seen highly controversial accounts of its future availability. Estimates of the remaining accessible P reserves gained from phosphate rock vary strongly according to projections on economic and technical usability of deposits and therefore must be regarded as snapshots (BGR 2013). Between 1988 and 2012, statistical projections indicated an availability of P between 84 and 396 years (BGR 2013). In one study, a Peak P in the year 2033 is estimated (Cordell et al. 2009). The concept of Peak P in general is open to question from a historical and economic point of view (Ulrich 2013; Heckenmüller et al. 2014). So, instead of putting the main focus on geological P availability, Ulrich and Frossard (2014) suggest greater focus on socio-economic and environmental vulnerabilities that may occur from current and future P production and demand, such as by establishing fertilizer access to smallholding farmers and avoiding pollution in soil and water by using P (ibid.). The latter perspective obviously aims towards more careful use of P in general. Because of that, and because there is no doubt about the fact that P deposits are finite, there is no genuine difference in policy implications from either perspective.

The possible environmental damage and scarce P reserves in particular raise the question as to the quality of the remaining resources with respect to the availability of P not increasingly contaminated with heavy metals such as uranium (U) or cadmium (Cd). Rock phosphates from sedimentary deposits represent most of the P reserves still accessible worldwide (BGR 2013). These in particular are relatively heavily contaminated with heavy metals due to geogenic influences in contrast to igneous deposits. Table 1 shows the average geogenic burden of rock phosphates with Cd and U according to origin.

Almost 90 % of P fertilizers produced worldwide are obtained from sedimentary rock phosphate (Van Kauwenbergh 1997). According to a study by Dittrich and Klose (2008), the average Cd concentration in mineral nitrogen, phosphorus and potassium-based NPK fertilizers per unit of P lies between 18 mg Cd/kg P_2O_5 and

Table 1 Uranium and cadmium concentration of rock phosphates of different origins (Mortvedt and Beaton 1995; Van Kauwenbergh 1997; Raven and Loeppert 1997; Kharikov and Smetana 2000; Gupta and Singh 2003 as resumed in Schnug and Kratz 2005)

Rock phosphate					
	Sedimentary		Igneous		
	USA	Morocco	China	Middle East	Russia (Kola)
Cadmium (mg kg^{-1})	6.1–92	15–38	<2–2.5	1.5–35	0.1–1.3
Uranium (mg kg^{-1})	65–180	75–155	23–31	40–170	10–28

over 60 mg Cd/kg P_2O_5 in triple superphosphate fertilizers. Results from 2003 showed that Cd concentrations of up to 20 mg Cd/kg P_2O_5 might not lead to long-term accumulation of Cd in the soil, whereas concentrations of 60 mg Cd/kg P_2O_5 and more would likely lead to a relatively high accumulation in the soil over a period of 100 years (EU Commission 2003). Apart from possible long-term accumulation, the main issue is that Cd ions in soils may become mobile and translocate easily into crops or groundwater through deeper soil layers. Cd already becomes mobile at pH values of 6.5 (Blume 2004). A tendency for soils with low pH values and/or humus levels towards high transfer rates into crops has been reported (Ruppe et al. 2009). Concentrations in the human body may reach critical levels caused by grain and vegetable consumption (BfR 2010). The European Food Savety Authority (EFSA 2011) established a tolerable weekly intake of 2.5 µg/kg body weight for Cd according to a recent risk assessment by the joint FAO/WHO Expert Commitee on Food Additives (WHO 2011). As a consequence, European Commission recently amended EC Regulation No. 1881/2006 for maximum Cd levels in foodstuffs in 2014.

Even though eliminating Cd and U from P is technically possible, economic factors have prohibited this from being practiced on a regular basis. Raising the price in the aftermath of the procedure and price-regulating speculation of remaining P reserves have encountered skepticism, as this raises concerns of possibly fostering unequal fertilizer distribution (Lange 2009). Ulrich et al. (2014) also anticipate considerable economic potential by using U from phosphate rock in the nuclear industry. New nuclear power plants under construction worldwide show that there most probably will be a demand for U in the future (www.world-nuclear.org). Issues facing permanent disposal, non-assessable uncertainties of disasters beyond all expectations and risks of terrorist attacks raise questions as to the suitability of nuclear energy as a future energy source (Welch 1980; McDaniels 2006; Kirpotin and Nemceva 2015).

The use of P and other mineral fertilizers such as nitrogen (N) is highly interlinked with the environmental problem of soil degradation as well as eutrophication in water bodies, biodiversity loss and global climate change. Continuous soil degradation worldwide ranges from soil sealing, desertification and salination to soil compaction, nutrient loss and soil acidification to soil contamination. According to the secretariat of the United Nations Convention to Combat Desertification, this accounts for losses of around 24 billion tons of fertile topsoil worldwide per year

(UNCCD 2012). Soil degradation may not only occur due to heavy metals introduced into the soil through mineral fertilizers; the amount of humus in soils and therefore also natural soil fertility may decrease where rapid-action mineral fertilizers are mainly applied to the soil and crop rotation and cover crops to preserve natural soil fertility no longer play a role in intensive farming in specialized agricultural businesses (LfL 2011, pp. 43–44). Indeed, especially high amounts of N may lead to soil acidification (Kotschi and Wetter 2015). Once degraded, soils with low pH levels may finally store less nutrients such as P for release into plants; nutrients may be washed out of the soil quickly by rain and flow into water bodies (Kotschi and Wetter 2015). Agriculture is a major contributor for most of the diffuse input of P into water (Withers et al. 2015; EEA 2012; Frossard et al. 2004). The anthropogenic increase in surface water nutrients leads to more restricted biodiversity, such as due to the possibility of excessive growth of algae in coastal regions and overturning waters. Over 400 "death zones" in water bodies have been identified worldwide (UBA 2014). Once in a river or sea, P is deposited in sediments in lakes or delivered to coastal waters where P recovery is virtually impossible (Lange 2009; Beusen et al. 2005).

At the same time, conventional agriculture in particular is still highly dependent on mineral NPK fertilizers. Yield increases that began in the 1950s in Europe have been directly connected to mineral fertilizers (Withers et al. 2015). The worldwide production of agricultural products has almost tripled since 1950, while agricultural area has increased only by 12 % (Beste 2015). We are still confronted with global population growth and growing demand for meat products (FAO 2012) as well as the never ending thirst for energy, which both depend on economic growth. This has led to an increasing demand for agricultural land and the respective fertilizers for food, animal feedstock and energy crop production (WWF 2014, p. 23 et seq.; Brot für die Welt 2014). The International Fertilizer Industry Assosiation (IFA 2014) expects a 2 % global increase in demand for P fertilizers up to 2018. Most of the demand is expected to come from Latin America, followed by Africa and Asia. Cultivated land areas and therefore demand for P-fertilizers has been expending steadily, especially in Latin American countries (IFA 2014). One reason is the increasing large-scale production of animal feed mainly exported to Europe and used for intensive livestock farming (Lymbery 2015; WWF 2014, p. 34 et seq.). P fertilizers are imported alongside crops for animal feedstock.

In contrast to the decreasing demand for P fertilizers in developing countries—and without taking into account P in imported food and animal feedstock—demand for P in countries such as Germany has stagnated at a relatively high level since 2012 (BMELV 2015a). The year 2014 saw 284,000 tons of P_2O_5 sold in Germany (IVA 2015). The P input is regularly higher than P withdrawn by harvesting crops in the past, resulting in a surplus of nutrients such as P (and N) in a large total area in Germany and throughout Europe. Nutrient balances between 2005 and 2008 indicate P surpluses of 2–3 kg/ha in Germany, up to 25.5 kg/ha in Malta and 20 kg/ha in the Netherlands (Eurostat 2014). Huge amounts of manure from industrial mass meat production adds considerably to the problem of over-fertilization (SRU 2008, p. 468). Surplus P discharge through erosion, leaching or run-off from soil into water bodies has serious consequences as previously mentioned.

As P from phosphate rocks is a finite resource more sound recycling-oriented management seems to be warranted due to the environmental issues. The same follows considering the distribution issues already described and the connection between P fertilization and the environmental impact of conventional agriculture as mentioned above. Cordell et al. (2009) assumed that demand for P could be cut by 50 %, initially only by changing dietary habits and efforts in food chain efficiency such as lowering food overproduction, reduced product chains and minimized food wastage. Several billion euros worth of food are wasted every year throughout Europe (ibid.). Secondly, recycling and recovery of P and the scarcer use of resources, but also organic farming merit more attention. Organic farming is preferable to conventional agriculture in several ways: cover crops that prevent erosion and improve P availability in soils are usually used more often (Pimentel et al. 2005), and mixed farming and less specialized farms are also more common compared to conventional farming, which is mostly monocultural (BMELV 2015b). This may close P cycles more easily, while dispensing with mineral fertilizers, many of which are strongly polluted with Cd or U, and chemical pesticides. One-third less fossil energy per hectare of farmland is used and approximately twice the carbon dioxide can be stored in the soil without the use of mineral fertilizers and improved soil quality (Beste 2015), a decisive advantage in combating global climate change. Apart from that, around 2 % of world energy is dedicated to processing N fertilizers in the Haber-Bosch process for energetic and material needs (Sutton et al. 2013, p. 8). For that reason fossil fuel and mineral N fertilizer prices are highly coupled. Indirectly, there may be also a connection to the price of mineral P fertilizers where mineral N, P and also Potassium (K) fertilizers are often sold in combination. Furthermore, fossil fuel price increases would increase the cost of mineral N fertilizers, placing organic farming that operates mainly without mineral fertilizers at an advantage compared to conventional farming. Moreover, fertilizers are only applied to ensure soil fertility in organic farming rather than to meet a presumed requirement for the highest yield. Remaining inputs of harmful substances due to fertilization are therefore lower overall with a heavy general reduction in the risk of over-fertilization (Pimentel et al. 2005).

The global trend, however, remains towards intensifying agriculture. The FAO (2012) predicts a production volume of 3 billion tons of grain (without rice) every year until 2050. Ninety percent of the necessary increase, starting at around 1.7 billion tons globally in 2010, is supposed to be attained through higher yields and intensified agriculture. Meanwhile, the amount of area used for agriculture has to be increased worldwide (BMELV 2010). The projected increase in intensity will inevitably lead to increased use of P fertilizers involving the impacts described above. Dietary habits and food handling as well as acreage competition between food and energy crops need to be placed on the agenda (see Ekardt and von Bredow 2012 for details).

It is interesting to note that there is obviously a call for action regardless of who is right in the discussion on Peak P. Economic and technical fields of action all point in a similar direction: as a resource, P needs a different economic value to engender a more sound use of rock phosphate while intensifying the utilization of alternative P sources.

EU Policy Approaches to Phosphorus

Politics at EU level has been paying increasing attention to P as scarce resource vital to humans, plants and animals on one hand, and P and related heavy metals as harmful pollutants in soil and water on the other. Current approaches and policy programs relevant to the P issue will be presented briefly in this section. To remain within the scope of this contribution, we will be focusing on EU levels while only touching on individual countries.

The European Union has created a flagship initiative under the Europe 2020 Strategy for "a resource-efficient Europe" that includes the target of transforming the European economy into a resource-efficient low-carbon economy. Other measures such as the Raw Materials Initiative and the European Partnership on Raw Materials point in the same direction (COM 2014a, 297, 2). These efforts are based on the necessity of securing Europe's supply with crucial resources (COM 2011b, 21, 2–3). In May 2014, the European Commission presented an updated list of critical resources, which now contains twenty resources (COM 2014a, 297). This list is an important instrument in the EU Raw Materials Strategy, and is intended to create incentives to produce, mine or recycle resources within Europe or conduct the respective research. Phosphate rock is among those six newly added resources classified as critical (COM 2014b, IP/14/599, 1–2). Including P as a resource on the list shows the increasing recognition of its economic relevance and importance for security of supply. However, there is no comprehensive legal regulation of P use so far.

The consultative communication of the European Commission (COM 2013b, 517) compiles all of the topics relevant to P as a resource so far. After a rather optimistic account of the current and future supply and demand situation, the communication highlights the environmental effects within the whole P cycle. Special focus lies on losses along the P chain, problems in mining, processing and conversion of P in fertilizers and animal feedstock, water pollution, and soil contamination caused by agricultural P use. The European Commission has pointed out that efforts need to be directed towards mining, processing, upgrading and transporting phosphate rock more efficiently. Increasing prices and the exhaustion of optimal deposits are the most probable driving factors for better use of P according to the Commission. Actual legal steps in this field have not so far been addressed.

Improvements in the inclusion of P in fertilizer security and transparency in fertilizer ingredients by labeling have been announced, however. Changes may start with the revision of the Fertilizer Regulation even though the time scheme of the revision is not yet clear. The use of phosphates and other P compounds in household detergents for textiles and dishwashers is limited by the Detergent

Regulation. A revised version of the regulation was passed in 2013, and is intended to limit P input from detergents further (COM 2013b, 517, 15).

Another important starting point for the EU lies in more efficient P application and reduction of P losses in agriculture. Various action programs within the scope of the Nitrates Directive have been drawn up to lower current and prevent future water pollution. The directive also requires member states to implement and enforce rules according to good agricultural practice. Good agricultural practice includes time periods in which fertilizers should not be applied to agricultural areas or provisions on procedures for fertilizer input (Annex II of the Nitrates Directive). Besides mandatory regulation of nitrate fertilization, some member states such as Ireland and the Netherlands have already introduced P limits at national level (COM 2013a, 683, 10). This has improved some aspects of fertilizing practices throughout the last decades. The report by the European Commission on the implementation of the Nitrates Directive has however found major differences regarding N intensities. For instance, elevated nitrate concentrations in waters are measured in Germany (COM 2013a, 683, 4–5), so Germany has been requested to tighten its fertilization regulation, the main instrument for implementing the EU Nitrates Directive. A corresponding infringement proceeding aimed against Germany was initiated at the end of 2013 (COM 2014c). There are also major differences among the European member states with regard to P intensities (COM 2013a, 683, 4).

Other legal acts on soils and water bodies in the EU indirectly address the P issue. The EU Nitrates Directive is an important instrument in implementing the EU Water Framework Directive, which aims at reaching a good ecological status in European waters. While a clear reference to phosphate fertilizers is also missing here, the directive points out that P compounds as harmful to waters and lists P as one of the agro-environmental indicators. The indicators describe the greatest burdens on water quality, but are not being applied yet due to limited available data and methodological difficulties (European Court of Auditors 2014, 46).

Even if there is an increased interest in soil protection as can be seen in the thematic Strategy for Soil Protection, an overarching European directive on soil is missing. Instead, the standards for good agricultural and environmental condition of land (GAEC), part of the Cross Compliance regulation within the common agricultural policy framework, encompass some soil-related requirements such as minimum soil cover or minimum land management reflecting site-specific conditions to limit erosion and thus contribute to the reduction of P losses in agriculture (Annex II of the Regulation (EU) No. 1306/2013). These regulations are not explicitly directed at P though, and there is still scope for improvement. For instance, obligations to limit P use by business owners are not covered by the Cross Compliance rules; the European Commission relies on national nitrate action programs to contain P-related regulations (COM 2013b, 517, 15).

"Greening" the common agricultural policy was introduced with the 2013 reform, and is intended to make the agricultural sector more environmentally friendly by linking direct payments to farmers to environmental performance. Disregarding the question of how ambitious these conditions actually are, the GAP remains focused on avoiding environmental harm through N surplus. Furthermore, the success of Greening depends on the member states, which have considerable leeway for implementation. The design of agro-environmental programs in rural development policy framework is also in the hands of the member states (Garske and Hoffmann 2015). Those programs could in fact contribute to decreasing P losses in agriculture with the focus set accordingly. So far, this focus is missing.

The EU's approach to increasing P efficiency in animal farming is based on increasing accreditation of new phytase enzymes as additives for animal feed for mono-gastric livestock as well as phosphorus contents in diets adapted to requirement during the different life stages of animals known as "phase feeding." This aims at decreasing P intensities in animal feed. The European Commission hopes for further approaches to solutions from the current EU Program for Research and Innovation Framework (Horizon 2020) and the European Innovation Partnership on Agricultural Productivity and Sustainability (COM 2013b, 517, 15–16). Here again, that actual approach is somewhat vague. Moreover, improvement in P efficiency in animal farming does not preclude rebound effects, and any reductions of P in animal food may be nullified by rising consumption and production of animal-based food.

The EU also recognizes the potential of food waste avoidance and recycling as spelled out in the European Resource Efficiency Roadmap, which aims at reducing the disposal of food waste still fit for consumption by 2020. An actual legal regulation is missing—yet again. Besides avoiding food wastage, the EU focuses on improving the use of food waste. The Landfill Directive requires member states to reduce landfilling with biodegradable waste to 35 % of the amounts of such waste produced in 1995 by 2016 (COM 2013b, 517, 17).

Additionally, the European Commission includes an improved legal framework on the usage of animal by-products in animal feed to recover higher amounts of P from agricultural wastage and agricultural by-products (COM 2013b, 517, 16–17). Similar efforts are envisaged for reusing P in organic fertilizers. The European Commission sees a lot of unused potential here. So far, there is no common strategy for promoting the recycling of P from manure, sewage or organically degradable waste while various techniques to eliminate hazardous materials from recycled P are not yet marketable. To make matters worse, the price of mineral phosphate fertilizers usually lies below the price of fertilizers produced from recovered P. This also renders efforts at increasing P recycling ineffective (COM 2013b, 517, 18–19).

Sewage treatment is another field of action in improving P management. The Urban Waste Water Treatment Directive provides for a removal of P from sewage. Sequestration of P into a usable form is not pursued, even though procedures are available such as from processing sewage sludge. Using sewage sludge for fertilization is regulated in the Sewage Sludge Directive. Harmonization of high quality standards for the usage of sewage sludge on agricultural areas is missing (COM 2013b, 517, 17–18). Accordingly, there may be a problem with contamination in sewage sludge—and little incentive to replace mineral fertilizer with sewage sludge. Member states have individually taken specific efforts to protect the P resource. Sweden, for example, has set a national intermediate target to recover at least 60 % of P compounds in sewage and use it in fertilizers by 2015. The Netherlands has formulated a target to increase the reuse P in different production processes in an agreement on the phosphate value-added chain. German legislation is also working on legal regulations to lower P wastage (COM 2013b, 517, 3). The German Federal Ministry for the Environment, Nature Conservation, Building and Nuclear Safety is planning binding regulations for phosphate recycling while revising the sewage sludge regulation, which may also include encouraging recycling for other materials. The necessary regulations are to be written down in a new regulation on regaining phosphate (AbfPhosV). Passing the regulation through parliament has been delayed, and its implementation is not yet in sight in the near future (Bundestag 2013, Drs. 17/12984, 54).

All this shows that the issue of P has increasingly gained a place on the European agenda while giving reason for political declarations of intent. However, little in the way of legal activity has emerged with actual requirements for improved handling of P. Far-reaching regulations such as limits for water and soil pollution have been met at member state level (see Ekardt et al. 2010b for more information on existing environmental regulatory law for P in Germany). There are still fundamental doubts as to whether traditional strategies of environmental law with individual command and control in matters such as fertilization and good agricultural practise is the best approach to P. This will be addressed in the next chapter.

Issues in Existing Legal Approaches

The situation described is usually answered with propositions aimed at changing details in regulatory law, such as imposing new obligations and restrictions for farmers what is referred to as the command and control approach. However, there are several reasons to assume that those or similar regulatory policy instruments will not ultimately be able to solve resource and environmental problems concerning P. The existing problem is of a more fundamental nature that goes far beyond the issue of P. Until today, environmental law most frequently consists of substantial legislation for specific products, crops or other topics such as vehicle emissions, building services and other clearly defined topics with specific sanctions in case of non-compliance. Issues of resources and environment are, however, a problem of quantity. In other words, the total quantity of resource usage needs to be taken into account—and oftentimes also the total quantity of environmental impact.

Concerning P fertilization, this implies that the accumulation of many fertilization procedures and the resulting over-fertilization and expenditure of a finite resource, not one individual fertilization procedure alone, causes environmental pollution and resource scarcity issues. This applies in a similar fashion to other environmental land use problems such as the contribution of agriculture to climate change through energy-intensive fertilization or methane-emitting livestock production.

Generally speaking, we are confronted with profound governance problems in environmental law. They arise where regulation is mainly aimed at single products, installations, industries or geographic spaces (see Ekardt 2011; Ekardt and Hennig 2015; Ekardt et al. 2010a for a more detailed account; see also Ekardt et al. 2010b).

- An effective policy approach has to be coherent between substantial strength and the set target. With respect to P, increasingly recycling P as a resource and promoting organic farming would help cover the need for P in an alternative, more cycling-orientated fashion using organic fertilizers, cover crops and so on. This may avoid long-term scarcity and the increasing usage of contaminated stock as well as other environmental problems discussed at the beginning of this article. So far, there is no clear legal guideline leading in that direction. It is hard to imagine this formulated by means of a command and control approach as this would require regulating many details without addressing the special needs of a specific farmer.
- Regulatory law is generally subjected to exceptions and margins of appreciation for individual cases as exceptions will not apparently change the outcome. These exceptions will at times undermine the purpose of the norm if frequently applied.
- An effective governance approach needs to be effectively enforced. Not all environmental policies are complied with in practice. Environmental problems in agriculture cause particular difficulties due to the infinitely high number of minor procedures requiring administrative monitoring. It is hardly realistic to place a "police officer on every tractor," which would be necessary to enforce legal fertilization restrictions for example.
- Regulating individual procedures, products or crops, as is typical in regulatory law, causes incentives for sectoral, geographic and resource-based shifting (or relocating) effects. This occurs when environmental damage or resource consumption is relocated by the enterprises or citizens concerned as a result of policy measures. Resource consumption may be shifted into other sectors of life, places or to more intensive use of other resources. As an example, the transition from fossil to bio-fuels has placed greater burden on soil, water and nature in general without any cap on energy consumption (sufficiency). Capping P consumption in the EU only would contribute towards minimizing ecological problems in waters and soil in Europe but the resource problem would still prevail with P deposits used elsewhere.
- Another issue often confused with shifting effects is the rebound effect: regulations in environmental law for specific products, houses, industrial plants or cars have made it possible to improve these particular issues. At the same time, however, the number of greenhouse gas-emitting sites or products has still increased due to increasing prosperity in modern life, with overall consumption growth partially or completely eliminating any gains in efficiency of a specific good or service. Another form of rebound effect occurs where increasing consumption is explicitly caused by the ecological improvement of a crop or service. Examples include raising room temperature due to a cleaner conscience from living in a low-energy house, or adopting a harder driving style in an energy-efficient car. More broadly, rebound effects include the phenomenon that a newer and more efficient appliances do not substitute the older ones, but are used in addition to them. This implies in terms of P that a cap on P application "per hectare" by regulatory law might seem like a good option at first glance,

but if areas so far unused are cultivated for animal feedstock to meet globally growing meat consumption or for bio-fuel crops, there will be no overall reduction in P usage.

- Furthermore, regulatory environmental regulations cannot reflect issues that are hard to specify precisely. Aspects such as validity of food security or long-term availability of non-contaminated P are difficult to translate into regulatory criteria for single plants as there is no concrete relationship to specific fertilizer use.

The reasons for ignoring these governance problems and the subordination of ecological and resource-political questions go deeper than explanations such as economic and administrative self-interest might indicate. Ultimately, it is a multi-layered vicious circle involving farmers, consumers, politicians, law applicants, fertilizer producers and others that mutually strengthens certain basic attitudes contributing to this context as all participants are jointly dependent on each other (Ekardt 2011). This is why agriculture in its current orientation towards increasing short-term profit in addition to economic self-interest is aligned to traditional values such as increasing production, illustrating the underlying concept of the omnipresent growth paradigm. Furthermore, anthropogenic constants such as the narrow space-time focus of human emotionality on the "here and now" as well as habits, denial, and convenience will presumably make it rather difficult for most of those involved to face a long-term and currently invisible P problem with any level of commitment. Moreover, there is a problem with public goods: All those stakeholders involved know that possibly the ecological problem dimension and definitely the resource-problem dimension with respect to P cannot be resolved by single individuals, which often makes action less appealing. These are the same general problems that prevail within every societal transition towards increased sustainability.

Phosphorus Governance Through Economic Instruments

The governance problems of traditional environmental policy as described in the above suggest that a command-and-control approach alone might not be the best legal option. However, this does not suggest a wait-and-see strategy either, as this would solve the issue of casual use of P even less. A governance approach which reflects the required holistic perspective would be more favorable—genuine quantity control and limitation would make mineral P scarcer and more expensive, giving incentives to increase recycling rates and resource protection while relieving ecological stress. Fertilization would mainly occur only when actually necessary, which might be assessed differently in different countries with different soil qualities. The same—or even stronger—effects may be reached in an indirect way, by limiting or pricing fossil fuels (see below). Further details on the approach outlined in the following will be subject to research at the Leibniz Science Campus Phosphorus Research at Rostock University during the next years.

Necessary quantity control (also called "pricing" in the following) may either explicitly limit the use of the P resource, and therefore also its ecological side effects, or put a price on the P resource. One possible way of pricing could be a reallocation of EU agricultural subsidies towards environmental services instead of supporting mass livestock production. This may also be advantageous from a fiscal point of view and in the interest of free global trade and fair market access for developing countries. Another sensible approach to pricing may be a tax on P. Alternatively, a cap-and-trade system could be implemented along the same lines as EU emissions trading. This would require the initial distributors of P to input only those quantities of P into the market for which they have previously purchased the certificates by auction. The amount of certificates would be limited and eventually discounted.

To avoid shifting effects between different countries, this approach should theoretically be adopted worldwide. This will not happen in the short-term; the EU should lead the way until then. It should invite other countries to join in putting a price on P. Revenues from pricing could go towards developing countries to compensate for rising food prices. Imported products (and maybe also exported goods) from countries not willing to participate in pricing could be taxed through border adjustments comparable to value-added tax (see Ekardt 2011 for legitimacy in international trade law). This governance approach—given that it is adopted on a geographically and substantially broad scale (including border adjustments)—has clear advantages over regulatory law (Ekardt 2011; Ekardt and Hennig 2015; Ekardt et al. 2010a).

- Quantity control is easier to implement and less likely to encounter enforcement problems than regulatory law.
- Quantity control prevents shifting effects as evasion is impossible—especially if the approach is applied comprehensively by a large number of countries.
- The approach eliminates rebound effects as it covers all forms of P usage.
- The approach could therefore also be unbureaucratic, supporting the principles of freedom and democracy as decision-making remains with the parliaments instead of administrative agencies.
- The approach gives incentives to increase recycling rates and resource protection while reducing ecological burden. Likewise, ecologically preferable rotation-oriented forms of land use such as organic farming would be supported by pricing P. Recycling P from residues of sewage treatment and waste business (sewage sludge and similar) into agriculture is another issue to address. This, however, would only be possible if the impending overload of soil with heavy metals and further organic constituents is countered by new treatment concepts.

Regarding P, as with other environmental issues like climate change, it is a matter of choice whether quantity control or taxes should be applied to the initial distribution or the end use. End use could mean for instance meat sold to a consumer. Taxing fertilizer producers would be easier to implement; higher fertilizer costs would be transferred to the consumer by the farmers purchasing the fertilizer.

The example of P also shows the importance of viewing different resources in their interdependencies. For instance, implementing quantity controls for P without corresponding greenhouse gas or fossil fuel quantity control including land use could lead to reduced fertilizing, and thus to lower crop yields in some circumstances; this may be compensated by higher land use for agriculture. Moreover, burning forests (fire clearing) and similar may increase as another means of fertilizing soil. A more detailed solution for the best economic governance approach to be analyzed in the Phosphorus Campus in the coming years will therefore need to address different resource and environmental issues in context. Besides the issue of P, problems such as climate change, loss of biodiversity, land use and N surplus problems require consideration. Pricing in one of these areas might lead to positive effects in many of the others. In any case, it will be essential to use governance units that are easy to measure to avoid implementation issues. The entrance point of a resource such as P fertilizer to the EU market can easily be monitored, but biodiversity and greenhouse gas emissions from a specific agricultural area are not as easy to estimate.

The governance unit most easily to measure would be fossil fuels. (Intensive) pricing on fossil fuels would not only strengthen climate protection in general, but also in terms of land use in particular, making N as a fertilizer component more expensive while reducing greenhouse gas emissions (carbon dioxide, methane, nitrous oxide), biodiversity loss and contamination and degradation of soils and groundwater. Pricing on fossil fuels would trigger resource efficiency in terms of fertilizers and food waste reduction, but also make organic farming much more attractive. Effects such as closed-loop cycles and organic farming would also indirectly lead to a reduction in mineral P overuse.

Quantity control of P or fossil fuels would also make the production of livestock in particular less attractive as the production of one calorie of livestock product currently requires a multiple of vegetable calories. Animal products would become more expensive and scarcer, food would be consumed more sparingly. Therefore, a price increase of P or fossil fuels would not endanger global food security, but trigger resource efficiency and a more ecologically sound lifestyle in Western countries. Moreover, this would lead to more lasting access to P. Explicitly pricing P or fossil fuels additionally provides the opportunity to reimburse developing countries for increasing food prices, if they participate directly or indirectly in the pricing system. The costs of this measure would therefore effectively lie with industrialized countries.

None of this necessarily implies that economic incentives will be sufficient in P policy. Even where economic incentives are created, additional regulatory measures may also be necessary for addressing issues such as appropriate sewage sludge treatment, which should be used more intensively while also being restricted to certain ecological and technical conditions as ecological damage in specific places (hot spots) is more effectively regulated by a command-and-control approach than by pricing.

P use and, in general, any administrative law or quantity control approach eventually has implications for social distributive justice. This not only refers to conflicts between economic freedom and the protection of physical preconditions of freedom—partly also guaranteed by fundamental rights and human rights—which are always present in environmental protection (Ekardt 2011). Rather, this refers to secondary effects that arise from the resulting compromises between these different rights in environmental law. In other words, harm and benefit arising from P application may not always align. This problem has a national and global dimension (Ekardt 2011). Eventually declining phosphate rock reserves are likely to result in higher prices and potential quality degradation due to higher heavy metal loads. While industrialized countries are still able to pay the price for high quality and fertilizer in general, developing countries may face severe availability, accessibility and quality issues. Moreover, soils in the southern hemisphere are currently exposed to substances such as heavy metals for production that is mostly consumed in industrialized countries. However, these particular questions on distribution speak for quantitative regulation rather than administrative regulation by legislation—it is not problematic to side with social adjustment payments in the former case, such as by paying higher prices for foodstuffs and other commodities. These compensation payments may for instance distribute the revenues arising from a charge or from a certificate system auctioning per capita to the citizens of every country. Another option would be partially or completely framing them as a north-south transfer.

Conclusion

Regardless of any individual estimation as to how scarce the resource P actually is, a call for action also emerges from the environmental impact of P. Specific, practical measures need to be taken to foster sound use of P; instead, the issue so far has remained in the sphere of non-binding proclamations. However, typical approaches in regulatory law—the command-and-control approach addressing individual activities—are not appropriate for addressing the characteristics of resource and environmental problems as quantity problems. In contrast, economic policy instruments are an effective supplement in avoiding compliance issues, rebound and shifting effects, and encourage more recycling of P as well as more sparing use. This also provides the chance of addressing different environmental issues concerning land use, resources, climate change and biodiversity at the same time.

Acknowledgments The major point of origin of the ideas of this paper is our participation in the Leibniz Science Campus Phosphorus Research of Rostock University and other institutions funded by Leibniz Gemeinschaft, the German Federal Government and other donors.

References

Beste A (2015) Intensivfeldbau: Industrielle Landwirtschaft mit Zukunftsproblemen. In: Heinrich-Böll-Stiftung, IASS, BUND/Le Monde diplomatique (ed.): Bodenatlas, Daten und Fakten über Acker, Land und Erde, pp 18–19, 3rd edition, Berlin

Beusen AHW, Dekkers ALM, Bouwman AF, Ludwig W, Harrison J (2005) Estimation of global river transport of sediments and associated particulate C, N and P. In: Global biochemical cycles, vol 19, GB4S05

BfR (2010) Aufnahme von Umweltkontaminanten über Lebensmittel (Cadmium, Blei, Quecksilber, Dioxine und PCB). Ergebnisse des Forschungsprojektes LExUKon, Rostock

BGR (2013) Phosphat mineralischer Rohstoff und unverzichtbarer Nährstoff für die Ernährungssicherheit weltweit. Online: http://www.bgr.bund.de/DE/Themen/Zusammenarbeit/TechnZusammenarbeit/Politikberatung_SV_MER/Downloads/phosphat.pdf?__blob=publicationFile&v=4. Accessed 01 Dec 2014

Blume HP (2004) Handbuch des Bodenschutzes. Bodenökologie und –belastung. Vorbeugende und abwehrende Schutzmaßnahmen, 3. Auflage, Augsburg

BMELV (2010) Die deutsche Landwirtschaft. Leistungen in Daten und Fakten. Edition 2010, Berlin

BMELV (2015a) Online: http://www.bmelv-statistik.de/index.php?id=139&stw=D%C3%BCngemittel. Accessed 08 Aug 2015

BMELV (2015b) Agrarpolitischer Bericht der Bundesregierung 2015, Berlin

Brot für die Welt (2014) Der Energiepflanzenanbau gefährdet das Recht auf Nahrung. Die aktuelle Biokrafstoffpolitk aus entwicklungspolitischer Perspektive, Berlin

Bundestag D (2013), 17. Wahlperiode, Drucksache 17/12984 vom 05.04.2013

Cordell D, Schmid-Neset T, White S, Drangert JO (2009) Preferred future phosphorus scenarios: a framework for meeting long-term phosphorus needs for global food demand. In: International conference on nutrient recovery from waste water streams, IWA Publishing, London, Vancouver, 10–13 May 2009

Dittrich B, Klose R (2008) Schwermetalle in Düngemitteln. LfULG-Schriftenreihe, Heft 3/2008, Dresden

EEA (2012) European waters—assessment and pressures. European Environment Agency, Copenhagen. Available at: http://www.eea.europe.ea/soer/synthesis/synthesis

EFSA (2011) Scientific opinion statement on tolerable weekly intake for cadmium: EFSA panel on contaminants in the food chain (CONTAM). EFSA J . 9(2):1975

Ekardt F (2011) Theorie der Nachhaltigkeit: Rechtliche, ethische und politische Zugänge [Theory of sustainability: legal, ethical and political perspectives]. Nomos, Baden-Baden

Ekardt F, Hennig B (2015) Ökonomische Instrumente und Bewertungen von Biodiversität. Lehren für den Naturschutz aus dem Klimaschutz? Metropolis Verlag, Marburg

Ekardt F, von Bredow H (2012) Extended emissions trading versus sustainability criteria: managing the ecological and social ambivalences of bioenergy. Renew Energies Law and Policy 49–68

Ekardt F, Hennig B, Hyla A (2010a) Landnutzung, Klimawandel, Emissionshandel und Bioenergie. LIT Verlag, Münster

Ekardt F, Holzapfel N, Ulrich AE (2010b) Phosphorus, land-use and absolute quantity reductions as a legal problem. J Eur Environ Plann Law 267–286

European Commission (2003) Draft proposal relating to Cadmium in fertilisers, Online: http://ec.europa.eu/enterprise/newsroom/cf/_getdocument.cfm?doc_id=2966. Accessed 21 Nov 2012

European Commission COM (2011b) 21 from 26.1.2011 [Communication from the Commission to the European Parliament, the Council, the European Economic and Social Committee and the Committee on the Regions: a resource-efficient Europe—flagship initiative under the Europe 2020 Strategy]

European Commission COM (2013a) 683 from 4.10.2013 (Report from the Commission to the Council and the European Parliament on the implementation of Council Directive 91/676/EEC

concerning the protection of waters against pollution caused by nitrates from agricultural sources based on Member State reports for the period 2008–2011)
European Commission COM (2013b) 517 final from 8.7.2013 (Communication from the Commission to the European Parliament, the Council, the European Economic and Social Committee and the Committee on the Regions: consultative Communication on the sustainable use of phosphorus)
European Commission COM (2014a) 297 final from 26.5.2014 (Communication from the Commission to the European Parliament, the Council, the European Economic and Social Committee and the Committee of the Regions: on the review of the list of critical raw materials for the EU and the implementation of the raw materials initiative)
European Commission COM (2014b) IP 14/599, Press Release from 26.5.2014 (20 critical raw materials—major challenge for EU industry)
European Commission COM (2014c) Nitratbelastung im Grundwasser: Kommission fordert Deutschland zum Handeln auf. Online: http://ec.europa.eu/deutschland/press/pr_releases/12542_de.htm. Accessed 20 Nov 14
European Court of Auditors (2014) Integration of EU water policy objectives with the CAP: a partial success. Special Report No. 4/2014
Eurostat (2014) http://epp.eurostat.ec.europa.eu/statistics_explained/index.php/Fertiliser_consumption_and_nutrient_balance_statistics/de%23Brutton.C3.A4hrstoffbilanzen. Accessed 01 Dec 2014
FAO (2012) World agriculture towards 2030/2050: the 2012 revision. Available at http://www.fao.org/docrep/016/ap106e/ap106e.pdf. Accessed 10 June 2015
Frossard E, Julien P, Neyroud J-A, Sinai S (2004) Phosphor in Böden, Düngern, Kulturen und Umwelt – Situation in der Schweiz. Schriftenreihe Umwelt Nr. 368
Garske B, Hoffmann K (2015) Reform der gemeinsamen Agrarpolitik, Beiträge zum transnationalen Wirtschaftsrecht 180, Halle
Gupta CK, Singh H (2003) Uranium resource processing. Secondary Resources, Berlin
Heckenmüller M, Narita D, Klepper G (2014) Global availability of phosphorus and its implications for global food supply. An Economic Overview, Kiel Working Paper No 1897, Jan 2014, Kiel
IFA (2014) Fertiliser outlook 2014–2018. In: Heffer P, Prud'homme M (eds), 82th annual conference Sydney, Australia, May 2014
IVA (2015) Der Düngemittelmarkt 2014/15. Starkes Inlandsgeschäft treibt Umsätze, Frankfurt
Kharikov AM, Smetana VV (2000) Heavy metals and radioactivity in phosphate fertilisers: short term detrimental effects. Online: http://www.fertiliser.org/en/ItemDetail?iProductCode=6869Pdf&Category=TECH. Accessed 09 June 2015
Kirpotin S, Nemceva G (2015) Environmental, economic and social risks of nuclear power enginieering: the case of the southern part of the Ob-river basin. Int J Environ Stud 72:1–12
Kotschi J, Wetter K (2015) Düngemittel. Zahlende Konsumenten, Intrigante Produzenten, in: Bodenatlas, Daten und Fakten über Acker, Land und Erde, pp 20–21
Lange J (2009) Phosphor – so wichtig wie Luft, so knapp wie Erdöl? Working paper as a basis for a workshop. 23.01.2009, Freiburg
LfL (2011) 20 Jahre Boden-Dauerbeobachtung in Bayern, Part 3, Entwicklung der Humusgehalte zwischen 1986 und 2007, Freising-Tüntenhausen
Lymbery P (2015) Futtermittel: Viel Land für viel Vieh, in: Bodenatlas, Daten und Fakten über Acker, Land und Erde, 3rd edition, Berlin
McDaniels T (2006) Chernobyls effects on the perceived risks of nuclear power: a small sample test. Risk Anal 8:457–461
Mortvedt J, Beaton J (1995) Heavy metal and radionuclide contaminants in phosphate fertilisers. In: SCOPE 54—Phosphorus in the global environment—transfers, cycles and management, Chap 6. Online: http://www.scopenvironment.org/downloadpubs/scope54/6mortvedt.htm. Accessed 09 June 2015
Pimentel D, Hepperly P, Hanson J, Douds D, Seidel R (2005) Environmental, energetic, and economic comparisons of organic and convetional farming systems. Bio Sci 55(7)

Raven KP, Loeppert RH (1997) Trace elements composition of fertilisers and soil amendents. J Environ Qual 26:551–557

Ruppe A, Bahr C, Pohl A (2009) Fortschreibung von Beurteilungsmaßstäben für den Wirkungspfad Boden-Pflanze: Methodik zur flächenrepräsentativen Erfassung pflanzenverfügbarer Stoffgehalte in unbelasteten Böden und Stoffgehalten in Nahrungs- und Futtermittelpflanzen. Forschungsbericht FKZ 206 74 200, Umweltbundesamt, Dessau-Roßlau

Schnug E, Kratz S (2005) Schwermetalle in P-Düngern, Landbauforschung Völkenrode, Special Issue 286

SRU (2008) Umweltgutachten 2008, Umweltschutz im Zeichen des Klimawandels, Berlin

Sutton MA et al (2013) Our nutrient world: the challenge to produce more food and energy with less pollution. Global overview of Nutrient Management, Centre für Ecology and Hydrology, Endinburgh

UBA (2014) Eutrophierung der Ostsee. Online: http://www.umweltbundesamt.de/daten/gewaesserbelastung/ostsee/eutrophierung-der-ostsee. Accessed 01 Dec 2014

Ulrich AE (2013) Peak phosphorus: opportunity in the making. exploring global phosphorus management and stewardship for a sustainable future. Dissertation, ETH Zürich

Ulrich AE, Frossard E (2014) On the history of a reoccurring concept: phosphorus scarcity. Sci Total Environ 490:694–707

Ulrich AE, Schnug E, Prasser HM, Frossard E (2014) Uranium endowments in phosphate rock. Sci Total Environ 478:226–234

UNCCD (2012) Desertification land degradation and drought—some global facts and figures. Online: http://www.unccd.int/Lists/SiteDocumentLibrary/WDCD/DLDD%20Facts.pdf. Accessed 01 Dec 2014

Van Kauwenbergh S (1997) Cadmium and other minor elements in world resources of phosphate rock. The Fertiliser Society, Proceedings No 400

Welch B (1980) Nucelar power risks: challenge to the credibility of science. Int J Health Serv 10:5–36

WHO (2011) WHO Food Additives Series 64, 73rd meeting of the JECFA. World Health Organisation, Geneva

Withers PJA, van Dijk KC, Neset TS, Nesme T, Oenema O, Schoumans OF, Smit B, Sylv P (2015) Stewardship to tackle a global phosphorus inefficiency: The case of Europe. Ambio 44 (2):193–206

WWF (2014) Fleisch frisst Land. Ernährungsweisen. Fleischkonsum. Flächenverbrauch. Online: https://www.wwf.de/fileadmin/fm-wwf/Publikationen-PDF/WWF_Fleischkonsum_web.pdf. Accessed 17 July 2015

Index

E. Schnug and L.J. De Kok (eds.), *Phosphorus in Agriculture: 100 % Zero*,
DOI 10.1007/978-94-017-7612-7

MIX
Papier aus verantwortungsvollen Quellen
Paper from responsible sources
FSC® C105338

If you have any concerns about our products,
you can contact us on
ProductSafety@springernature.com

In case Publisher is established outside the EU,
the EU authorized representative is:
Springer Nature Customer Service Center GmbH
Europaplatz 3, 69115 Heidelberg, Germany

Printed by Libri Plureos GmbH
in Hamburg, Germany